Technical Writing Style

THE ALLYN AND BACON SERIES IN TECHNICAL COMMUNICATION

SERIES EDITOR: SAM DRAGGA, TEXAS TECH UNIVERSITY

Thomas T. Barker
Writing Software Documentation: A Task-Oriented Approach

Dan Jones
Technical Writing Style

Charles Kostelnick and David D. Roberts
Designing Visual Language: Strategies for Professional Communicators

Carolyn Rude
Technical Editing, Second Edition

Technical Writing Style

Dan Jones
University of Central Florida

Allyn and Bacon
Boston • London • Toronto • Sydney • Tokyo • Singapore

Series Editor: Eben W. Ludlow
Production Coordinator: Christopher H. Rawlings
Editorial Production Service: Omegatype Typography, Inc.
Composition and Prepress Buyer: Linda Cox
Manufacturing Buyer: Suzanne Lareau
Cover Administrator: Suzanne Harbison

Copyright © 1998 by Allyn & Bacon
A Viacom Company
160 Gould Street
Needham Heights, MA 02194

Internet: www.abacon.com
America Online: Keyword: College Online

All rights reserved. No part of the material protected by this copyright notice may be reproduced or utilized in any form or by any means, electronic or mechanical, including photocopying, recording, or by any information storage and retrieval system, without written permission from the copyright owner.

Library of Congress Cataloging-in-Publication Data

Jones, Dan, (Dan Richard), 1952–
    Technical writing style / Dan Jones.
      p. cm.
    Includes bibliographical references (p.) and index.
    ISBN 0-205-19722-1 (pbk.)
    1. English language—Technical English. 2. English language—
Style. 3. Technical writing. I. Title.
PE1475.J57 1998
808'.0666—dc21   97-14813
                                                                                    CIP

Printed in the United States of America
10 9 8 7 6 5 4 3 2 1   02 01 00 99 98 97

Acknowledgments:
Tom McArthur, Editor. From *The Oxford Companion to the English Language.* Copyright © 1992. Reprinted by permission of Oxford University Press.

(Acknowledgments continued on page 302, which constitutes a continuation of the copyright page)

For Gloria Restivo Jones, 1926–1995

Contents

Foreword to the Series xi

Preface xiii

1 Understanding Technical Writing Style 1
*Chapter Overview 1
The Problem with Defining Style 2
A Definition of Style 3
Style in Technical Writing 3
The Domain of Technical Writing Style 7
Why Technical Writing Is Often Ineffective 10
Technical Writing as a Craft 11
Chapter Summary 11
Case Study 1 Three Mile Island and Ineffective Communication 12
Questions/Comments for Discussion 14
Exercises 14
Notes 17*

2 Reflecting the Style of a Discourse Community 19
*Chapter Overview 19
What is a Discourse Community? 20
How Does a Discourse Community Affect Style? 22
Questions to Consider Before Writing to a Discourse Community 25
Case Study 2 The Space Shuttle* Challenger *and Different
   Discourse Communities 26
How Does Style Relate to Subject, Purpose, Scope, Audience, and Other Elements of
   Writing? 34
Chapter Summary 38
Questions/Comments for Discussion 38
Exercises 39
Notes 40*

3 Choosing an Appropriate Style 42
*Chapter Overview 42
On Style and Grammar 43*

Prescriptive and Descriptive Approaches to Style 43
Good and Bad Writing 44
Good and Bad Style 45
Appropriate and Inappropriate Styles 46
Kinds of Styles 46
Chapter Summary 56
Questions/Comments for Discussion 57
Exercises 57
Notes 59

4 Persuading through Style 62
Chapter Overview 62
Technical Prose as Persuasion 63
The Rhetoric of Science 64
Some Practical Consequences of the New Rhetoric of Science 67
Persuasion and Technical Prose Style 70
Rhetorical Choices and Style 70
Persuasive Strategies and Stylistic Choices 71
Case Study 3   The Exxon Valdez *and Persuasion* 75
Chapter Summary 75
Questions/Topics for Discussion 79
Exercises 80
Notes 82

5 Choosing Appropriate Words—Diction 85
Chapter Overview 85
On Diction 86
Levels of Diction 87
Denotation and Connotation 90
Challenges Presented by Specialized Diction 90
Major Diction Strategies in Technical Prose 91
Common Diction Faults 97
Technical Prose and a Literary Style 106
On Euphemisms and Dysphemisms 109
Chapter Summary 110
Case Study 4   The Ruby Ridge Incident and Diction 110
Questions/Topics for Discussion 113
Exercises 113
Notes 114

6 Handling Technical Terms and Jargon 118
Chapter Overview 118
Using Technical Language 119
Words, Terms, and Terminology 120
Technical Terms as Jargon 120
The Challenges of Using Technical Terms 122
The Challenges of Using Jargon 125
Other Specialized Languages 129
Abbreviations, Acronyms, Initialisms 133
Tips for Using Clear Technical Language 134
Chapter Summary 134
Case Study 5   Apollo 13 and Jargon 135

Contents   ix

              Questions/Topics for Discussion   138
              Exercises   139
              Notes   141

7   Creating Sentences with Style   144
       Chapter Overview   144
       Sentences in Technical Prose   145
       A Brief Review of Sentence Basics   145
       Sentence Combining Techniques   150
       Emphasis   153
       Rhythm   155
       Sentence Variety   155
       Sentence Elegance in Technical Prose   156
       Sentence Faults   156
       Chapter Summary   158
       Questions/Topics for Discussion   159
       Exercises   159
       Notes   161

8   Structuring Paragraphs and Other Segments   163
       Chapter Overview   163
       Writing Technical Paragraphs   164
       Writing on One Topic   164
       Developing the Paragraph Topic   167
       Achieving an Effective Flow   174
       Providing Adequate Details   177
       Larger Segments in Technical Prose   179
       Common Paragraph Faults   179
       Chapter Summary   180
       Questions/Topics for Discussion   181
       Exercises   181
       Notes   184

9   Establishing an Appropriate Tone   187
       Chapter Overview   188
       Tone in Technical Prose   188
       The Key Emphasis of Tone   190
       Strategies for Developing an Appropriate Tone   202
       Humor in Technical Writing   204
       Chapter Summary   207
       Case Study 6   Computer Viruses and Tone   208
       Questions/Topics for Discussion   211
       Exercises   211
       Notes   213

10   Considering Style and Bias   215
       Chapter Overview   215
       Bias in Technical Prose   216
       Kinds of Bias   218
       Political Correctness   228
       Chapter Summary   229
       Case Study 7   Gender-Neutral Language   230

Questions/Topics for Discussion   231
Exercises   232
Notes   233

11 Determining the Ethics of Style   236

Chapter Overview   236
What is Ethics?   237
Ethics and Technical Prose   238
Ethics and the Professions   238
Criteria for Judging Ethical Actions   241
Unethical Language   242
The Ethics of Style   245
Chapter Summary   246
Case Study 8   Deception and the Exxon Valdez   247
Questions/Topics for Discussion   249
Exercises   249
Notes   252

12 Editing for Style   253

Chapter Overview   254
Misconceptions about Editing   255
Approaches to Editing   256
The Challenges of Editing for Style   258
Strategies to Help You Edit for Style   259
Copyediting for Style   264
Chapter Summary   265
Questions/Topics for Discussion   265
Exercises   266
Notes   269

Glossary   270

Bibliography   280

Index   290

Foreword by the Series Editor

The Allyn and Bacon Series in Technical Communication is designed for the growing number of students enrolled in undergraduate and graduate programs in technical communication. Such programs offer a wide variety of courses beyond the introductory technical writing course—advanced courses for which fully satisfactory and appropriately focused textbooks have often been impossible to locate. This series will also serve the continuing education needs of professional technical communicators, both those who desire to upgrade or update their own communication abilities as well as those who train or supervise writers, editors, and artists within their organization.

The chief characteristic of the books in this series is their consistent effort to integrate theory and practice. The book offers both research-based and experience-based instruction, describing not only what to do and how to do it but explaining why. The instructors who teach advanced courses and the students who enroll in these courses are looking for more than rigid rules and ad hoc guidelines. They want books that demonstrate theoretical sophistication and a solid foundation in the research of the field as well as pragmatic advice and perceptive applications. Instructors and students will also find these books filled with activities and assignments adaptable to the classroom and to the self-guided learning processes of professional technical communication.

To operate effectively in the field of technical communication, today's students require extensive training in the creation, analysis, and design of information for both domestic and international audiences, for both paper and electronic environments. The books in the Allyn and Bacon Series address those subjects that are most frequently taught at the undergraduate and graduate levels as a direct response to both the educational needs of students and the practical demands of business and industry. Additional books will be developed for the series in order to satisfy or anticipate changes in writing technologies, academic curricula, and the profession of technical communication.

Sam Dragga
Texas Tech University

Preface

Many introductory technical communication textbooks tell you almost everything you need to know about the basics of the documentation process. Excellent introductory texts include Jimmie Killingsworth's *Information in Action* (Allyn and Bacon, 1996); Paul Anderson's *Technical Writing: A Reader-Centered Approach* (3rd edition, Harcourt, 1995); Elizabeth Tebeaux, Tom Pearsall, and Kenneth Houp's *Reporting Technical Information* (9th edition, Allyn and Bacon, 1998); and Mary Lay, Billie Wahlstrom, Stephen Doheny-Farina, Ann Hill Duin, Sherry Burgus Little, Carolyn Rude, Cynthia Selfe, and Jack Selzer's *Technical Communication* (Richard Irwin, 1995). In these texts you will find valuable advice on planning, scheduling, organizing, writing, designing, illustrating, editing, revising, and producing technical information.

Increasingly, advanced technical communication textbooks are becoming available, with each textbook solely devoted to a specialized topic such as technical editing, design, illustration, usability testing, online documentation, or software documentation. Two excellent advanced textbooks are Gerald Alred, Walter Oliu, and Charles Brusaw's *The Professional Writer: A Guide for Advanced Technical Writing* (St. Martin's Press, 1992) and Carolyn Rude's *Technical Editing* (2nd edition, Allyn and Bacon, 1998).

Yet, despite all of these introductory and advanced texts, no one text is available which is specifically devoted to the challenges of style in technical communication. Many companies have style guides specifying the writing standards for all documents produced by the company, and many of these guides offer a great deal of information about style. Yet most also include information about design, illustrations, and page layout. Similarly, Philip Rubens' *Science and Technical Writing: A Manual of Style* (New York: Henry Holt, 1992) covers topics ranging from audience analysis and planning to matters of style in the text to indexing, illustrating, and designing documents. As for texts on general style, Joseph Williams's *Style: Ten Lessons in Clarity and Grace* (4th edition, Harper Collins, 1994) remains the best textbook available on the characteristics of a complex prose style, but Williams devotes little attention to the particular challenges of writing technical prose.

Some of the introductory and advanced texts have good chapters on style topics, but style in technical communication is such a complex subject that much more needs to be covered than what is covered in these texts. As you will see, style concerns more than knowing the difference between active and passive voice, avoiding unnecessary technical terms, achieving an effective tone, or using acronyms and initialisms appropriately. And style, as it is discussed here, has little to do with punctuation, spelling, grammar, design, or illustration.

This book has one aim: to make you a better writer of technical prose. It focuses in depth on the many elements of technical writing style, and it demonstrates the many stylistic strategies you must consider for every technical document you write.

Audience

For this text to be useful to you, you should already have some basic knowledge about technical communication. You should know how to assess an audience, plan, state a purpose, organize, and outline for a technical document. You should have some experience in writing various kinds of technical proposals, reports, or manuals. Having some knowledge of illustrating and designing technical documents would also be helpful. *Technical Writing Style* is aimed primarily at those of you who want to build on this basic knowledge to improve your technical prose.

You may be a college student, a practicing technical communicator, an engineer, a scientist, a programmer, or a member of any other technical profession. You may be a college junior majoring in forensic science or a senior or graduate student majoring in technical communication. You may be a biology professor working on a grant proposal or an anthropology professor preparing a journal article. You may be an engineer faced with your first major technical report or you may be an experienced engineer about to write your 100th report. You may be a computer programmer writing a travel report or an architect writing a feasibility report. This book will help you develop effective stylistic strategies no matter what your technical subject or technical knowledge is. You will study many different stylistic strategies for writing to a variety of audiences, ranging from an unknown or phantom audience; to a lay person, paraprofessional, executive, technician, or expert audience; or, most commonly, to a mixture of audiences or a complex audience. Although the strategies and challenges of providing an effective technical writing style for these audiences vary a great deal, there are also many similar strategies and challenges for writing to any audience. Also, while experts often resent having to write to a lay audience, learning the strategies necessary for doing so can be beneficial for writing in other situations, to other experts. In brief, experts can learn to write more effectively to each other by learning to write clearly to less knowledgeable audiences.

Organization

This book is organized according to twelve strategies for achieving an effective technical writing style. These basic strategies correspond to the twelve chapters of the textbook. To be a more effective writer, you need to know

- what technical writing style is
- what discourse community you are writing in
- what kinds of styles are available to you

- what the best tactics for persuasion are
- what the best words for your purpose are
- how to handle technical terms and jargon
- how to write effective sentences
- how to write appropriately detailed paragraphs
- how to achieve the best tone
- how to avoid bias
- how to consider ethics
- how to edit your style.

Accordingly, the twelve chapters of this book are:

1. Understanding Technical Writing Style
2. Reflecting the Style of a Discourse Community
3. Choosing an Appropriate Style
4. Persuading Through Style
5. Choosing Appropriate Words—Diction
6. Handling Technical Terms and Jargon
7. Creating Sentences With Style
8. Structuring Paragraphs and Other Segments
9. Establishing an Appropriate Tone
10. Considering Style and Bias
11. Determining the Ethics of Style
12. Editing for Style

Features

Technical Writing Style offers many features to help those who teach or work with technical prose.

- A helpful twelve-point approach to the complex topic of technical prose style including a closing chapter offering a useful approach for editing for style.
- A chapter overview at the beginning of each chapter that provides students with learning objectives for each chapter.
- Six case studies of recent interesting events to illustrate the principles of effective or ineffective technical prose style and two case studies asking students to critique published documents on the basis of style.
- Numerous style examples presenting all kinds of effective and ineffective technical prose styles.
- Challenging questions and topics for discussion at the end of each chapter.
- Exercises at the end of each chapter to help students apply the principles of style they have studied.
- A detailed glossary of terms related to technical writing style.
- A detailed bibliography bringing together in one place some of the most important recent work on technical prose style.

Acknowledgments

Eben Ludlow, Doug Day, and Liz Egan of Allyn and Bacon helped me to complete this book with their advice, encouragement, and patience. The final manuscript benefited from the expertise of Christopher Rawlings at Allyn and Bacon and Barbara Cook, Tim Barnes, and Mary Young at Omegatype Typography, Inc.

Sam Dragga of Texas Tech University first suggested two years ago that I write a textbook on technical writing style, and he graciously invited me to submit the manuscript for consideration for the new series of technical communication textbooks now being published by Allyn and Bacon. Sam offered valuable advice and support with this undertaking. His comments on an early draft of the manuscript were particularly helpful.

My reviewers offered many constructive suggestions to help me revise and prepare the final manuscript: Jimmie Killingsworth of Texas A & M University, Sherry Burgus Little of San Diego State University, and Henrietta Nickels Shirk of North Texas University.

Area technical communication professionals Dan Voss, John T. Oriel, and Dick Hughes provided or pointed me to some examples or topics. My colleagues at the University of Central Florida—John Schell, Madelyn Flammia, Gloria Jaffe, and David Gillette—offered much encouragement throughout the project.

Like all teachers, I am indebted to the students in all of my classes over the years, but especially students in my undergraduate technical writing style classes who, for the past twelve years, have helped shape this book in countless ways. Their ideas and the ideas of the graduate students in my class on technical prose style have been invaluable.

The following students in these classes helped identify good prose examples or contributed to some of the exercises: Sharon Anderson, Laurie Benson, Shannon L. Berry, Mike Boyce, Marilyn Chastain, Grace M. Edwards, Jennifer H. Geaslen, Rebecca Gillespie, Mary Ellen Gomrad, David Haas, Leighan Hanson, Karen Lane, Jennifer Lenz, John Miner, Laura Matherson, Joanne Muratori, John T. Oriel, Christina Richie, Rosa Sobhraj, Melanie Woods, Debra Winter, and Sue Wrzesinski.

Eugene G. Gray, Cynthia A. Legg, Nadejda Bojilova, and Jennifer Jones contributed greatly to some of the case studies. Lori Brodkin, Joy Friday, Molly McCoy, and other current or former students helped to shape some of the ideas in some of the chapters.

This book would never have been completed without the extensive help of Karen Lane who, in addition to providing examples and ideas, helped in many ways with the glossary, the bibliography, the exercises, some of the case studies, the index, and with copy-editing.

Finally, my wife, Carol, provided valuable editing comments on many drafts of various chapters, and, along with the help of our son, Sam, gave me the time necessary to undertake and finish this project.

Dan Jones
djones@pegasus.cc.ucf.edu

Technical Writing Style

Chapter 1

Understanding Technical Writing Style

Everybody talks about style, but almost nobody understands the meaning of the word in the business environment. And this lack of understanding hurts both those who write letters for another's signature and those who write for themselves. Neither knows where to turn for help.[1]
—JOHN S. FIELDEN

Style can mean a great many things.[2]
—JACQUES BARZUN

Aphoristic definitions abound: style is the man; it is proper words in proper places; the dress of thoughts; the way you write; etc. But what does the writer aim for?[3]
—JOHN WALTER

Chapter Overview

The Problem with Defining Style
A Definition of Style
Style in Technical Writing
The Domain of Technical Writing Style
Why Technical Writing Is Often Ineffective
Technical Writing as a Craft
Chapter Summary
Case Study 1: Three Mile Island and Ineffective Communication
Questions/Topics for Discussion
Exercises
Notes

The Problem with Defining Style

Texts without Definitions

It's surprising how many books about writing discuss style without defining it. The authors assume for one reason or another you know what they are talking about when they offer advice about how to address problems of style. For example, three of the best known books on nontechnical prose style—William Strunk and E. B. White's *Elements of Style,* Joseph Williams's *Style: Ten Lessons in Clarity and Grace,* and William Zinsser's *On Writing Well*—offer no precise definitions of style.[4] It seems everyone has something to say about style but no one wants to define it.

James Kilpatrick's *The Writer's Art,* another valuable book on the craft of writing, offers all kinds of good advice on how to write well, but he too avoids defining style: "Whatever it is, style provides the individual hallmark that writers stamp upon their work—but that metaphor is inapt, for it suggests that style is something you put onto a piece after you have finished it, as if it were ketchup on chili or lemon on fish. Style doesn't work that way. It's more of a marinade, permeating the whole composition."[5]

Some authors of fine texts on the craft of writing think it's unimportant to define *style*. In *Write to the Point,* Bill Stott believes "the ultimate product that writers sell is what they say, not who they are."[6] He even admits, "I'm no longer sure I know what style is. I don't know that anybody knows. A magic word indeed."[7] For Stott, the most specific meanings are organization and tone. He prefers using these words in place of the word *style*.

Of course, many people believe it's fruitless to define style in any precise way. H. L. Mencken points to the nebulous quality of style, the part that's difficult to pin down:

> The essence of a sound style...is that it cannot be reduced to rules—that it is a living and breathing thing, with something of the demoniacal in it—that it fits its proprietor tightly and yet ever so loosely, as his skin fits him. It is, in fact, quite as securely an integral part of him as that skin is. It hardens as his arteries harden. It is gaudy when he is young and gathers decorum when he grows old.... In brief, a style is always the outward and visible symbol of a man, and it cannot be anything else. To attempt to teach it is as silly as to set up courses on making love.[8]

Texts with Good Definitions of Style

To be fair, there are some good definitions of style in books about writing. In *Technical Editing,* Carolyn Rude defines style as "the cumulative effect of choices about words, their forms, and their arrangement in sentences."[9] In *Developing a Written Voice,* Dona Hickey suggests that style is "a writer's habitual language choice" or "a writer's habits of language use."[10] To Hickey, "A writer can be said to have a style when he or she intentionally and habitually relies on particular verbal patterns."[11]

A good definition of style, however, must reflect that style is more than just a matter of habitual language choice or tone. Style is also a matter of the ethos or character of the writer. Nevin Laib suggests this aspect of style in his *Rhetoric and Style:* "Style is the external expression of your values and beliefs, your personal or social rhetoric. It is your persona or role as an author—a public image you project, a mask you wear, or a stance you take toward the subject and audience."[12]

A Definition of Style

Style affects or influences almost all other elements of writing. Style is your choices of words, phrases, clauses, and sentences, and how you connect these sentences. Style is the unity and coherence of your paragraphs and larger segments. Style is your tone—your attitudes toward your subject, your audience, and yourself—in what you write. Style is who you are and how you reflect who you are, intentionally or unintentionally, in what you write. And all of these defining characteristics of style are true for both nontechnical and technical prose.

This one paragraph definition is more complex than it appears. In fact, one of the major purposes of this textbook is to discuss this definition in depth. For example, your choice of words is the focus of Chapter 5: Choosing Appropriate Words—Diction and Chapter 6: Handling Technical Terms and Jargon. As for the elements of a sentence and how these elements may be connected, Chapter 7: Creating Sentences with Style, provides many strategies. For the important style elements of unity and coherence in your paragraphs and larger segments, Chapter 8: Structuring Paragraphs and Other Segments, addresses not only these issues but also many different ways to develop technical paragraphs. Chapter 9: Establishing an Appropriate Tone discusses the many different attitudes you may express toward your readers and your subject in technical prose. This chapter also addresses the important concept of ethos or the writer's character and how this is revealed in prose. Finally, other important elements of style closely related to tone and ethos are covered in Chapter 10: Considering Style and Bias and Chapter 11: Determining the Ethics of Style.

Style in Technical Writing

Confusion Caused by Style Guides

Part of the confusion about defining technical prose style is caused by a broad view of what encompasses style in this area. Corporate style guides used in many industries have helped to create the impression that all of the rules and conventions to be followed by a company in creating its documents are matters of style. This kind of style guide typically covers agreed upon conventions for format, punctuation, spelling, grammar, illustration, design, and tone. As long as the writers at the company follow these conventions, the company's documents will be consistent in content, format, and tone. These style guides leave us with the impression that everything in the documentation process—from planning to production—is a matter of style.

Confusion Caused by Misconceptions

The idea that technical prose has a style, or many styles, should be obvious (Chapter 3: Choosing an Appropriate Style discusses the wide range of styles in technical prose). However, when you bring up the topic of style, people tend to think more of their favorite novels, newspaper columns, or collections of essays than a recent technical document they read. If they discuss the technical document at all they tend to complain about its strange terminology, confusing sentences, impersonal tone, and so on. Of course, these issues and many others all concern style, in this case, the ineffective style of the feasibility report, VCR manual, software user guide, or whatever the seemingly indecipherable technical document is. The following description of a Department of Transportation project offered for university research is a typical example:

> The nature, extent, and consequences of carlessness and operational and institutional means of improving mobility for this diverse group (the elderly, poor, young, etc.) is to be identified and analyzed. Data has been collected and disaggregated to identify the groups affected by carelessness and attitudinal surveying techniques utilized to determine the groups' travel needs and priorities (actual and latent) allowing the development of a trip priority matrix. Changes in demand for transit by the carless resulting from changes in transit policies and operations (routes, fares, amenities, etc.) will be analyzed. Disaggregated demand and behavioral models will be applied to develop demand elasticities for travel by the carless in terms of modal choice (based on trip priority, opportunity, and accessibility and socio-economic and demographic characteristics) and trip purpose.

Weaknesses in this example include the failure to identify clearly who the agent of the action is (who is doing the identifying) in the opening sentence, poor sentence emphasis in most sentences, passive voice ("Data has been collected…"), wordiness, poor diction choices ("carlessness," "demand elasticities," "modal choice"), pretentiousness ("disaggregated," "disaggregated demand"), and an overly impersonal tone. A reader must reread this paragraph many times to understand what the writer is trying to express.

So the confusion about style in technical prose—yes, it has a style because all prose does, but it seems to lack style because it's often handled so poorly—is a problem. This confusion about style in technical prose is compounded by the fact that many people have a limited view of the role of style in this kind of writing. They believe that the only essential style for technical prose is a transparent style. It should be completely inconspicuous, unobtrusive, and invisible. After all, these people argue, the most important part of a technical document is the subject. Anything which detracts from our understanding of the subject is unnecessary. For example, in *How to Write and Publish a Scientific Paper,* Robert Day argues against literary qualities in scientific prose: "I take the position that the preparation of a scientific paper has almost nothing to do with literary skill. It is a question of *organization.* A scientific paper is not 'literature.' The preparer of a scientific paper is not really an 'author' in the literary sense."[13]

There are important differences in the aims and genres of science, scientific, and technical writing, but the primary difference is one of intended audience. *Science* writing is written by either scientists or science journalists to explain to the layperson various scientific discoveries or other news about science. The chief audience is not an expert audience. *Scientific* writing is primarily expert-to-expert writing. Whether this kind of writing is in *The New England Journal of Medicine* or the *Journal of Immunology,* experts, in general, are explaining their findings to their peers. *Technical* writing most often aims for a wide range of audiences, from lay person to expert, but the most common audience in technical writing is the combined or complex audience, consisting of people with all kinds of backgrounds and levels of knowledge about the subject.

Despite these essential differences in audiences, you are still free to choose from a wide range of styles in science and technical writing, and, to a degree, even in scientific writing. However, the view that technical prose in technical documents should somehow be transparent is understandable. Yes, it is true that readers of technical prose don't prefer an ornate or literary style in many different types of technical documents. If they are trying to install a modem, for example, they want clear, accurate, and concise instructions. They usually don't want the author's stylistic idiosyncrasies detracting from the necessary information. *But every decision the author makes to keep the style inconspicuous is a stylistic decision.*

Some readers prefer that the instructions offer some encouragement or even humor to help them overcome their fears and anxieties. Many types of printed and online instructions use these motivational strategies and many others. But there's an important point that's often overlooked here. Whether the set of instructions is clear, accurate, and concise, or clear, accurate, concise, and motivational, the document has a definite style. How does the author (or team of authors) achieve clarity, accuracy, and conciseness? And, if motivational factors are used, how does the author motivate you? The answers to these questions are in large part matters of style.

Whether you are writing a science article for a lay audience, a scientific journal article for peers, or a technical document for home users of a computer program, you have a style in your writing. Whether you are writing the lead story for your local newspaper or a business letter, you have a style. Whether you write fiction or nonfiction, you have a style. Some styles are more conspicuous than others, but all writers have a style. As long as writers write with words, they adopt a style whether they do so consciously or subconsciously, conspicuously or inconspicuously, competently or incompetently.

Stylistics and Technical Prose

Some of the bias against style in technical prose can be seen in definitions of stylistics. Definitions of stylistics reflect a preoccupation with literary rather than technical prose. A typical dictionary definition of *stylistics* is "an aspect of literary study that emphasizes the analysis of various elements of style (as metaphor and diction)."[14] In *A Dictionary of Stylistics,* Katie Wales defines *stylistics* as "the study of style,"[15] but she notes a confusion of *stylistics* with *literary stylistics,* "literary because it tends to focus on literary texts."[16] Wales offers the helpful distinction of *general stylistics:* "a cover term to cover the analyses of non-literary varieties of language."[17]

Perhaps the major reason stylistics traditionally has been preoccupied with literary rather than scientific and technical prose is that literary prose appears to be more emotive or more expressive. However, this view simplifies a complex issue. Scientific and technical prose is often very expressive of the individual self. In *Science as Writing,* David Locke offers a compelling argument in defense of the expressiveness of scientific prose: "…I vigorously deny that traditional assessment in literary and scientific camps that one, if not *the,* discriminating feature between literary language and scientific language is that the former is expressive and the latter is not."[18] Locke is careful not to suggest "that all scientific language is highly expressive or that all scientific language is equally expressive."[19] What Locke argues is "that scientific language can be expressive...of the individual self and expressive of ideas and of powerful feelings."[20] Locke knows that the writing in most scientific journals shows little expressiveness, but he believes there is a residue even in these publications. Within the discourse community of a particular scientist, the expressiveness is there for all to see: "Scientists who know well the workers in their particular field will know precisely who wrote a given paper, whether or not the name appears on it, and they will know equally well what the author's feelings are about the work reported."[21]

The Tradition of Style in Technical Prose

The ignorance about technical writing style arises in part from an ignorance concerning the history of scientific and technical communication. Even a casual look at some of the major scientific and technical literature over the past 400 years shows a wide variety of expressive styles in technical prose. Francis Bacon, William Harvey, Charles Darwin, Thomas Huxley, Michael Faraday, Thomas Edison, and many others have distinct styles as they discuss topics ranging from the circulation of blood in the body to the invention of motion pictures. Over the past few decades there has been a renaissance in science writing by scientists and science journalists. Lewis Thomas, Oliver Sacks, Stephen Jay Gould, and Peter Medawar are just some of many fine stylists of technical prose writing today.

The Old Paradigm for Technical Prose

Much of the confusion about style in technical communication also arises from a misunderstanding about what technical communication is. Many people still believe in the old paradigm which says technical writing is or should be objective (in the sense of being neutral), and, therefore, good technical writing should be impersonal, clinical, functional—anything but stylish. This old paradigm is sometimes referred to as the conveyor belt metaphor because writers are seen as neutrally passing along technical information to readers without shaping or creatively interpreting the material.

The New Paradigm for Technical Prose

A new paradigm concerning technical writing emerged in the 1970s and has become firmly entrenched. This new paradigm rejects the conveyor belt view and recognizes technical writing as not objective but persuasive and therefore fundamentally rhetorical. (The persuasive nature of technical writing will be discussed in more detail in Chapter 4: Persuading through Style.) This new paradigm recognizes the creative and dynamic role that all writers of technical prose play when they write. Whether you are explaining how to use banking software or discussing Einstein's theory of relativity, you are advocating a point of view. Whether you are discussing an idea, product, organization, or service, you are an interpreter of technical information. Whether you are explaining a complex subject to peers or you are making a complex and unfamiliar subject simpler and more familiar to a mixture of audiences, you do so with much creativity, many strategies, and a constant manipulation of style.

The Elements of a Technical Writing Style

Is the style of technical writing different from the style of other kinds of writing? Yes and no. Technical writing style is somewhat different because technical writing is chiefly expository (its purpose is to inform) and mainly functional (it provides useful information). It reflects an extreme awareness of a specific audience. It is often characterized by a specialized vocabulary, a virtual shorthand for discussing a technical subject. It relies on certain sentence types (for example, *subject–verb–direct object* order) and paragraph patterns (for instance, *general to specific*) more commonly than other types of writing. But technical writing is also writing, and to be effective all writing has to take into account purpose, audience, and numerous other elements. In these respects and others, technical writing has much in common with other kinds of writing. As Chapter 5: Choosing Appropriate Words—Diction will demonstrate, technical writers face many of the same general problems faced by writers of any kind of prose.

Many people who write technical prose will no doubt continue to misrepresent and misunderstand the nature of technical writing. That's why it's even more important to recognize that technical writing is first and foremost writing, and that writers of technical prose face many of the same difficulties as do writers of any other kind of writing. For this reason, and for the reason that technical writing is fundamentally persuasive, technical writing style is far from uncreative or unimaginative.

The Diversity of Technical Writing Styles

As you will see in this textbook, there is a wide range of technical writing styles. A technical prose style may be formal, conversational, or chatty. It may be transparent or ornate; plain, complex, or overly complex. The level of formality and the level of difficulty of the technical writing depend on many factors, especially on the level of technicality of the subject. The level of technicality may be technical, semitechnical, or nontechnical. And, as with other writing styles, technical writers may use a variety of strategies for concealing or revealing their style.

The Domain of Technical Writing Style

Technical Writing and Rhetoric

As we discussed in the previous section, some people want to limit the possibilities of style in technical writing. In effect, they want to limit the domain—the sphere of influence or territory—of style in technical writing. That's why it's so important to understand how technical writing is grounded in rhetoric. As Daniel Marder reminds us: "We fail to realize that in technical reporting, only the subject is technical. Despite all the razzle-dazzle in our most technical of technological times, the substance of technical reporting remains a formal discipline of the mind: a rhetorical discovery of the best ways and means to persuade an audience."[22] Understanding the rhetorical nature of technical writing broadens your understanding of the domain of style in technical prose. In other words, accepting the fact that technical writing is fundamentally rhetorical is to have a better understanding of the possibilities of style in technical writing.

Defining Rhetoric

What is *rhetoric?* The word has several meanings. Aristotle's definition is "the faculty of observing in any given case the available means of persuasion."[23] Stated another way, rhetoric is the art of speaking or writing effectively. This art can be learned by studying the rules and principles of composition formulated by classical or modern rhetoricians or, more generally, by studying writing or speaking as a means of communication or persuasion. Rhetoric also refers to your skill in the effective use of speech, even including insincere or grandiloquent language. Finally, rhetoric broadly refers to verbal communication.

Every endeavor, every discipline, every technology, every science has its own rhetoric. Rhetoric in this sense is the way things are known and expressed in an area. Nevin Laib supports this broad view of rhetoric when he comments that "Rhetoric is the means through which we support, explain, debate, and even discover ideas."[24] Rhetoric is our way of knowing what we know. Rhetoric isn't the opposite of truthful; it's the very essence of truthful inquiry.

Kinds of Rhetoric

Richard VanDeWeghe expresses this view of rhetoric when he argues for the fundamentally rhetorical nature of technical communication: "Because its domain is information and the work of human beings, technical communication has its roots in rhetoric, a comprehensive field that has gotten a bum rap ever since the Greek sophists gave it a bad name and Plato questioned its integrity."[25] VanDeWeghe clarifies the kind of rhetoric technical communication is by discussing four kinds of rhetoric: (1) sub-rhetoric "or words used deliberately to deceive people or to obscure issues"; (2) mere rhetoric "where language is used for sincere, not deceptive selling of a cause"; (3) Rhetoric B, "or language used in value disputes where the agenda calls for an open-minded 'discussion of the issues' but where you and I know all people have their own private visions to promote"; and (4) Rhetoric A, "words used for true inquiry—no hidden agendas, no predetermined point of view; it is the search for answers in honest pursuit of discovering and appraising our values and views."[26] Rhetoric A is the "highest form" of rhetoric. For VanDeWeghe, accepting that technical communication is concerned with Rhetoric A helps us have "a much healthier and more sophisticated image of the profession of technical communication."[27] Technical communication is a far more creative and artistic activity than many people realize.

Two Theories of Style

Richard Lanham makes a similar argument about style in general, not just technical writing style, in *Analyzing Prose*. He challenges what he calls the C-B-S Theory of Style. This theory builds on the three central values of clarity, brevity, and sincerity and "argues that prose ought to be maximally transparent and minimally self-conscious, never seen and never noticed."[28] According to Lanham, this view holds that *rhetoric* "becomes a dirty word, pointing to superficial ornament on the one hand and duplicity on the other. It becomes, that is, everything which interferes with the natural and efficient communication of ideas."[29]

However, Lanham argues that there are some problems with this C-B-S Theory of Style. Lanham believes that the terms *clarity, brevity,* and *sincerity* are confusing and unhelpful at the very least. There are many ways to be clear. As for brevity, messages vary in length. And sincerity is a matter of determining the right kind of sincerity. Lanham believes "the C-B-S theory of prose style seems not only unhelpful but a violation of our common sense."[30] Lanham suggests that if we were always clear, brief, and sincere "we would not last long in society; in fact we'd probably be locked up."[31]

For Lanham the C-B-S theory of style also suggests a gap between "literature" and "ordinary prose."[32] For instance, the ambiguities of poetry are tolerated; the ambiguities of prose are not. Yet this view is limited. Lanham believes "the history of prose style shows clarity, brevity, and sincerity to be not only rare attainments but even rarer goals. People want to do all kinds of other things through their prose—show off, fool you, fool themselves, run through all the feints and jabs of human sociality."[33]

In contrast to this theory of style, Lanham offers his rhetorical view. For Lanham rhetoric

> is everything in a message which aims not to deliver neutral information but to stimulate action. And rhetoric is the whole domain of ornament, of verbal play, that impulse which always seems to move in on purposive communication like an ornamental border gradually taking over the page in a medieval manuscript.[34]

Lanham suggests that *rhetoric* has been unfairly treated by what he calls the "Newtonian Interlude," the view that dominated from the seventeenth through the nineteenth centuries suggesting "the world was clearly 'out there' and all of us clearly 'in here.'"[35] Writing teachers are still influenced by this Newtonian ideal of verbal behavior.

As an alternative Lanham suggests we expand our notions of the domain of style to include not only purpose but also play and competition. We will recognize more readily how style, which "usually means the game-and-play part of the message" sometimes "really *is* the message and so style becomes content."[36]

Lanham's analysis of style shows us how rhetoric is central to all kinds of writing and that understanding prose style is a way to understand the full range of human behavior. A good grasp of prose style helps us to understand how to be persuasive in all kinds of situations: "The really persuasive people have an instinctive grasp of purpose, game, and play. And it is precisely this mixture that prose style in its fullness always expresses, and that prose analysis can teach."[37] Lanham believes a study of prose style helps us to know "*how* to use information" and to know "*what kind of message* a message is."[38] For Lanham the study of prose style "shows us what is most centrally 'humanistic' about the humanities, what is truly 'liberal' in a liberal education."[39]

Of course, the views of VanDeWege and Lanham about the domain of prose style are supported by many others.[40] To study prose style is to study people and how they comprehend and use information. To study prose style is to map out the strategies necessary to get people to understand or do what you want them to. To study prose style and appreciate fully its domain is to have the most fundamental knowledge about communication.

Why Technical Writing Is Often Ineffective

Because of lingering misconceptions of what technical writing is and because so much writing is done so poorly, many people expect technical writing to be something difficult, if not impossible, to understand, something mechanical and uncreative, in brief, something totally uninteresting. Many people are understandably frustrated with such prose. As consumers, they would rather install or assemble a product on their own than consult the instructions because they have often been victimized by poorly written and overly complex technical documents. Whether these documents are instructions for installing ceiling fans or using a software program, many fail to meet the needs of their readers. Despite numerous complaints, many companies continue to produce totally inadequate documentation for the products they sell. Quality documentation has been and continues to be a low priority for many companies. Yet other companies are also seeing that quality documentation is an essential part of their product. They realize they can dramatically decrease service calls if the documentation is well written (and, of course, if the product is well designed).

Why Some Technical Writing Fails

Lots of technical writing fails for the same reason other kinds of writing fail. Many writers of technical prose do not clearly define their purpose, assess their audience, plan, or know their subject. Many writers lack a command of the basics of grammar, punctuation, and spelling. And many writers who have strengths in all of these areas still write ineffective technical prose because they don't understand the intricacies and demands of style. Effective writing, regardless of the kind of writing, is a complex and difficult skill to master. It simply isn't true that anyone can write, and it is especially not true that just anyone can write clear technical prose.

Problems Readers Have

In a study of writing in organizations, Janice Redish notes five categories of problems that readers have with corporate prose:

1. The document is not organized so that readers can find what they want in the time they are willing to spend looking for it.
2. The content of the document is not what readers need.
3. Readers cannot picture themselves in the text.
4. People have to read a sentence several times to understand what it means.
5. Readers do not know the words the document uses, or the words mean different things to the reader and the writer.[41]

All of these problems are complex and have many causes. However, they all have one thing in common: writers failing to consider the reader. Writing technical prose is a social activity. Effective technical writing should be audience-based rather than subject-based; too much technical prose ignores the audience.

Concerning the fault of poorly organized documents, people read for different purposes at different times. Sometimes people read to learn. Sometimes they read to do. Writers of technical prose must consider how people are using their documents and organize the information in such a way that readers can find the necessary information easily. As for content not being what readers need, writers too often focus on what they know rather than on what the reader needs to know. On readers not picturing themselves in the text, a style can be so formal and abstract that it completely fails, in the reader's terms, to clarify how or when to do a task. About readers having to reread sentences to understand them, too many writers put too many ideas in one sentence or create unnecessarily long and convoluted sentences. Finally, as for readers not understanding the words writers use, writers of technical, bureaucratic, or business prose forget they are often writing to a complex audience, use inappropriate strategies for writing to multiple audiences, and often even assume their peers are familiar with the same jargon when they aren't.[42]

Technical Writing as a Craft

Technical prose, like any other kind of prose, is often ineffective because writers not only fail to remember they are writing for people, no matter how technical the subject, but also because writers take too much for granted about the craft of writing. There's a common notion that anybody can write, but writing is an art, a craft, a skill, a technique. It's hard to do well. As with any other craft, there are many things to learn, and it takes time to become an expert. And writing technical prose presents special challenges.

Chapter Summary

To write well, you must understand the craft of writing. Understanding the craft of writing means, first of all, understanding its basic terminology. Writers should be careful about how they define the terms of their craft, and *style* is one term, in particular, which must be used carefully. Style is our choices of words, clauses, phrases, and sentences. Style is our paragraph pattern choices. Style is our tone—our attitudes toward our subject, our audience, and ourselves—in what we write. Style is who we are and how we reflect who we are, intentionally or unintentionally, in what we write. Technical writing, like all writing, consists of many styles, and choosing a style is closely related to a writer's purpose, subject, audience, context, and many other factors.

Accepting technical communication as a social and rhetorical activity broadens the domain of style in technical writing. It is a domain which allows for freedom, creativity, play, and many other seemingly nontechnical factors. Of course, writers of technical prose may write poorly just as writers of any other type of prose may, but accepting the fact that style in technical writing can be controlled gives technical writers a way to write appropriately for a wide range of audiences on a wide range of subjects for all kinds of purposes and contexts.

Case Study 1: Three Mile Island and Ineffective Communication[43]

Early in the morning of 28 March 1979, Unit 2, a nuclear reactor at Three Mile Island (TMI) in central Pennsylvania, suffered a loss of coolant, severely damaging the reactor and, until it was brought under control, threatening the surrounding population with a potentially fatal radiation disaster. The immediate crisis took five days to spin out and the cleanup took many years, but what was evident from the start was that the officials charged with the oversight of the plant and the safety of the public were ill-prepared to deal with the disaster.

Communication problems had long burdened the Nuclear Regulatory Commission (NRC), one of the federal agencies with jurisdiction over nuclear utility plants in the United States. Poorly written manuals, ignored and misplaced memoranda, and ineptly conveyed technical jargon compounded the problems of working with a dangerous technology. When the system broke down, plant operators made decisions that resulted in a loss-of-coolant accident (LOCA) and partial core meltdown of the nuclear reactor.

Perhaps calling the event a crisis in communication may seem to belittle the very real public safety disaster that unfolded, but in reality the danger was significantly exacerbated by the ineffective and misleading communications that surrounded the accident. Because plant operators did not realize the nature of the events that were playing out, their actions were inappropriate for the situation, and because of their misinterpretation of the problem, their remedies only intensified it. Their training did not address the specific configuration that led to the partial meltdown of the core, so their actions were uninformed, with disastrous and nearly catastrophic results.

The Accident

4:00 A.M.: The night shift workers at Three Mile Island nuclear power plant are unaware that sludge blocking the demineralizing system in Unit 2 (TMI-2) has clogged the pipes, shutting off the feedwater pumps that feed water needed to cool the unit.

4:04 A.M.: Plant operator Craig Faust notices an anomaly on the control panel that seems to indicate that the nuclear core is being flooded with too much water. In fact, emergency valves inadvertently left closed after routine maintenance are actually blocking the flow of water through the system. Based on his misunderstanding of the situation, Faust instructs the system to curtail the flow, allowing the core's coolant water to boil away.

4:08 A.M.: Eight minutes into the accident, operators identify the problems with the feedwater valves and promptly open them only to discover another obstacle. Water flowing through the two newly opened valves bursts a disk in the overflow tank. More than eight thousand gallons of radioactive water and steam escape the containment building and flood into an auxiliary building.

5:18 A.M.: The first series of hypersensitive radiation alarms sounds, indicating excessive levels of radiation in the air of the containment building.

6:48 A.M.: Nearly three hours after the start of the accident, shift supervisor Bill Zewe declares a Site Emergency at TMI-2. All personnel are ordered to leave the auxiliary building.

The Response

In the days following the incident, public officials tried to make sense of the ongoing crisis and formulate a response. Particularly crucial was the decision whether or not to evacuate the population from the surrounding area. Governor Thornburgh finally announced the evacuation of young children and pregnant women but never agreed to a general evacuation.

John F. Ahearne (National Research Council Committee on Risk Perception and Communication chair), seeking to identify the cause of breakdowns in communicating risk to the public, points to several factors that operate to impede open communication.

- Legal issues—Public officials give evasive or dissembling answers on issues that affect the public. Admitting responsibility for hazards or risks may open the directing agencies to lawsuits by concerned citizens.
- Jurisdiction—Regulatory oversight by several agencies makes a determination of responsibility difficult. For example: both the Environmental Protection

Agency (EPA) and the Nuclear Regulatory Commission (NRC) have control over nuclear safety.
- Ambivalence toward the role of the public—Public agencies often discourage and even exclude the public from contributing its input, while claiming to encourage active participation.
- Discourse community expectations—Disparities between scientists and government or industry officials' communication styles compound the problem and almost guarantee miscommunication as an outcome.

All of these factors were at work in the communication disaster that was Three Mile Island. Technicians were uncertain as to the relevance of the questions they were being asked. Metropolitan Edison's weekly press releases were written by engineers in technical and complex terminology. Government sources were in conflict over what information to release and what actions to take. Representatives of industry issued misleading statements or none at all. Reporters, charged with conveying information to the public, were unable to judge adequately the validity of the information they received.

Officials making public pronouncements often tried to put a more benign spin on their bulletins in order to minimize the apparent severity of the crisis.

- An official from the Department of Environmental Resources tried to allay concerns expressed after the utility dumped 400,000 gallons of radioactive waste water into the Susquehanna River by likening the radioactive waste to a carbonated beverage: "Most of it will be dissipated like bubbles in soda pop."
- The announcement of lethal radiation in the plant's turbine room was couched in expressly non-threatening terms: "Plutonium has taken up residence in the building." The rhetoric is meant to suggest a house guest moving in, albeit not a welcome one.
- The accident was not initially called an accident, but rather an "incident," a much less severe event.
- The public was treated to information about the "abnormal evolution" at the power plant, rather than its "breakdown."
- Company officials, unable to say for certain that no human harm would develop from the radiation leak, fell back on the assertion that "we didn't seriously contaminate anybody," somehow not as reassuring as no contamination at all.

Euphemism and understatement are commonplace in an industry whose mandate includes reassuring the public that nuclear power, the same force that leveled Hiroshima and Nagasaki, could be harnessed safely in close proximity to human population centers. So a previous occurrence at a similar plant is referred to as a "transient," accidents are "abnormal events," and incidents that pose a threat to the integrity of the plant have "safety significance."

With the omission of background or contextual information in particular we see the importance of communication during a crisis. Metropolitan Edison declared a "general emergency" in the early morning several hours after the accident began. A "general emergency" is "the highest level of radiation emergency, ...defined as an incident that has the potential for serious radiological consequences to the health and safety of the general public.... [T]he main decision that has to be made after declaration of a general emergency is whether to evacuate the local population."[44] For fear of creating panic in the population or perhaps to put a milder light on the crisis, officials did not offer, nor did reporters request, a clarification of the expression. Without a shared understanding of the message intended by the term, the force of the warning was dissipated.

The inability of industry and government officials to speak with a single voice, to give the public the information it needed to understand the crisis and make informed decisions based on realistic options crippled the communication process. Various groups were operating under different mandates: technicians and scientists tried to understand the technical situation; company officials tried to avoid legal repercussions; public officials tried to avert panic; news media tried to distill the news and sell newspapers or air time; the public tried to cope with an avalanche of conflicting, confusing technical information and advice. The result of the communication breakdown was a loss of confidence in the government and the nuclear power industry, a loss that in some ways has never been recouped.

Questions

1. Defend or refute the following statement: In the case of Three Mile Island, the effects of communication

mismatches among engineers, company officials, government agencies, the news media, and the public superseded the effects of any equipment problems that the plant experienced.

2. How did the declaration of a "general emergency" enable officials to fulfill their responsibilities to inform the public while averting panic? Do you think they handled the announcement appropriately?

3. If you were preparing a plan to be followed in the event of a nuclear accident, what recommendations would you offer that would enable the clear and accurate flow of information among discourse communities?

Questions/Comments for Discussion

1. What is *technical writing* and how does it differ from other kinds of writing?
2. How would you describe your prose style? What habitual choices do you make in your writing?
3. Why is it important to know what style is to achieve an effective style?
4. Why is it important to understand that technical writing is persuasive writing? What does it mean to say technical writing is persuasive?
5. What does the *domain of style* mean? What does the domain of technical writing style mean? What is important for us to recognize about the domain of technical writing style?
6. You've just read about some of the reasons so much technical prose is ineffective. What ineffective technical document have you read recently, and why was it ineffective?
7. Do you think writing is a craft? Why or why not?
8. What do you find to be the most difficult thing about writing?
9. What do you think are some of the special challenges presented by writing technical prose?
10. What do you think are the best ways for you to improve your technical writing skills?
11. Richard Lanham suggests that we need to expand our notion of the domain of style to include not only purpose but also play and competition. In what ways can an expanded notion of style include play and competition?

Exercises

1. Find two definitions of *style* and two definitions of *stylistics*. Also find at least one definition of *expressive* or *emotive* prose. Present these definitions for class discussion.
2. Consider the following two prose examples. Identify the qualities that make them technical. What prose qualities do they share? In what prose qualities do they differ?

Example 1

Defragment your hard disk at regular intervals. MS-DOS 6.0 comes with a defragging utility (called DEFRAG), as do most popular disk utility packages. Third-party and shareware defraggers are also available.

There are various defragging methods to choose from. A full-optimization—one that moves all files to the beginning of the disk and stores them in contiguous sectors—could take a half hour or more, depending on the PC's speed, the hard disk's size, and the fragmentation level. A files-only optimization, which places files only in contiguous areas, is faster, but also slightly less effective. Most defragging utilities will analyze your disk and recommend which form of optimization to use. Defragment your hard drive every couple of months and you'll retain peak performance.[45]

Example 2

To prevent a decrease in your computer's speed, you should defragment (reorganize) your hard disk every couple of months. MS-DOS 6.0 (with a command called DEFRAG) and other popular disk utility packages (programs used for routine operations such as sorting and dumping data) are available to defrag your computer's hard drive. This defragging can be done in one of two ways. One method is full-optimization, which moves all files to the beginning of the disk and stores them in neighboring sections. This method could take a half hour or more, depending on the computer's speed, how scattered the files are, and the hard disk's size. The second method, files-only optimization, which is faster but less effective, places files only in neighboring areas. Most defragging utilities will recommend which method of optimization to use.

3. Choose an essay from one of Lewis Thomas's collections of essays—*The Lives of a Cell, The Medusa and the Snail, Late Night Thoughts on Listening to Mahler's Ninth Symphony,* or *The Fragile Species*—and discuss some of the characteristic stylistic elements of the essay.
4. Find a science book or essay from the sixteenth-, seventeenth-, eighteenth-, or nineteenth-century. Select one section consisting of at least three paragraphs and discuss the expressive elements of the style. Discuss your impressions of the author of this example.
5. Consider this excerpt from a scientific article. What makes the paragraph scientific writing?

    The amount of heterogeneity between studies for the association of prognostic factors with mortality was quantified using the Q-statistic for RDs and the Mantel-Haenszel test for homogeneity for ORs. These two statistics were calculated for individual prognostic factors when the association between mortality and a given factor was reported in two or more studies. A P of less than .05 associated with either statistic indicated that a significant amount of heterogeneity existed among studies. Visual displays of the data were used to assess heterogeneity in mortality rates for specific etiologic agents and the frequency of reported morbid complications. When study outliers were identified, sensitivity analyses were performed to determine whether exclusion of the outlier estimates changed the summary estimate in a clinically significant way.[46]

6. Provide one example each of a science, scientific, and technical writing document. What are some of the elements these three documents have in common? How do they differ?
7. Authors may choose from many possible style elements when crafting nonfiction prose. What are some of the characteristic style elements in this selection?

    If it is true as one prominent Canadian public figure once stated that "the government has no business in the bedrooms of the nation" it is clear that government has surely found a place in every other room in the house. Picture for a moment a middle-aged political scientist standing in front of his bathroom mirror about to scrape the excess hair from his haggard features. "In here," he might be overheard to say with satisfaction, "the government certainly has no place."

But then he might pause and ponder his immediate surroundings. He notices a little note etched on the corner of the mirror, which indicates that this particular piece of glass meets certain government standards. The label on the aerosol can containing his shaving cream warns him, "Do not puncture or incinerate." A government agency somewhere has decided that such warnings are necessary. The same label tells him that the can contains "350 ml" (however much that is!), because still another agency has abolished fluid ounces.... As he brushes his teeth, our hero, by now on the verge of paranoia, remembers that the electricity and the water in the small room are supplied by the public utilities, and that the municipality puts various chemicals in the water to protect him from typhoid, dysentery, tooth decay, and sundry and other public health horrors. The point of this little vignette is that governments in the 1980s touch upon literally every aspect of our day-to-day existence.[47]

8. These two selections, one scientific and one fictional, tell us something about the human heart. Compare the styles of the two paragraphs to explain what makes them so different.

Example 1

This seemingly tireless pump is about the size of a clenched fist and weighs slightly more than half a pound. Lying nearly on its side beneath the breastbone, it rests upon the dome-shaped diaphragm below, and between the right and left lungs, which fold partially around its front side. It has a truncated conical shape, somewhat bulging at one end like a pear, and its tip lies to the left of the breastbone, directed toward the front of the chest. The large vessels, which carry blood to and from the heart, emerge from the portion away from the tip.[48]

Example 2

In the silence that followed, Baby Suggs, holy, offered up to them her great big heart....
"...This is flesh I'm talking about here. Flesh that needs to be loved. Feet that need to rest and to dance; backs that need support; shoulders that need arms, strong arms I'm telling you. And O my people, out yonder, hear me, they do not love your neck unnoosed and straight. So love your neck; put a hand on it, grace it, stroke it and hold it up. And all your inside parts that they'd just as soon slop for hogs, you got to love them. The dark, dark liver—love it, love it, and the beat and beating heart, love that too. More than eyes or feet. More than lungs that have yet to draw free air. More than your life-holding womb and your life-giving private parts, hear me now, love your heart. For this is the prize." Saying no more, she stood up then and danced with her twisted hip the rest of what her heart had to say while the others opened their mouths and gave her the music. Long notes held until the four-part harmony was perfect enough for their deeply loved flesh.[49]

9. Find what you consider to be a weak set of instructions on some type of assembly or installation. Discuss what makes the language of these instructions ineffective. Discuss what improvements you would make in the prose.
10. Consider the following excerpt from a technical document. Why do you think it fails to communicate? Discuss what makes it ineffective.

TACTech's REAL-TIME component information software is able to automatically identify what part is being discontinued, what configuration or assemblies in the system utilize the part, identify the common utilization across multiple equipment impacted and identifies equivalent parts by a secondary source of supply. TACTech is also able to provide all sources of supply that are in production for each microcircuit and discrete component listed on a programs Provisioning Parts List (PPL) and isolates any and all procurement problems in addition to printing out why the line item is a procurement problem. TACTechs' software tools provide component life cycle trend analysis information for parts on a specific bill of material. All faulty part numbers are also identified and are isolated and flagged for assisting in the configuration management process improvement. The software tools of TACTech bridges the gap between logistics and design engineering for new hardware designs, by providing component selection criteria to ensure that the most state-of-the-art type of technology is being designed in. The TACTech software tools can also eliminate obsolete parts from being designed in and assists with the efforts of standardization under QPL and QML qualification levels. Any engineer that uses TACTech is able to obtain a detailed design analysis at the configuration level for whatever program they are currently working on. If utilized properly, the software tools of TACTech can eliminate hundreds of engineering hours in manual effort and supplier research.[50]

Notes

1. John S. Fielden, "'What Do You Mean You Don't Like My Style'?" *Harvard Business Review* 60.3 (May–June 1982): 129.

2. Jacques Barzun, *Simple and Direct: A Rhetoric for Writers,* rev. ed. (New York: Harper & Row, 1985) 67.

3. John Walter, "Usage and Style in Technical Writing: A Realistic Position," *The Practical Craft: Readings for Business and Technical Writers,* eds. W. Keats Sparrow and Donald Cunningham (Boston: Houghton Mifflin, 1978) 94.

4. William Strunk and E. B. White, *The Elements of Style,* 3rd ed. (New York: Macmillan, 1979); Joseph Williams, *Style: Ten Lessons in Clarity and Grace,* 4th ed. (New York: HarperCollins, 1994); and William Zinsser, *On Writing Well: An Informal Guide to Writing Nonfiction,* 5th ed. (New York: Harper Perennial, 1994).

5. James Kilpatrick, *The Writer's Art* (New York: Andrews, McMeel & Parker, 1984) 53.

6. Bill Stott, *Write to the Point: And Feel Better About Your Writing* (Garden City, NY: Anchor, 1984) 131.

7. Stott, p. 131.

8. As quoted in Kilpatrick, p. 52.

9. Carolyn D. Rude, *Technical Editing* (Belmont, CA: Wadsworth, 1991) 222.

10. Dona Hickey, *Developing a Written Voice* (Mountain View, CA: Mayfield, 1993) 1, 23.

11. Hickey, p. 23.

12. Nevin Laib, *Rhetoric and Style: Strategies for Advanced Writers* (Englewood Cliffs, NJ: Prentice Hall, 1993) 21.

13. Rob Day, *How to Write and Publish a Scientific Paper,* 3rd ed. (Phoenix, AZ: Oryx, 1988) 5.

14. *Merriam Webster's Collegiate Dictionary: Deluxe Electronic Edition* CD-ROM, 1995. This source also more broadly defines *stylistics* as "the study of the devices in a language that produce expressive value."

15. Katie Wales, *A Dictionary of Stylistics* (New York: Longman, 1989) 437.

16. Wales, p. 438.

17. Wales, p. 438.

18. David Locke, *Science As Writing* (New Haven: Yale UP, 1992) 86.

19. Locke, p. 85.

20. Locke, p. 85.

21. Locke, p. 85.

22. Daniel Marder, "Technical Reporting Is Technical Rhetoric," *Technical Communication* 25.4 (1978): 11.

23. J. Barnes, ed. *The Complete Works of Artistotle: The Revised Oxford Edition,* vol. 2, trans. W. R. Roberts (Princeton, NJ: Princeton UP, 1984) 2155.

24. Laib, p. 1.

25. Richard VanDeWeghe, "What is Technical Communication? A Rhetorical Analysis," *Technical Communication* 38.3 (1991): 296.

26. VanDeWeghe, p. 296.

27. VanDeWeghe, p. 295.

28. Richard Lanham, *Analyzing Prose* (New York: Scribner's, 1983) 2.

29. Lanham, p. 2.

30. Lanham, p. 3.

31. Lanham, p. 3.

32. Lanham, p. 5.

33. Lanham, p. 4.

34. Lanham, p. 7.

35. Lanham, p. 7.

36. Lanham, p. 9.

37. Lanham, p. 9.

38. Lanham, p. 10.

39. Lanham, p. 10.

40. See, for example, Carolyn R. Miller, "A Humanistic Rationale for Technical Writing," *College English* 40 (1979): 610–17, and Russell Rutter, "History, Rhetoric, and Humanism: Toward a More Comprehensive Definition of Technical Communication," *Journal of Technical Writing and Communication* 21.2 (1991): 133–53.

41. Janice Redish, "Writing in Organizations," *Writing in the Business Professions,* ed. Myra Kogen (Urbana, IL: NCTE/ABC, 1989) 104.

42. Redish, p. 104–114.

43. This case study is based on the following sources: John F. Ahearne, "Telling the Public About Risks." *The Bulletin of the Atomic Scientists* (September 1990): 37–39; Mitchell Stephens and Nadyne Edison, "Coverage of Events at Three Mile Island." *Mass Comm Review* 7.3 (Fall 1980): 3–9; United States, President's Commission on the Accident at Three Mile Island, *Report of the President's Commission on the Accident at Three Mile Island: The Need for Change,* by John G. Kemeny, (Washington: GPO, 1979).

44. Stephens and Edison, p. 7.

45. From Jeff Prosise, "Help: Working Smarter—Care and Feeding of Your Hard Disk," *PC Computing* June 1995: 194–198.

46. Michael J. Fine, Melanie A. Smith, et al., "Prognosis and Outcomes of Patients with Community-Acquired Pneumonia." *JAMA* (10 January 1996): 135.

47. Michael Whittington, *The Canadian Political System: Environment, Structure, and Process.* 4th ed. (Toronto: McGraw, 1987).

48. *Collier Encyclopedia,* p. 743.

49. Toni Morrison, *Beloved* (New York: Knopf, 1987).

50. Michael C. Galuzzi, 77"Post-Production Support through Real-Time Information Software and Design Process Reform." Paper delivered at a conference for the Department of Defense, military personnel, and defense contractors. The conference was provided by U.S. Army Missile Command, Redstone Arsenal, Alabama.

Chapter 2

Reflecting the Style of a Discourse Community

> *Researchers...suggest that style is more than decoration applied to content, even more than a way of shaping meaning. It is a way of belonging to a community, and each community has its own style, not simply its own diction and format but...its own values, emphases, ways of seeing and thinking. And within large communities are smaller units, each with its own perspectives. Do our current texts help students learn how to belong to these communities? Do they teach them how to discover styles and to adjust styles to audiences?"[1]*
> —MARY ROSNER

Chapter Overview

What Is a Discourse Community?

How Does a Discourse Community Affect Style?

Questions to Consider Before Writing to a Discourse Community

Case Study 2: The Space Shuttle **Challenger** and Different Discourse Communities

How Does Style Relate to Subject, Purpose, Scope, Audience, and Other Elements of Writing?

Chapter Summary

Questions/Topics for Discussion

Exercises

Notes

What Is a Discourse Community?

A *discourse community,* simply defined, is a group of people who commonly associate with each other. Typical discourse communities are students in the same class, students in the same major, members of a committee, members of a professional society, members of a local civic or business organization, members of a particular profession, or practitioners of a hobby. People in the same discourse community share a special language, experiences, basic knowledge, expectations, values, methodologies, idiosyncrasies, and goals. Discourse communities may be described broadly or narrowly, globally or locally. Perhaps at the broadest level are the discourse communities of different cultures, for example, middle eastern or third world. Each nation and each ethnic group within these nations are other examples of broad discourse communities.

The Professions and Trades as Discourse Communities

Other broad discourse communities are those of the professions and trades. Doctors, nurses, dentists, lawyers, architects, accountants, engineers, scientists, programmers, educators and many other professionals have specific rhetorics and styles, and there are many rhetorics and styles within each of these professions. The same is true of any of the trades such as machinists, automobile mechanics, plumbers, or electricians.

Fields of Study as Discourse Communities

All fields of study—from astronomy to zoology in the sciences or from art to Zen in the humanities—are specialized communities as well. Each field of study has a complex set of criteria for determining what constitutes knowledge in that field. College students are, of course, a discourse community distinct from others in society and represent para-professionals in the process of learning about their specialized communities. As they study their disciplines, they acquire the particular ways of knowing and performing in their fields. Students majoring in literature acquire the terminology—from *alliteration* to *zeugma*—to discuss literary style intelligently. They learn how to discuss a text—its author, themes, imagery, characters, irony, symbolism, structure, and figurative language. They learn how to take various critical approaches to texts ranging from aesthetic to linguistic to psychological to philosophical to rhetorical. Similarly, science majors acquire their terminology, study the major theories and laws of their particular science, learn and practice the rigorous methodology of the scientific method, and, depending on the science, conduct laboratory experiments. Law students study case histories, acquire the specialized language of the legal community, study the techniques of their particular legal specialty, learn research skills, and come to understand the complex strategies for effectively practicing law.

Hobbyists as Discourse Communities

Hobbyists are another broad area of discourse communities also important for considerations of style. Hobbyists range from experts to people who merely dabble. Aircraft enthusiasts, amateur astronomers, bird watchers, card players (from bridge to poker), chess players, gardeners, hackers, motorcyclists, painters, photographers, potters, stamp collectors—these are just some of the thousands of hobbyists. These hobbyists, perhaps more easily than the other groups mentioned, illustrate the demands of knowing a subject. Anyone who has seriously pursued a hobby knows it often takes many years of practice, reading, and discussion to know a hobby well. And many things are taken for granted by hobbyists in terms of technical vocabulary, basic knowledge and skills, common assumptions and values. To be knowledgeable about a hobby is to devote a lifetime to becoming a member of a distinct discourse community.

See the following description of the characteristics of hackers:

Hackers is a term applied to broad group of people many of whom have shared interests.

General Appearance

Intelligent. Scruffy. Intense. Abstracted. Surprisingly for a sedentary profession, more hackers run to skinny than fat; both extremes are more common than elsewhere. Tans are rare.

Dress

Casual, vaguely post-hippie; T-shirts, jeans, running shoes, Birkenstocks (or bare feet). Long hair, beards, and mustaches are common. High incidence of tie-dye and intellectual or humorous 'slogan' T-shirts (only rarely computer related; that would be too obvious).

Hackers dress for comfort, function, and minimal maintenance hassles rather than for appearance (some, perhaps unfortunately, take this to extremes and neglect personal hygiene). They have a very low tolerance of suits and other 'business' attire; in fact, it is not uncommon for hackers to quit a job rather than conform to a dress code.

Female hackers almost never wear visible makeup, and many use none at all.

Reading Habits

Omnivorous, but usually includes lots of science and science fiction. The typical hacker household might subscribe to *Analog, Scientific American, Co-Evolution Quarterly,* and *Smithsonian.* Hackers often have a reading range that astonishes liberal arts people but tend not to talk about it as much. Many hackers spend as much of their spare time reading as the average American burns up watching TV and often keep shelves and shelves of well-thumbed books in their homes.

Other Interests

Some hobbies are widely shared and recognized as going with the culture: science fiction, music medievalism (in the active form practiced by the Society for Creative Anachronism and similar organizations), chess, go, backgammon, war games, and intellectual games of all kinds. (Role-playing games such as Dungeons and Dragons used to be extremely popular among hackers but they lost a bit of their luster as they moved into the mainstream and became heavily commercialized.) Logic puzzles. Ham radio.

Other interests that seem to correlate less strongly but positively with hackerdom include linguistics and theater teching.

Education

Nearly all hackers past their teens are either college-degreed or self-educated to an equivalent level. The self-taught hacker is often considered (at least by other hackers) to be better motivated, and may be more respected, than his school-shaped counterpart. Academic areas from which people gravitate into hackderdom include (besides the obvious computer science and electrical engineering) physics, mathematics, linguistics, and philosophy.

Communication Style

Though hackers often have poor person-to-person communication skills, they are as a rule quite sensitive to nuances in language and very precise in their use of it. They are often better at writing than at speaking.[2]

To succeed in any profession, trade, field of study, or hobby, you have to know the special subject matter, jargon, expected manner of presentation, and methods of inquiry associated with the area. Knowing the rhetoric and style of an area is to know how knowledge is acquired in the area, the ways things are done, how to talk to others to coordinate efforts, learn new skills, and make decisions. Technical knowledge takes a long time to acquire, and knowing how to express this technical knowledge appropriately is a complex skill. Considering the difficulties, no one should be surprised that technical people resent outsiders misusing technical terminology or misunderstanding basic assumptions.

Because rhetorical situations vary so much from culture to culture, profession to profession, trade to trade, field of study to field of study, or hobby to hobby, considerations concerning style must vary a great deal too. The point here is more than the old adage, "Consider your audience before you write." The point here is that the discourse community heavily influences technical writing style just as it does any other kind of writing style.

How Does a Discourse Community Affect Style?

Politics and Style

Styles are shaped in large part by the discourse communities in which and for which documents are written. Recognizing this fact is to recognize the reality of writing about subjects for people. Whether or not to put something in writing, what to write, how to write it effectively—all of these decisions are part of the politics of style. Richard Lanham tells us: "Learn the rules and then, through experience, train your intuition to apply them effectively. *Stylistic* judgment is, last as well as first, always *political* judgment."[3] Technical writing is not done in a vacuum. It's a social activity involving interpreting complex information for people with all kinds of knowledge levels, who need to know, use, or act on this information. As a social activity, it's subject to all kinds of influences, biases, and distortions.

On the Preferences of Naval Officers

Several recent surveys of people writing in various professions offer strong support for the view that politics may take precedence over rules in various discourse communities. For example, in one study 262 naval officers were asked to read two memos with similar content, one written in a high-impact style and the other in a bureaucratic style.[4] The high-impact memo used concrete language, active verbs, first- and second-person pronouns, and stated the bottom line up front. The bureaucratic memo used long and complex sentences, passive verbs with implied subjects, abstract language, no personal pronouns, and buried the bottom line.

Although the officers read the high-impact memo more efficiently, they preferred the bureaucratic style because it reflects the language customs of the Navy. Also, many of the officers felt that bureaucratic "businesses" can be advantageous. One officer commented that the bureaucratic style helps you "to cover your stern, and that's smart, shrewd writing if you want to survive in the Navy."[5]

On the Writing of Scientists and Engineers

Another study shows that many scientists prefer a simpler style of writing but are reluctant to write more simply for fear of losing credibility or esteem in the eyes of their peers. Although scientists may prefer to read a simplified style of writing, they will have a higher regard for the author of a more complex style. In crafting a more complex style of writing, the scientist sometimes ignores audience needs in exchange for personal or political considerations.[6]

A study of the prose of engineers and scientists found at least twelve common characteristics: long sentences, passive voice, a preference for nouns instead of verbs, a preference for *which* instead of *that,* under-punctuated technical prose, underplayed key ideas, a tendency to personify the inanimate ("the missile's nose cone"), a preference for inflated or wordy prose, multiple adjectives, a preference for a Latinized vocabulary ("It would have been *difficult* to *accomplish* a *similar objective* with the older techniques.") instead of the Anglo-Saxon ("It would have been *hard* to *do* the *same thing* with the older techniques."), inapplicable logic, and verbosity and conciseness.[7] As for the last quality of verbosity and conciseness, paradoxically, technical prose of scientists and engineers may be verbose and concise at the same time. Multiple adjectives, for example, are a form of concise, science shorthand. The conciseness of much technical prose may be seen by simply trying to rewrite in lay terms a short passage of technical prose. As George Schindler notes in his study, "any rewrite for a nonspecialist reading audience will, on the average, demand more words.... It is quite apparent that, despite occasional wordiness, much technical prose is really an extreme form of shorthand developed by specialists for communicating with other specialists."[8]

Schindler believes the specialist often confronts the problem of intelligibility. Sometimes being misunderstood is not a fault; sometimes it is. If a layperson has trouble with a specialist's writing

> Such writing is not necessarily poor merely because it is so judged by people for whom it was not intended. It becomes poor when the writer intends to write for a broader

audience, but does not depart from the habits that have served him well within his field of science or engineering. Long sentences, heavy qualifications, "shorthand" terms, excess verbiage, Latinized words, and passive constructions become insuperable obstacles to the reader. The prose may possibly be very good for the specialist, but the general reader makes no allowances. He is usually not required to read it, so if he cannot read it easily, he simply quits.[9]

Schindler notes that the twelve characteristics he discusses can become a fault in the writing within a single profession: "The writer quite properly reacts to the pressure toward conformity with the writing practices of his group, but he errs if he succumbs abjectly."[10] This is an important point. We have to recognize that writing does not exist in a vacuum. All kinds of factors—specifically, all kinds of politics—influence what writers write. Yet in reacting to the politics of the writing situation, a writer can overreact and violate sensible points of style. For example, writers can spend so much time qualifying and defining that their sentences become so loaded with subordinate phrases and clauses that even their peers cannot read them. Specialists may also develop a style so spare, relying heavily on figures and tables, for instance, that no one can understand it. Schindler cautions us about judging these stylistic preferences. His point is that these preferences are commonplace among engineers and scientists, and we must have a better understanding of why these preferences are so widespread.

On Academic Prose

Of course, scientists and engineers hold no special claim to overqualifying what they write. Far too many in the discourse community of academicians overqualify, hedge, or hide what they want to say and in turn create largely dismal writing. Patricia Limerick argues, "Colleges and universities are filled with knowledgeable, thoughtful people who have been effectively silenced by an awful writing style, a style with its flaws concealed behind a smokescreen of sophistication and professionalism."[11] Limerick comments, "When you write typical academic prose, it is nearly impossible to make a strong, clear statement. The benefit here is that no one can attack your position, say you are wrong or even raise questions about the accuracy of what you have said, if they cannot tell what you have said."[12] Limerick believes academicians are creating a "cult of obscurity" and hiding "behind the idea that unintelligible prose indicates a sophisticated mind."[13]

As an example of obscure prose, she cites the following sentence:

> If openness means to "go with the flow," it is necessarily an accommodation to the present. That present is so closed to doubt about so many things impeding the progress of its principles that unqualified openness to it would mean forgetting the despised alternatives to it, knowledge of which makes us aware of what is doubtful in it.[14]

Limerick indicts the poor quality of writing of many academicians, not the complexities of their disciplines and specialized terminology.

While we must follow the style practices of our discourse community, doing so should not be seen as a license to write obscurely. Limerick uses an analogy pertaining to carpenters to suggest what all of us must keep in mind. If a carpenter makes a door for a cabinet and it does not fit, the carpenter "removes the door and works on it until it fits. That attitude, applied to writing, could be our salvation. If they thought more like carpenters, academic writers could find a route out of the trap of ego and vanity. Escaped from that trap, we could simply work on successive drafts until what we have to say is clear."[15]

Whether or not you agree completely with Limerick's judgement of most academic prose (there are pressures within the academic community to be obscure), her critique and the studies summarized here show that, stylistically, people in many discourse communities must weigh many factors before determining how they communicate to others. And, more important, writers in all discourse communities often decide to use whatever strategies are most commonly accepted by their peers rather than be governed by any universal prescription for clarity, accuracy, and brevity. Stated another way, what constitutes good style is relative to every discourse community.

Communications fail when writers within a discourse community misunderstand the stylistic demands of that community. Writers often fail to use words with the proper nuances or may use an arrangement deemed unacceptable by a community that demands a certain approach or may not measure up to a subject's complexity due to a naive use of the expected technical terms.

Perhaps one of the biggest challenges for writers is writing to audiences in another discourse community. This is why, for example, many experts have so much difficulty explaining a complex subject to a lay audience. These experts cannot take for granted the many factors they take for granted in addressing an audience of their peers. Technical language enables these experts to take shortcuts with other experts, but when communicating with a lay audience many of these shortcuts are no longer available. Technical communicators often find themselves caught between the subject expert and the lay audience. A special skill of the technical communicator is the ability to shape prose so that a combination of audiences can understand it, use it, or act on it.

Questions to Consider Before Writing to a Discourse Community

Before writing your first word in a document intended for a member of your discourse community or another discourse community, you need to ask some basic questions. Some of these questions are, of course, basic audience analysis questions. Other questions remind you that you are not writing about a subject in a vacuum. Your technical prose is for other people and will have consequences.

Questions to help you consider your discourse community include:

- What is your relationship to the audience: Peer? Subordinate? Supervisor?
- Is the audience expecting your document? Motivated to read it?
- What will your audience use the document for?
- How knowledgeable is your audience about the subject?
- Are you using any technical terms or other words the audience may not know?
- Are you certain all of your technical terms are used correctly?

- Have you fully considered the connotations as well as the denotations of the nontechnical terms you are using?
- Are you assuming any basic knowledge (theories, assumptions, ideas) you should not assume the audience knows?
- Are there special preferences in this discourse community you need to give particular attention to?
- Is your audience comfortable with the level of formality in your document?
- Is your audience comfortable with your other stylistic preferences?
- Are there particular political elements (tensions, relationships among people, histories, backgrounds) you need to pay special attention to?
- Is this the best time to present this document or would a later time be better?
- What will the results be if your document fails? Succeeds?
- Have you considered all of the other important elements necessary for an effective document in this particular discourse community?

This list is only a beginning to aid you in preparing to write for your community or another community. Many other questions must also be asked and answered if you are to communicate effectively.

Case Study 2: The Space Shuttle Challenger and Different Discourse Communities

Industrial and corporate history is full of incidents in which members of one discourse community have failed to consider the special considerations of another community. Companies are continually marketing products that do not meet the needs of consumers, are poorly designed, or have poor documentation. The poor quality of so much computer software documentation is one example of a company and its programmers and managers totally ignoring or misjudging the needs of the small business or home personal-computer user.

The space shuttle *Challenger* accident of 28 January 1986 has been used to demonstrate many problems from the political to the rhetorical to the technological. There are many possible explanations for what finally caused the disaster that killed seven people only 73 seconds after liftoff. One key explanation is that a major communication gap occurred between management at one level and engineers at another. In effect, this incident shows dramatically what can happen when two very different discourse communities fail to communicate with each other and when one community distorts the values of another. We will look at a case study about this disaster in detail to demonstrate how members of two discourse communities can misunderstand each other with, in this instance, horrible consequences.

Scenario

On the morning of 28 January 1986, the Space Shuttle *Challenger* sat poised on Kennedy Space Center's (KSC) Launch Complex 39B ready for her tenth flight into space, designated Flight STS-51L.[16] Just three miles away, a large crowd of distinguished visitors and anxious 51L family members watched patiently as the large digital clock methodically ticked down the remaining minutes before launch.

Inside the *Challenger* were Commander Francis Scobee; Pilot Michael Smith; mission specialists Judith Resnik, Ellison Onizuka, and Ronald McNair; payload specialist Gregory Jarvis; and Christa McAuliffe, a high school teacher chosen for the flight in a nationwide competition. The banter between them was calm and professional. *Challenger* was ready to fly them safely into space and they were ready to go.

The early morning of 28 January was a cold one. The temperature had not risen above 36°F; a full 15 degrees colder than any previous Shuttle launch (V1, 19). The NASA Ice Team was concerned about the ice that had accumulated on the launch pad over the blustery 17° night. The engineers of Morton Thiokol Inc. (MTI), the makers of *Challenger's* Solid Rocket Boosters (SRBs), were equally concerned about the low early morning temperature.

In Utah, MTI Staff Engineer Roger M. Boisjoly was attending a pre-launch teleconference the evening before the launch with MTI representatives and NASA officials at both Marshall Space Flight Center, Huntsville, Alabama and KSC. Boisjoly was the most vocal opponent of the *Challenger* liftoff on this chilly January morning. From tests conducted at MTI's Brigham City Booster Plant (Utah), the MTI engineer knew the facts were on his side. He would later state:

> The conclusion was we should not fly outside of our data base, which was 53 degrees. Those were the conclusions. And we were quite pleased because we knew in advance, having participated in the preparation, what the conclusions were, and we felt very confident with that. (V1, 90)

Boisjoly was right. The recommendation from the Brigham City engineering team was included in a chart (see Figure 2.1) presented by MTI Vice President of Engineering Robert K. Lund during the early morning teleconference. However, a few hours later, KSC officials received a telefax from MTI's Vice President of Space Booster Programs Joe C. Kilminster clearing *Challenger* for launch (see Figure 2.2). Much to the dismay of Roger Boisjoly, MTI management gave the OK to launch *Challenger*. He later testified about the teleconference and MTI's final decision: "There was never one positive, pro-launch statement ever made by anybody. There have been some feelings since then that folks have expressed that they would support the decision, but there was not one positive statement for launch ever made in that room" (VI, p.90).

At 11:38 A.M. EST, after more than two hours of hardware related delays, *Challenger* lifted off from Pad 39B. After just 73 seconds, a fireball erupted to engulf and destroy *Challenger* and her crew of seven.

MTI Assessment of Temperature Concern on SRM-25 (51L) Launch

- Calculations show that SRM-25 O-rings will be 20° colder than SRM-15 O-rings
- Temperature data are not conclusive on predicting primary O-ring blow-by
- Engineering assessment is that:
  - colder O-rings will have increased effective durometer ("harder")
  - "Harder" O-rings will take longer to "seat"
    - More gas may pass primary O-ring before the primary seal seats (relative to SRM-15)
      - Demonstrated sealing threshold is 3 times greater than 0.038
  - If the primary seal does not seat, the secondary seal will seat
    - Pressure will get to secondary seal before the metal parts rotate
      - O-ring pressure leak check places secondary seal in outboard position which minimizes sealing time
- MTI recommends STS-51L launch proceed on 28 January 1986
  - SRM-25 will not be significantly different from SRM-15

Joe C. Kilminster

Joe C. Kilminster, Vice President
Space Booster Programs
MORTON THIOKOL, INC.
Wasatch Division

Recommendations:

- O-ring temp must be ≥ 53°F at Launch
  Development motors at 47° to 52°F with putty packing had no blow-by SRM 15 (the best simulation) worked at 53°F
- Project ambient conditions (temp & wind) to determine launch time

FIGURE 2.1   Recommendations from the Engineering Team

FIGURE 2.2   MTI Assessment of Temperature Concern

In Utah, Roger Boisjoly sat stunned. At KSC's viewing stand, Christa McAuliffe's parents searched in vain for the orbiter to break free of the huge white cloud. *Challenger* never reappeared.

Background

In June of 1986, the President's commission investigating the *Challenger* accident concluded that below-freezing temperatures affected the reliability of 51L's SRB field joints. The field joints are the points where SRB segments are connected—five per SRB. Each field joint contains two rubber O-rings which seal the joint from the high pressure generated by the solid rocket fuel as it burns within the SRB. The commission determined that the cold weather caused both of the O-rings in the right aft SRB field joint to become hard and brittle. This condition allowed hot gases to escape the right SRB and ignite the highly combustible liquid oxygen and liquid hydrogen fuels within the Shuttle's External Fuel Tank.

The problems with SRB O-rings were well documented long before the liftoff of *Challenger*. Both MTI and NASA engineers were well aware of the potential for loss of life and vehicle should one of these field joints fail. Previous missions had experienced what is called "blow-by" when hot gases had blown through the O-rings sealing a field joint. Flight STS-51C (21 January 1985) sustained blow-by on two different field joints (V2, H-19). And as far back as STS-2 (12 November 1981), erosion of field joint O-rings was a documented problem (V2, H-19).

In 1979, a full two years prior to the first Shuttle mission (STS-1, 12 April 1981), the NASA Chief of Marshall's Solid Motor Branch John Q. Miller questioned the reliability of the SRB field joints (V1, 236). See Miller's memo (Figure 2.3).

The most damning evidence uncovered by the presidential commission confirming that MTI was aware of the problems with the field joint O-rings came with the release of Roger Boisjoly's memo to MTI's Vice President of Engineering in July of 1985. He makes it known that if the situation is not soon remedied "we stand in jeopardy of losing a flight along with all the launch pad facilities" (V1, 249–250). See Boisjoly's memo (Figure 2.4).

Dated 9 August 1985, and in total agreement with Boisjoly's memo, another MTI memo from Brian G. Russell, MTI Manager of the Solid Rocket Motor (SRM) Ignition System, discusses the suspect field joints. Unlike Boisjoly's internal memo, Russell's memo is addressed to a senior NASA official at Marshall Space Flight Center, James W. Thomas, Jr. (V5, 1568). See Russell's memo (Figure 2.5).

Boisjoly was also concerned with O-ring failures at low temperatures. According to the Flight Readiness Review (FRR) for STS-51E, dated 31 January 1985, evidence of cold weather effects on the performance of field joint O-rings was well documented (V2, H-19, H-31, H-59).

If evidence existed of faulty SRB field joints, why were the memos from Miller, Boisjoly, and Russell ignored by both NASA and MTI management? (These memos are not the only evidence presented during the presidential commission, but they are the most damning.)

Investigation

A close inspection of each of the three memos reveals clear and concise language concerning the problems with the SRB field joints.

The Miller Memo

Speaking as the Chief of the NASA Solid Motor Branch, John Q. Miller states that MTI's "position regarding design adequacy of the clevis [the aft side of the field joint] joint to be completely unacceptable." He further states the reasons for this inadequacy: "violates industry and Government O-ring application practices"; "excessive tang [the forward side of the field joint]/ clevis movement"; "the clevis joint secondary O-ring seal has been verified by tests to be unsatisfactory." There are no ambiguous words here. Miller is thorough with his analysis of the problems. By choosing the words *unacceptable, violates, excessive,* and *unsatisfactory,* he is clearly negative in his message.

The Boisjoly Memo

From the opening paragraph, Roger Boisjoly's intent is clear: "This letter is written to insure that [MTI] management is fully aware of the seriousness of the current O-ring erosion problem...." He also states that the MTI position to fly while studying the O-ring problem has been "mistakenly accepted" and that the "result [of a field joint failure] would be a catastrophe of the highest order—loss of human life." During the commission testimony, Boisjoly admitted that he personally believed what he was

National Aeronautics and Space Administration

NASA

George C. Marshall Space Flight Center
Marshall Space Flight Center, Alabama
35812

EP25 (79-13) January 19, 1979

TO: EE51/Mr. Eudy
FROM: EP25/Mr. Miller
SUBJECT: Evaluation of SRM Clevis Joint Behavior

As requested by your memorandum, EE51 (79-10), Thiokol documents TWR-12019 and letter 7000/ED-78-484 have been revaluated. We find the Thiokol position regarding design adequacy of the clevis joint to be completely unacceptable for the following reasons:

    a. The large sealing surface gap created by excessive tang/clevis relative movement causes the primary O-ring seal to extrude into the gap, forcing the seal to function in a way which violates industry and Government O-ring application practices.

    b. Excessive tang/clevis movement as explained above also allows the secondary O-ring seal to become completely disengaged from its sealing surface on the tang.

    c. Contract End Item Specifications, CPW1-2500D, page I-28, paragraph 3.2.1.2. requires that the integrity of all high pressure case seals be verifiable; the clevis joint secondary O-ring seal has been verified by tests to be unsatisfactory.

Questions or comments concerning this memorandum should be referred to Mr. William L. Ray, 3-0459.

John Q. Miller

John Q. Miller
Chief, Solid Motor Branch

cc:

SA41/Messrs. Hardy/Rice
EE51/Mr. Uptagrafft
EHO2/Mr. Key
EPO1/Mr. McCool
EP42/Mr. Bianca
EP21/Mr. Lombardo
EP25/Mr. Powers
EP25/Mr. Ray

FIGURE 2.3 The Miller Memo

Company Private

MORTON THIOKOL

Wasatch Division

Interoffice Memo

31 July 1985
2870: FY86:073

TO: R. K. Lund
Vice President, Engineering

CC: B. C. Brinton, A. J. McDonald, L. H. Sayer, J. R. Kapp

FROM: R. M. Boisjoly
Applied Mechanics—Ext.3525

SUBJECT: RM O-Ring Erosion/Potential Failure Criticality

This letter is written to insure that management is fully aware of the seriousness of the current O-ring erosion problem in the SRM joints from an engineering standpoint.

The mistakenly accepted position on the joint problem was to fly without fear of failure and to run a series of design evaluations which would ultimately lead to a solution or at least a significant reduction of the erosion problem. This position is now drastically changed as a result of the SRM 16A nozzle joint erosion which eroded a secondary O-Ring with the primary O-Ring never sealing.

If the same scenario should occur in a field joint (and it could), then it is a jump ball as to the success or failure of the joint because the secondary pressurization. The result would be a catastrophe of the highest order—loss of human life.

An unofficial team (a memo defining the team and its purpose was never published) with leader was formed on 19 July 1985 and was tasked with solving the problem for both the short and long term. This unofficial team is essentially nonexistent at this time. In my opinion, the team must be officially given the responsibility and the authority to execute the work that needs to be done on a non-interference basis (full time assignment until completed).

It is my honest and very real fear that if we do not take immediate action to dedicate a team to solve the problem with field joint having the number one priority, then we stand in jeopardy of losing a flight along with all the launch pad facilities.

R. M. Boisjoly
R. M. Boisjoly

Concurred by:

J. R. Kapp
J. R. Kapp, Manager
Applied Mechanics

COMPANY PRIVATE

FIGURE 2.4 The Boisjoly Memo

MORTON THIOKOL

Wasatch Division

9 August 1985
E150/BGR-86-17

Mr. James W. Thomas, Jr., SA42
George C. Marshall Space Flight Center
National Aeronautics and Space Administration
Marshall Space Flight Center, AL 35812

Dear Mr. Thomas:

Subject: Actions Pertaining to SRM Field Joint Secondary Seal

Per your request, this letter contains the answers to the two questions you asked at the July Problem Review Board telecon.

1. *Question:* If the field joint seconday seal lifts off the metal mating surfaces during motor pressurization, how soon will it return to a position where contact is re-established?

Answer: Bench test data indicated that the o-ring resiliency (its capacity to follow the metal) is a function of temperature and rate of case expansion. MTI measured the force of the o-ring against Instron plattens, which simulated the nominal squeeze on the o-ring and approximated the case expansion distance and rate.

At 100°F. The o-ring maintained contact. At 75°F. The o-ring lost contact for 2.4 seconds. At 50°F the o-ring did not re-establish contact in ten minutes at which time the test was terminated.

The conclusion is that secondary sealing capability in the SRM field joint cannot be guaranteed.

2. *Question:* If the primary o-ring does not seal, will the secondary seal seat in sufficient time to prevent joint leakage?

Answer: MTI has no reason to suspect that the primary seal would ever fail after pressure equilibrium is reached, i.e., after the ignition transient. If the primary o-ring were to fail from 0 to 170 milliseconds, there is a very high probability that the secondary o-ring would hold pressure since the case has not expanded appreciably at this point. If the primary seal were to fail from 170 to 330 milliseconds, the probability of the secondary seal holding is reduced. From 330 to 600 milliseconds the chance of the secondary seal holding is small. This is a direct result of the o-ring's slow response compared to the metal case segments as the joint rotates.

Please call me or Mr. Roger Boisjoly if you have additional questions concerning this issue.

Very truly yours,

Brian G. Russell

Brian G. Russell, Manager
SRM Ignition System

BGR/co

cc:
L. Wear, SA42, E. Skrobiszewski, SA49, I. Adams, MTI/MSFC

bcc:
J. Kilminster, A. McDonald, R. Ebeling, J. Elwell, B. Brinton, A. MacBeth, R. Boisjoly, A. Thompson, S. Stein

FIGURE 2.5 **The Russell Memo**

writing when he typed the final sentence: "We stand in jeopardy of losing a flight along with all the launch pad facilities." Unlike Miller's more technical memo, Boisjoly's memo reflects a personal approach to the problem. His appeal is humanistic and attempts to garner a sympathetic response from MTI management. By stressing the loss of human life, Boisjoly includes real, live humans—a point often overlooked in the aerospace world when dealing with hardware problems.

The Russell Memo

Brian G. Russell's memo attempts to answer questions presented by NASA's James W. Thomas Jr. concerning the secondary O-ring sealing surface (the outer of the two field joint O-rings). Although technical in nature, two of Russell's statements represent what should be MTI's position concerning field joint erosion.

First, Russell states that "secondary sealing capability in the SRM field joint cannot be guaranteed." This statement is especially relevant due to his finding that "at 50°F the O-ring did not re-establish contact in ten minutes…" Second, Russell states that "from 330 to 600 milliseconds [after ignition], the chance of the secondary seal holding is small." In other words, if there is initial blow-by of the secondary sealing O-ring, the chance of the O-ring resealing is small. It is ironic that Russell mentions the low temperature problem and the resealing of the secondary O-ring problem in the same memo. These two factors are exactly the cause of the failure of *Challenger's* left aft field joint.

Conclusion

There are three facts that are certain when looking at the events of the *Challenger* disaster:

1. MTI knew of the field joint problems, both with design and temperature;
2. NASA also knew of the field joint problems;
3. The warnings were ignored and *Challenger* and her crew were lost.

The presidential commission concluded that the low temperature was the ultimate cause for the loss of *Challenger*. Hence, the field joints have been redesigned with an added heater element to keep the field joints a constant 80°F.

The commission also concluded that communication, both internal and external, was faulty. They wrote: "Communication during the launch decision process was inadequate"; furthermore, "Key individuals' objections to launch were not registered to the top program officials" (V2, I-17).

Some Possible Explanations

Why were key individuals' objections not registered to the top program officials? The three memos seem clear enough, but why did a communication problem occur despite these and other communications warning about problems?

One possible explanation is that the writing style of the engineers' memos discussed here was not framed properly for higher level management. Dorothy Winsor discusses Brian Russell's memo to James Thomas. Winsor agrees the memo clearly shows "that the time necessary for the secondary O-ring to return to sealing position was a function of temperature. The lower the temperature, the longer the secondary O-ring took. Moreover, at temperatures below 50 (degrees) F, the O-ring did not return to position at all."[17] Winsor wonders if the experts were simply ignored by bureaucrats or if some other explanation is possible. Winsor suggests "that the meaning of these data was not obvious for those who read this memo when the memo was first written."[18] Winsor says the writer of this memo "passed on the information as though the meaning of the data was obvious and thus gave his reader no help in interpretation."[19]

Winsor critiques the memo's lack of impact. For example, it has a weak closing sentence: "The conclusion is that secondary sealing capability in the SRM field joint cannot be guaranteed." The writer could have written something like "The conclusion is that the secondary seal is not effective at temperatures below 50 (degrees) F, and thus the joint is highly vulnerable to catastrophic failure at such temperatures." Or the writer could have framed the memo differently, emphasizing that the temperature data was his reason for writing the memo. Perhaps framing the information differently would have helped the reader to interpret the information differently. The question and answer format implies that the writer produced the data only because the reader asked for them, not because the data were important in themselves.

Yet we can go beyond the wording of these memos to a deeper problem. The communication problems here are chiefly the result of one discourse community—management (in this case, consisting of many levels of management)—manipulating the knowledge framework of another discourse community—engineers.

Paul Dombrowski identifies two ways management reconceptualized the O-ring problem. First, an anomaly—charring of some O-rings—gradually was considered no longer an anomaly. Charring became acceptable after a succession of completed space shuttle flights prior to the *Challenger* flight. "The very success of these flights was taken to demonstrate that exposure of the O-rings to the exhaust gases was not a serious concern and could and should be tolerated."[20] If the mission was successful, and if an O-ring broke but the break was not bigger than before, then this became evidence that charring should be tolerated. "Thus what was never to happen came to be permissible, even being taken as an indication of safety rather than danger." Stated in the report: "But rather than identify this condition as a joint that didn't seal, that is, a joint that had already failed, NASA elected to regard a certain degree of erosion or blow-by as 'acceptable'"[21]

The second major reconceptualization occurred when management increasingly questioned engineers at both MTI and NASA "pressing them to prove that their reservations involved certain peril to the mission."[22] Engineers were forced finally to perform a complete flip-flop on the way they normally handled assumption and the burden of proof. According to Dombrowski, engineers at both MTI and NASA "found themselves in the situation of being asked by management to prove absolutely and certainly that the *Challenger* flight would end in disaster."[23] Engineers "were totally thrown into confusion and frustration by the change of assumption and conceptualization."[24] Traditionally, the job of these engineers was "to prove that the vehicle *would* fly safely." Traditionally, management is supposed to scrub the flight if engineers could not assure shuttle safety. At the teleconference, however, the evening before the launch, "When engineers could not prove that *Challenger would not* certainly go up in flames, management took their implicit assumption that *Challenger* would fly as thus confirmed and unqualified."[25]

Boisjoly comments on how the situation was reversed from what engineers were normally expected to do: "This was a meeting where the determination was to launch, and it was up to us to prove beyond a shadow of a doubt that it was not safe to do so. This is in total reverse to what the position usually is in a preflight conversation or flight readiness review. It is usually exactly opposite to that."[26]

Even though they were ignored, these engineers continued to object to approving the launch. A casual observer may think the engineers made their objections all the way to the top. Their objections were stated clearly, but, significantly, their objections were interpreted differently by others. The prior conceptualizations of management show how the meaning of evidence can be interpreted differently by another discourse community, and that this discourse community of managers can even "refuse to recognize conflicting evidence."[27]

The space shuttle *Challenger* incident dramatically illustrates how reading and writing styles are entrenched in discourse communities. Simply stated, engineers have their way of knowing and acting on that knowledge. Managers have a different way of knowing and acting on that knowledge. You might think the memos provided here would be equally clear to everyone, interpreted the same way by everyone who reads them. But how we read as well as how we write is shaped by the communities we belong to. This brief overview of the *Challenger* episode should remind us of how, as writers, we are tied to our communities.

Questions

1. In what ways is this case study a lesson in how different discourse communities—in this case engineering and upper management as well as government agencies and contractors—interpret information differently?
2. Do you think the three memos by Miller, Boisjoly, and Russell in Case Study 2 are clear in their warnings about the O-rings? Why or why not? What do you think is the major reason the warnings in these memos were ignored by higher level management?
3. What are some of the style features of each of the three memos? Which one of the three do you think is the most effective and why?
4. After reading this case study and the three memos, do you think it's fair to say that the *Challenger* disaster is as much a communication disaster as it is a technological disaster? Why or why not? What advice would you give to the parties involved with this case to help them avoid or at least minimize future communication problems of this sort?

How Does Style Relate to Subject, Purpose, Scope, Audience, and Other Elements of Writing?

As you have seen, the discourse community shapes the way a writer writes. In addition to adjusting your style to a discourse community, you must adjust your style to many other elements as well. Your style influences all of these elements, and these elements influence your style. How you approach your subject, limit your scope, state your purpose, assess your audience, as well as how you punctuate, spell, and adhere to the traditional rules of grammar, are decisions that influence your style and which are shaped by your style.

Your Subject

Your **subject** and its complexity play a large role in determining which style you will use. Of course, one of the major factors often distinguishing technical writing from other kinds of writing is the subject matter. Technical writing is predominantly about a scientific or technological subject—a user's guide on how to use an e-mail program, an engineering report on a traffic intersection, a lab report on blood test results, a status report on the condition of O-rings, a proposal to obtain funding for a research grant, an installation guide to setting up your VCR—all of these and many others are examples of technical subjects. That's obvious.

What's less obvious is the best way to write about each of these topics. For example, what you discuss in a user's guide on how to use an e-mail program depends on your understanding of the program, the complexity of the program, the background and knowledge of the audience, how much of the e-mail program you want your audience to be able to use, and numerous other factors. The complexity of the e-mail program is one particularly significant issue. Some e-mail programs are well designed and easy to use, and some are poorly designed and cumbersome.

The complexity of a subject can be caused by poor design, but complexity can also be caused just by the nature of the subject matter. Many subjects are highly technical, some are moderately technical, and some have a low technicality. Each demands its own stylistic choices.

One way to look at level of technicality is to consider the audience. The level of technicality of a subject is often in direct proportion to the audience's familiarity with the subject. Experts and technicians, of course, will find many topics in their field much less technical than people outside of their field. Most disciplines are so complex, however, that it's almost impossible for anyone to be an expert about everything in a particular discipline. So many topics in a particular area can be highly technical even to insiders. It's also true that many people find some subjects to be far more technical, complex, or intimidating than other people find these subjects to be. Level of technicality can be difficult to establish clearly and objectively for all audiences.

Highly Technical Prose

In technical prose the highest level of technicality most often appears in writing by experts for other experts in a particular profession, field of study, or hobby. Take, for example, the following selection from *New Generation Computing*:

> To illustrate how the multi-stop bounding algorithm provides lower-bound probabilities for states in D_i, we describe a sequence of transformations of the rate matrix G such that for each state space transformation, state probabilities for states in D_i in the original model are individually bounded from below by the state probabilities for in the transformed model. During this sequence of state space transformation, we use the basic aggregation/disaggregation technique described in [Courtois, 1988]. Of course, just as in the one-step bounds procedure, exact aggregation is not actually required in the computation of the steady state system availability bounds. We merely use the existence of an exact aggregation in the intermediate steps of the development. In the end, we only need bounds on transition rates out of the aggregate states.[28]

The first sentence loses anyone unfamiliar with computer programming. The sentence contains fifty-eight words, a complex sentence structure, and a passive voice with many prepositional phrases. The prepositional phrases alone are foreign to most laypeople. Additionally, the example has words, symbols, and phrases unfamiliar to most people outside of the field.

Moderately Technical Prose

A moderately technical prose may be seen in the next examples. The sentences are less complex and shorter, and there is less technical jargon. Some definitions of terms are also offered in the examples. The first sentence in the first example gives us a definition of the word *nodes:*

> The term hypertext (coined by Ted Nelson in 1965) is commonly used to describe an electronic text composed of nodes (blocks of text) which are linked together in a non-linear web. When viewing each node, certain words are highlighted to indicated that they 'yield' to a separate node; these are anchors. By tracing from node to node, readers create their own paths in the textual network. The traditional dominance of a single author-fixed reading is overturned. Readers shape their own experience of the text not only at the subjective level of interpretation, but at the objective level of words on a page. The provision for shared authoring, links between previously distinct works, and innovative access methods all greatly alter the concept of the book. Issues such as copyright, the canon, the Barthes' writerly versus readerly text take on a renewed significance.[29]

The first sentence in the second example first mentions "using a traditional programming language," goes on to give an example of a traditional programming language as being "BASIC or C," and even goes further to explain that it is the programming language in which application software is written:

> More sophisticated multimedia authoring software requires using a traditional programming language, most often either some form of BASIC or C, the language in which Unix and much applications software is written. Pascal, a programming language written to teach programming rather than to be a working language, is also used

as the basis for programming some authoring software. Windows-based presentations might require you to write custom drivers or call dynamic link libraries (DLLs).[30]

Although the definitions or explanations help us to understand more of the text, the paragraphs still contain words or phrases unfamiliar to people who don't know anything about the field. The phrase "non-linear web" in the first example or "authoring software" in the second example still keep the person unfamiliar with the field from understanding what the author is talking about.

Slightly Technical Prose
The final example offers a look at the lowest level of technicality:

> There are several ways to include text in a PageMaker document, and we might as well start using the correct terminology; PageMaker calls documents publications, and so we will, too. Each publication can comprise one or more stories. A story is a continuous block of text that PageMaker recognizes as a single unit. For example, each article in a newsletter might be a separate story. A story can be as short as a single headline or as long as a chapter in a book.[31]

At this level the sentence structure is less complex and the sentence length shorter. But like the examples at the medium level, the longer sentences at this level are caused by explanations or definitions. Although the example contains technical advice or information, the writer keeps the technical terms to a minimum. The writer could easily have used the term *desktop publishing* instead of the plain word *publishing*. Anyone not familiar with programs that do desktop publishing would wonder what it meant. Words or concepts that may be unfamiliar are either not used or they are defined.

The complexity of the subject, general diction, technical terms, explanation of technical terms, the number of words per sentence, sentence structure, sentence length, level of formality, tone, and many other factors help determine the level of technicality in a document. Writers of technical prose must be aware of the level of technicality they are trying to achieve and of the elements that will help them achieve that level. Many of these other necessary elements are discussed in later chapters.

Your Purpose

Your **purpose**—what you want your readers to do, know, believe or accept—is another big factor determining your style. Are you, for example, trying to inform, persuade, dissuade, delay, misinform, or entertain your technical reader? All of these purposes and many others are valid purposes in technical communication. If you are chiefly trying to inform, your general diction, technical terms, sentence structure, examples, analogies, paragraph length, and many other stylistic choices will aim for simplicity and clarity. A lot of software documentation both informs and entertains. Humor, a stylistic choice discussed in more detail in a later chapter, becomes a key strategy in writing an effective document.

Your Scope

The **scope**—the boundaries of your subject and the depth of detail you plan to discuss for your subject—of your technical document also influences the style. How much of the subject you decide to cover can shape many choices concerning your style. Are your providing a general overview or an in-depth analysis in your report, a quick-start guide or a reference guide in your software document, or an online reference guide or a printed reference guide? What style you use is directly related to the challenges you set forth in your scope.

Your Audience

As the discussion about discourse communities earlier in this chapter shows, perhaps no other single factor is more important than **audience** in determining the style you will use in a document. For a document to be effective, every word, phrase, clause, paragraph, or larger section must be chosen with the reader in mind, and many of these choices are stylistic choices. Conducting a careful audience analysis is essential. Whether you classify your audience by categories (lay person, para-professional, executive, technician, or expert) or use a heuristic approach (gather information through interviews, questionnaires, or other research), you must have a very specific image of who your audience is in technical prose. Numerous questions about your audience's discourse community, identity, background, knowledge level, and motivations have to be answered before you can determine the style you will use. As we have discussed, too much technical prose fails to be effective because it completely or even partially ignores the audience. Writing for an audience effectively is difficult because getting people to do, know, believe, or accept something is difficult. Many writers of technical prose forget this basic principle by concentrating far too much on their knowledge of the subject and what they want to say about the subject.

On Punctuation, Spelling, and Grammar

Finally, as was mentioned at the beginning of this section, even your punctuation, spelling, and grammar choices, are influenced by your stylistic choices, or may even become stylistic choices themselves. You may, for instance, sometimes choose to rely more on semicolons than commas, dashes more than colons, or parentheses more than dashes. These punctuation marks have important differences, but in some instances what you use is a matter of preference. You may choose to rely on one spelling more than another (*judgment* or *judgement, eyeing* or *eying, online* or *on-line*). You may violate a rule of grammar if you think it will gain you an advantage with your audience. Plural subjects may not always take plural verbs; pronouns sometimes don't agree with their antecedents; and using *hopefully* at the beginning of a sentence may be the effect you are striving for. When it comes to achieving your purpose in a document, you may make many controversial choices about elements of writing other than elements of style.

Chapter Summary

Discourse communities are groups of people who associate with each other. They may be international, national, regional, statewide, local, or personal. They may be described culturally, ethnically, professionally, religiously, personally, and in many other ways. These communities have ways of knowing, ways of doing, and ways of succeeding or failing. To live in a discourse community is to accept its rules, and these rules cover how members of the community communicate with each other and with those outside of the community. As you would expect, your writing style as a member of one of these communities is influenced, as is everything else, by the community. These communities also shape style at a more fundamental level. What is valued as knowledge in each community in part determines if and how a subject is discussed. Writers of technical prose are also members of their own discourse community. They communicate with each other professionally using a specialized vocabulary, a kind of shorthand.

Technical writing is, like any other kind of writing, a political judgement. Many factors shape what is written. And, of course, writers can go to extremes within a discourse community, for example, writing highly obscure prose when the community values obscurity. Writers of technical prose, like writers of anything else, must weigh the demands of politics in a discourse community with common sense judgments about style.

Finally, to write effective technical prose, you must be aware of how much your subject, scope, purpose, and audience (as well as many other elements) influence your style and, in turn, are shaped by your style.

Questions/Topics for Discussion

1. What are some of the major challenges presented to a member of one discourse community writing to a member of another discourse community?
2. In what ways do technical communicators deal with many different discourse communities? In what ways must other professionals deal with many discourse communities?
3. How many discourse communities do you think you belong to? Choose one of these communities and discuss what you think some of the chief values of this community are.
4. After reviewing the list of questions writers should consider before writing to someone in a discourse community, what other questions do you think writers should keep in mind?
5. In the discussion of politics and style, Richard Lanham writes, "Learn the rules and then, through experience, train your intuition to apply them effectively. *Stylistic* judgment is, last as well as first, always *political* judgment." In what ways is this statement true about the academic writing you have done for various college professors? If you have worked part- or full-time, in what ways is this statement true about the writing you have done for various employers?
6. Why is it necessary to distinguish subject, purpose, scope, and audience from style?
7. Why is it helpful to distinguish levels of technicality in technical prose? What do you think most determines a level of technicality—the complexity of the subject, the style of the writing, or the knowledge level of the audience?
8. What do you think are the most important questions to consider about audience in a thorough audience analysis?

9. Is it possible for a member of one discourse community to understand and communicate fully with a member of another discourse community without sharing a similar background? Why or why not?
10. Patricia Limerick discusses some of the faults of academic writing. Do you agree with her observations about the poor quality of much academic writing? If so, what do you think are some of the causes? If you don't agree with her observations, what do you think are the faults, if any, of academic writing?
11. In what ways do you think college students are hindered by academic prose in textbooks, journal articles, and lectures?

Exercises

1. Many strategies are available for learning more about a discourse community. If the discourse community is a profession or a field of study, the obvious answer for learning more about the community is to study formally the profession or field. Most of us, however, are only able to know one profession or field well. We are outsiders or laypeople when it comes to knowing someone else's specialty. As outsiders we talk to people in the profession or field, or we can do library research, or we can read some of the principle introductory texts in the area.

   One great source for finding out about discourse communities is the news groups on the Internet. There are well over 6000 of these groups, with all kinds of special interests—computer software, gardening, wine, photography, and chess, to mention only a few examples—and from all kinds of professions. Many technical communicators, for instance, subscribe to the TECHWR-L. By subscribing to this group, you can get a good idea of the particular interests of educators, students, and practicing technical communicators. Most hobbies, interests, and professions are represented by discussion groups. Many of these groups make a Frequently Asked Questions (FAQ) list a part of the group. You should read this FAQ and monitor the group for a few weeks before posting your first message. This way you'll have a good idea of the particular rules and discussed topics for the group.

   Increasingly, as the Internet becomes more sophisticated and more organized, it will be possible for a layperson to learn a great deal about almost any subject by accessing readily available online reference works, reading FAQs, monitoring news groups, using search engines, and many other tools for quick and helpful information.

   For this exercise, locate a news group and read its FAQ (if one is available). Then read the e-mail discussions in this group to get an idea of what kinds of topics are discussed on a typical day. Make a list of the characteristic concerns of this discourse community and report these findings to class during the next class meeting.
2. Find one technical document with a low level of technicality in its subject or style, one with a moderate level of technicality, and one with a high level of technicality. In what specific ways do the levels of technicality differ in these documents?
3. Identify your hobbies and tell what you think qualifies as basic knowledge in one of these hobbies. What are some particular technical terms a novice pursuing this hobby should know?
4. Review the description of the discourse community of hackers (pages 21–22). Choose one discourse community you know particularly well and create a similar description of this community. Try to use at least four of the same headings.
5. In this passage, a doctor describes his sensations as his leg heals from a serious injury. Identify any elements that indicate the writer's awareness that his readers do not belong to a medical

discourse community. In other words, how does this passage differ from what you would expect to read on a similar topic in a medical journal?

> I had very much the feeling of an electrical storm—of lightning flashes jumping from one fiber to another, and an electrical muttering and crackling in the nerve-muscle. I could not help being reminded of Frankenstein's monster, connected up to a lightning rod, and crackling to life with the flashes. I felt, then, on Saturday, that I was "electrified" or rather, that a small and peripheral part of the nervous system was being electrified into life: not me, *it*.... I played no part in these local, involuntary flashes and spasms. They were nothing to do with *me*, my will. They did not go with any feeling of intention or volition, nor with any idea of movement. They neither stimulated, nor were stimulated by, idea or intention; thus they conveyed no *personal* quality; they were not voluntary, not *actions*—just sporadic flashes at the periphery—none the less a clear and crucial and most welcome sign that whatever had happened, or had been happening, peripherally, was now starting to show some return of function—abnormal, flash-like, paroxysmal function, but any function was better than no function at all.[32]

6. Critique the style of the three memos provided in the case study of the *Challenger* disaster. Are there any similar stylistic elements in the three memos?
7. Find a technical document which has a clear statement of purpose. Identify how other elements in the document support the clear purpose of this document.
8. Find a technical document which has a well defined scope. Discuss why you think this scope is well defined.
9. Interview a student in a major completely different from your major. Ask the student to describe some of the important basic knowledge of his or her discipline and to review a dozen or so sample technical terms.
10. Using a dictionary, find at least five words which have alternate spellings. Provide both spellings in your list.
11. Words can mean different things when used within different discourse communities. For example, *driver, green,* and *hook* are terms whose everyday meanings do not apply when used by golfers. Can you think of five other words that take on different meanings depending on the discourse community?

Notes

1. Mary Rosner, "Style and Audience in Technical Writing: Advice from the Early Texts," *The Technical Writing Teacher* 11.1 (1983): 43–44.
2. "A Portrait of J. Random Hacker," *The New Hacker's Dictionary* online, http://www.tpconsultants.com/ENHD/portrait.htm
3. Richard Lanham, *Revising Prose,* 3rd ed. (New York: Macmillan, 1992) 96.
4. James Suchan, and Robert Colucci, "An Analysis of Communication Efficiency Between High-Impact and Bureaucratic Written Communication," *Management Communication Quarterly* 2 (1989): 454–84.
5. Suchan and Colucci, p. 474.
6. LaRae Donnellan, "Technical Writing Style: Preferences of Scientists, Editors, and Students." Conference on College Composition and Communication, 1985 ERIC ED 258 262.
7. George Schindler, "Why Engineers and Scientists Write As They Do—Twelve Characteristics of Their Prose," *IEEE Transactions on Professional Communication* 18.1 (March 1975): 5–10.
8. Schindler, p. 10.
9. Schindler, p. 10.
10. Schindler, p. 10.
11. Patricia Limerick, "Dancing With Professors: The Trouble With Academic Prose," *The New York Times Book Review,* 31 October 1993: 24.
12. Limerick, p. 3.

13. Limerick, p. 3.
14. Limerick, p. 3.
15. Limerick, p. 24.
16. All references found within this case study are taken from the *Report to the President by the Presidential Commission on the Space Shuttle Challenger Accident, Volumes 1–5; William P. Rogers, Chairman;* published 6 June 1986 by the Government Printing Office, Washington, D.C. All text references include the appropriate volume and page number (example: V, 100).
17. Dorothy A. Winsor, "The Construction of Knowledge in Organizations: Asking the Right Questions About the Challenger," *Journal of Business and Technical Communication* 4.2 (September 1990): 14.
18. Winsor, p. 15.
19. Winsor, p. 15.
20. Paul Dombrowski, "*Challenger* and the Social Contingency of Meaning: Two Lessons for the Technical Communication Classroom," *Technical Communication Quarterly* 1.3 (1992): 78.
21. Dombrowski, p. 78.
22. Dombrowski, p. 79.
23. Dombrowski, p. 79.
24. Dombrowski, p. 79.
25. Dombrowski, pp. 79–80.
26. Dombrowski, p. 80.
27. Dombrowski, p. 81.
28. Jonas Barklund, "Bounded Quantifications for Iteration and Concurrency in Logic Programming," *New Generation Computing* 12.2 (July 1994): 684.
29. Robin, Laura "Hypertext Authoring Environment: A Critical Review," *Ejournal* 3.3 (November 1993), online http://carbon.cudenver.edu/~mryder/itcdatajournals.html
30. John McCormick, *Create Your Own Multimedia System* (New York: Windcrest, 1995) 120.
31. Rebecca Bridges Altman and Rick Altman, *Mastering PageMaker 5.0® for Windows*™ (San Franciso: Sybex, 1993) 37; and Cyndy Legg, "The Different Faces of Multimedia," unpublished essay (1994): 10–11.
32. Oliver Sacks, *A Leg to Stand On* (New York: Summit, 1984) 117–118.

Chapter 3

Choosing an Appropriate Style

> *Someone, somewhere, must be making decisions about "correct English" for the rest of us. Who?.... The legislators of "correct English," in fact, are an informal network of copy-editors, dictionary usage panelists, style manual and handbook writers, English teachers, essayists, columnists, and pundits. Their authority, they claim, comes from their dedication to implementing standards that have served the language well in the past, especially in the prose of its finest writers, and that maximize its clarity, logic, consistency, conciseness, elegance, continuity, precision, stability, integrity, and expressive range."*[1]
>
> —STEVEN PINKER

Chapter Overview

On Style and Grammar
Prescriptive and Descriptive Approaches to Style
Good and Bad Writing
Good and Bad Style
Appropriate and Inappropriate Styles
Kinds of Styles
Chapter Summary
Questions/Topics for Discussion
Exercises
Notes

On Style and Grammar

In Chapter 1: Understanding Technical Writing Style we defined style as your choices of words, phrases, clauses, and sentences, and how you connect these sentences; the unity and coherence of your paragraphs and larger segments; your tone; and your persona in your writing.

For a better understanding of style, you need to understand the relationship of style to grammar. Grammar has two broad meanings: "A set of rules and examples dealing with the syntax and morphology of a standard language"; and "In linguistics, a term for the syntactic and morphological system which every unimpaired person acquires from infancy when learning a language."[2] In brief, grammar has two broad concerns: syntax and morphology. *Syntax* focuses on the structure of sentences and concerns chiefly the rules governing the order in which words and clusters of words can appear. For example, in discussing the syntax of technical prose, you may focus on the common sentence order of subject, verb, and direct object in much technical prose. In contrast, *morphology* focuses on the structure of words and concerns such matters as inflectional endings and the way words can be built up out of smaller units. A morphological approach to technical prose would, for example, focus on the qualities of technical diction and the origins of technical terms, topics discussed later in Chapter 5 and Chapter 6.

A sentence is *grammatical* if it is "constructed according to a system of rules, known by all the adult mother-tongue speakers of the language, and summarized in a grammar."[3] In a strict sense, a sentence is *ungrammatical* if it violates what is an acceptable word order: writing or saying, for example, "I to town walked" instead of "I walked to town," or "She switched on it," rather than "She switched it on."[4]

Style, like grammar, is, of course, concerned with issues of both syntax and morphology. The habitual word orders or stylistically surprising word orders you choose in a sentence are syntactical choices. Your habitual diction choices—technical and nontechnical—are morphological choices. Style, then, is just one of many concerns in the grammar of a language, but style is not the same as grammar. Grammar depends on rules; style is more a matter of choice. There are many rules you must follow to make a sentence grammatical. However, you have limitless stylistic choices in what words you use and how you choose to order them. It's helpful to keep this distinction in mind in debates about correctness, grammar, and style.

Prescriptive and Descriptive Approaches to Style

Some people think "correct" English can be achieved by following a rigid list of rules: don't split an infinitive, don't use *hopefully* at the beginning of a sentence, don't end a sentence with a preposition, don't begin a sentence with *because,* don't write or say *everyone has their,* and don't confuse *who* and *whom.* Follow these rules, formulae, or prescriptions, you're told, and you can't go wrong.

You should know that these rules and many others about correctness are hotly debated by those who study language. These contrasting views can be simplified by referring to them as prescriptive or descriptive approaches.

A Prescriptive Approach

A *prescriptive* approach to language is a traditional, authoritarian approach. Rules or directions for good English are prescribed. Prescriptivists warn us about our "language crisis" because of all of these lapses in our language. Some well-known critics of language and how we use it include John Simon, Edwin Newman, and William Safire.[5]

These writers and others are often referred to as "pop grammarians" because they use their columns and books as a public forum for chastising us on how language is abused. Geoffrey Nunberg, typical of most linguists who have given this issue their attention, argues against the prescriptivists for regarding grammar "as a mandarin code that requires only ritual justification."[6] Nunberg reminds us that linguists "have been arguing for a long time that the rules of traditional grammar have no scientific or logical justification...."[7] Other linguists have coined even better labels for these commentators on language. In *The Language Instinct* Steven Pinker refers to these guardians of the language as the language "mavens," a term he borrows from William Safire, a maven who writes the column "On Language" for *The New York Times Magazine* (*maven* comes from the Yiddish word meaning expert).[8] Another linguist, Dwight Bolinger, refers to this group as the *shamans* because of their verbal wizardry.[9]

A Descriptive Approach

A *descriptive* approach to language is less rules-oriented than a prescriptive approach, more concerned with how language is spoken and written in various communities. A descriptivist is more tolerant of differences in dialect. A descriptivist pays more attention to context than to rules. Many linguists support this descriptivist approach to language and see themselves as scientists who study language.

Of course, prescriptivists and descriptivists will continue to debate the merits of all kinds of rules concerning correctness. Most important, as you study technical writing style in this textbook, you need to recognize that the English language is constantly changing. New words are coined because of fads, trends, and technologies; established words change in their meaning. No person or language academy has yet succeeded in suspending a language in time and protecting it from how people use it.

Good and Bad Writing

These arguments over correctness overshadow the more important issue concerning what good or bad writing is. Of course, everyone who teaches writing or who otherwise writes for a living claims to know what good writing is. (Many go even further to claim that the best writers are born with the skill, that a talent for writing cannot be taught.) Ask any two teachers of writing or professional writers what good writing is and you're likely to receive very different responses. Generally speaking, these disagreements are understandable. People who have been practicing a craft, any craft, for 20 or 30 years, will have different opinions about the fine points of that craft. Yet they will also agree on many of the basics. Teachers of writing do agree on many of the basic characteristics of good and bad writing, and of good and bad writers.

Good writing reveals, among other qualities, a thorough knowledge of the subject; a clear sense of purpose; a keen awareness of audience; a strong control of style; a command of detail; clear organization; and an exceptional understanding of grammar, punctuation, and spelling. Good writers revise, revise again, and revise still more. Good writers of technical prose also have a broad knowledge of the principles of design, page layout, and graphics.

Bad writing reveals, among other qualities, an inadequate knowledge of the subject; a vague sense of purpose; an ignorance of audience considerations; a weak understanding of style; a poor command of detail; illogical organization; and, at best, a superficial knowledge of grammar, punctuation, and spelling. Bad writers seldom revise adequately. And poor writers of technical prose lack the essential knowledge of design, page layout, and graphics.

Yet even these commonsense descriptions of good and bad writing are simplifications. Sometimes, for example, good writers may want to hide their extensive knowledge of a subject; obscure the purpose of their writing; confuse or mislead the audience; manipulate the style; be general and abstract rather than specific and concrete in the details; organize their thoughts in an apparently random matter; and, yes, even be ungrammatical, and ignore the conventions of punctuation and spelling.

Teachers of writing don't like to call attention to these strategies, but they are often legitimate strategies nonetheless. (Chapter 11 will discuss in detail whether these legitimate strategies are always acceptable or ethical.) Some would have you believe this kind of writing is always due to incompetence or laziness (and a great deal is), but all kinds of academic and industry writing have these characteristics. William Lutz's *Doublespeak: From Revenue Enhancement to Terminal Living* is a good collection of euphemisms, jargon, gobbledygook or bureaucratese, and inflated language (Lutz's four kinds of doublespeak).

Good and Bad Style

What then constitutes a good style? A bad style? As you read this textbook, you need to develop a healthy skepticism about rules, assumptions, and ideologies concerning writing and style. Many texts about writing will tell you that a plain style makes for good writing or that it's good to write clearly. These claims are easy to support, but it's important to recognize these statements as claims, and moral claims at that. Yes, most readers prefer prose to be straightforward rather than ambiguous, deceptive, or pretentious. And writers who are honest and clear with their readers appear to be more ethical than writers who are not. Yet there are many instances when a writer must be insincere or indirect, ambiguous, deceptive, or pretentious. And there are many occasions when writers are all of these things ethically or unethically.

So to argue that a plain style is better than a complex style is to stake out an ideological position. Ian Pringle comments on this point in his review of the first edition of Joseph Williams's *Style: Ten Lessons in Clarity and Grace:* "Williams' stance is the claim that the kind of style he considers good is good because it is judged to be good by those whose education, taste, and sensibility make them the best judges."[10] In effect, Williams' position on good writing is a prescriptive position. You might think the issue is simple. A plain style *is* better than a complex style, right? Ironically, many teachers of writing don't agree with this observation. Pringle cites a study showing how teachers of writing often prefer a complex style to a plain style and often penalize students more for writing in a plain style.[11]

This textbook offers a descriptive approach to style. However, the guidelines discussed here are not neutral by any means. Bad grammar, bad writing, and bad style do exist. This text attempts to identify those stylistic strategies that work and those that don't work.

Appropriate and Inappropriate Styles

Perhaps a better way to look at this issue is not to focus on good and bad styles, but rather on appropriate and inappropriate styles. Style is less concerned with rules and more concerned with guidelines concerning what is appropriate, what is best for the purpose, for a particular audience, on a specific subject, in a specific context, in a certain discourse community (See Chapter 2). Whether or not you use a formal, informal, or colloquial style depends on these factors. So too does your decision to use a plain or complex style, a simple or big word, a non-technical or technical term, and so on. Choices concerning style are contextual and situational, not fixed. If it suits your purpose, you may use slang, write loose sentences, write long paragraphs, or use confusing words.

Technical communication presents special problems to those who insist on seeing only good and bad styles. Much research is available showing that so-called bad elements of style are appropriate in many contexts. Jo Allen argues, for example, "clarity may be one of the essential components of a definition of technical writing, but a scientist's means of producing clarity may depend on field-specific conventions—appropriate uses of jargon, passive voice, and so on—that violate the technical-writing teacher's prescriptions."[12] Many others defend passive voice, noun strings or phrases, the anticipatory "it" in scientific writing, the unattended "this," long sentences, multiple adjectives, a Latinized vocabulary, and other controversial elements of style.[13]

Kinds of Styles

At least three kinds of styles have always been recognized. In classical rhetoric the three ways of speaking or writing were the **high** or **grand** style, the **middle** style, and the **plain** or **low style**. In literature the grand style was associated with epic poetry in the works of Homer, Virgil, or Milton, for example. ("Sing, goddess, the wrath of Pheleus' son Achilles, a destroying wrath which brought upon the Achaeans myriad woes, and sent forth to Hades many valiant souls of heroes"—Homer, *The Iliad*). The middle style was for education and edification, for instance, the writings of Sir Thomas More in the sixteenth century ("They have no lawyers among them, for they consider them as a sort of people whose profession it is to disguise matters"—*Of Law and Magistrates*). The plain or low style ("Let sleeping dogs lie—who wants to rouse 'em?"—Dickens, *David Copperfield*) was the language of popular entertainment in ballads and folktales. Of course, many writers make use of all three styles in their works—Chaucer, Shakespeare, and Dickens, for example.[14]

Today these styles are perhaps best typified by the solemn style of ceremonious occasions, the studied (or at least complex) style of much academic and industrial prose, and the conversational style of many popular magazines and newspapers, as well as the colloquial language of most college students. However, thinking only in terms of these broad styles is too limiting. You actually have many kinds of formal and informal styles available to you when you speak or write.

For those who must write letters, John Fielden suggests there are at least six major styles: a forceful style, a passive style, a personal style, an impersonal style, a colorful style, and a less colorful style.[15] According to Fielden, a forceful style uses active voice, is direct, uses primarily a subject-verb-object sentence order, avoids relegating a point or the action to a subordinate clause, and has a confident tone. A passive style avoids the imperative, uses passive voice heavily, avoids taking responsibility for negative statements, uses noncommittal words, and uses long sentences and heavy paragraphs. A personal style uses active voice, persons' names, personal pronouns, short sentences, contractions, and directs questions to the reader. An impersonal style avoids using persons' names, avoids using personal pronouns, uses passive voice, and makes some sentences complex and some paragraphs long. A colorful style relies on adjectives and adverbs as well as imaginative metaphors to make a point. Finally, a less colorful style avoids adjectives, adverbs, metaphors, and figures of speech, blends the impersonal and passive styles, and uses chiefly mundane diction.

Fielden emphasizes that these styles overlap and that there is no such thing as a universal style suitable for all occasions. He reminds us that every style has an impact on the reader, style can communicate almost as much as content, style cannot be isolated from the communication situation, and style must be suited to the circumstances.[16]

Although almost any style can be defended (depending upon the writer's purpose and many other factors already discussed), some styles are more effective than others.

Of course, some people are less optimistic about the kinds of styles available. Richard Lanham, for example, believes that the high, middle, and low styles are slowly becoming one: "The low style has dissolved, the high style has hardened and dehydrated, and the middle style has simply evaporated. The Official Style threatens to replace all three."[17] Lanham's Official Style is an important and genuine style. Prescriptivists would like to banish this style altogether for its lack of clarity and other problems. But as Lanham reminds us, there are legitimate reasons this style will always be used: "Clarity is often the last thing The Official Style really wants to create and, if you find yourself in a bureaucratic context, often the last thing you want to create."[18]

Just as writers of nontechnical prose have many styles to choose from when they write, writers of technical prose also have many choices. The following discussion is a brief overview of some of the principle styles available: a primer style, a telegraphic style, a passive style, a verb style, a nominal style, an affected style, a plain style, a complex style, and a literary style. Of course, these styles often overlap in a piece of writing. A document, for example, may be half verb style and half nominal style. As you plan your technical document, you must decide which style or combination of styles best suits your particular purposes.

A Primer Style

A *primer style* is an elementary writing style that consists of a series of short sentences. Although this style is fine for teaching children how to read, it is seldom an appropriate style for most other occasions. The sentences and paragraphs lack coherence, and the resulting choppy style is often difficult to read. Yet the style is used for purposes other than teaching children. Some military training manuals, for example, use this style if the intended audience has poor reading skills and limited education.

An example of the primer style:

> The process is very simple. Each file is described in terms of its attributes. Any attribute can be changed. There is a design command for each attribute. To change a file, change one or more attributes.

A Telegraphic Style

A *telegraphic style* omits articles, conjunctions, and transitional phrases. The style causes confusion because readers must fill in necessary words and because the style is also usually too choppy. Sentence rhythm and balance are usually missing.

Many of us have encountered instructions that read like the following:

Burner Assembly for Grill with Double or Single Valve

1. Find gas burner, venturi assembly with air shutter(s), gasket and four self tapping screws bagged separately with gasket. Inspect gasket before assembly. Determine that it has not been damaged.
2. Lay burner down with attach holes up. Place gasket on burner and venturi assembly on gasket. Assemble self tapping screws through plate on venturi assembly through gasket holes and into burner. Downward force on screws is required while turning screws clockwise.

Although this style has many disadvantages, it can be useful for taking technical notes, making lists, jotting down reminders, and for reducing a complex passage to its key points (see, for example the second exercise at the end of this chapter).

A Passive Style

The *passive style* is another very common style. In this style it is often difficult to determine who is doing what. A reader must read the sentence several times to know who the agent of the action is. The following selection from a memo is a typical example of passive voice:

> As you know, new scanners were delivered to each department by Office Warehouse. It has recently been brought to my attention that the people who are employed by this company have taken advantage of their positions by using the new scanner for personal business.
>
> Weaknesses in the current company policy were recently discovered. To discourage personal use of the scanners, a copy of the new policy will be placed by each scanner by Department Managers. In the attached pages the new policy is explained in detail.
>
> If there are any questions about this policy, please feel free to contact this office.

Several different kinds of passives are combined in this memo selection. The first sentence is an example of passive with *by*. The sentence would be improved by rewriting: "As you know, Office Warehouse delivered new scanners to each department." The second sentence is an example of a passive with *it*. Who brought the issue to the writer's attention? The next sentence is an example of a passive with no actor. Who discovered the weaknesses in the current company policy? The third sentence is another example of a passive with *by*. The sentence could be rephrased to say: "Department Managers will place copies of the new policy by each scanner." The last sentence is not passive voice, but it is impersonal. How are the readers to contact an office?

A Verb Style

A *verb style*, effectively written, is a style based on verbs, on action. A passive style is closely related to a verb style, but a verb style can be written completely in active voice. If you decide to create a verb style, your decision will affect many of your other choices in your prose. For example, strings of prepositional phrases, which are so common in noun or nominal style, practically disappear in a verb style.

In commenting upon this major style, Richard Lanham points to a selection by Gilbert White, an eighteenth-century English country clergyman and amateur naturalist. Lanham admires the strong verb style as well as the literary qualities of White's description of the flight of birds in his *Natural History of Selborne:*

> Hen-harriers fly low over heaths or fields of corn, and beat the ground regularly like a pointer or setting-dog. Owls move in a buoyant manner, as if lighter than the air; they seem to want ballast.... Rooks sometimes dive and tumble in a frolicsome manner; crows and daws swagger in their walk; wood-peckers fly *volatu undoso,* opening and closing their wings at every stroke, and so are always rising or falling in curves.... Parrots, like all other hook-clawed birds, walk awkwardly and make use of their bill as a third foot, climbing and descending with ridiculous caution. All the *Gallinae* parade and walk gracefully and run nimbly; but fly with difficulty with an impetuous whirring, and in a straight line. Magpies and jays flutter with powerless wings, and make no dispatch; herons seem encumbered with too much sail for their light bodies.... pigeons...have a way of clashing their wings, the one against the other, over their backs with a loud snap; another variety, called tumblers, turn themselves over in the air.[19]

White effectively relies on many verbs—*beat, want, dive, tumble, swagger, walk, climbing, descending, parade, flutter, clashing, turn*—to create an effective rhythm and parallelism throughout the passage. The verb style offers a vivid impression of the differences in bird flight that other styles could not capture as clearly.

A Nominal Style

The *nominal style,* or *noun style,* is another common style. The chief characteristics of this style are the abundant use of nominalizations, nouns, noun strings, or noun adjectives (See Chapter 5).

A noun style, like a verb style, can be quite effective and is one of the primary styles of writing. In fact, most writers rely on noun styles, verb styles, and a combination of both. But a noun style can be excessive, too, with too many nominalizations, noun strings, or noun adjectives.

When a verb is turned into a noun that noun is a nominalization (*compile* to *compilation, calculate* to *calculation*). When this nominalization is combined with a weak verb, the result is often an unnecessarily long sentence. For example:

The editor will make a compilation of the drafts.

Instead of

The editor will compile the drafts.

The following passage in noun style has many faults, including too many nominalizations:

A perception of the two extreme poles of this spectrum of language is derived from a conception of the range of language under consideration. The unliterary quality of the kinds of discourse found on each pole of the spectrum is characterized by the presence of the babble associated with children and the presence of children's nursery rhymes; in terms of the attitudes associated with this pole of the spectrum, there is a general preference for the derivation of pleasure from language. Language on the left pole of the spectrum is characterized by the reification of words; in other words, words are given a tangible quality. In accordance with this theory, words are found by the reader to be the objects of play. A categorization of the properties of this kind of language reveals in it the existence of a principle of contour, an audible quality, and a sense of physical corporeality. On the pole to the extreme right of the spectrum, there is an abundance of symbols, notably the equations associated with the practice of mathematical inquiry, which appear to exist for the sake of their denotative quality; transparent signifiers and their readily comprehensible signifieds are the body of ideas from which pleasure is derived. From this discussion it is apparent that the verbal surface of types of prose like nonsense rhymes is characterized by opacity; transparency is more characteristic of the surface texture of mathematical equations, however.[20]

Noun strings are a common device in many kinds of writing. Noun strings are groups of two or more nouns placed together:

personal-computer market base

software engineer

Noun adjectives are also commonly used:

White House spokesperson

Pentagon official

real-time user

daisywheel printer

As in the case of noun strings, noun adjectives can be overused.

An Affected Style

An *affected style* is perhaps best described as a style that combines the passive and nominal styles, throws in pretentiousness and gobbledygook, and otherwise strives for complexity. (See Chapter 6.) If a plain style has an opposite, it's an affected style.

Affected means assumed artificially or pretended, so it seems a good choice as a label for this kind of style. This style knows no boundaries. Industry, business, government, the military, and academe are all firmly entrenched in this style.

Sometimes this style may become quite convoluted, as in the following example from a philosophy text:

> With the last gasp of Romanticism, the quelling of its florid uprising against the vapid formalism of one strain of the Enlightenment, the dimming of its yearning for the imagined grandeur of the archaic, and the dashing of its too sanguine hopes for a revitalized, fulfilled humanity, the horror of its more lasting, more Gothic legacy has settled in, distributed and diffused enough, to be sure, that lugubriousness is recognizable only as langour, or as a certain sardonic laconicism disguising itself in a new sanctification of the destructive instincts, a new genius for displacing cultural reifications in the interminable shell game of the analysis of the human psyche, where nothing remains sacred.[21]

The author of this sentence is so preoccupied with embellishing the prose that the subject of the sentence is completely lost.

Russell Baker effectively parodies the affected style in his retelling of "Little Red Riding Hood":

> Once upon a point in time, a small person named Little Red Riding Hood initiated plans for the preparation, delivery, and transportation of foodstuffs to her grandmother, a senior citizen residing at a place of residence in a forest of indeterminate dimension.
>
> In the process of implementing this program, her incursion into the forest was in midtransportation process when it attained interface with an alleged perpetrator. This individual, a wolf, made inquiry as to the whereabouts of Little Red Riding Hood's goal, as well as inferring that he was desirous of ascertaining the contents of Little Red Riding Hood's foodstuffs basket, and all that.[22]

While it is easy to make fun of the affected style, it is a style that has a legitimate place in academic and corporate cultures. John Barry sees *technobabble* (his label for the affected style in the computer industry) as "a manifestation of a tendency to call a spade not a spade, but rather a 'wood- and steel-based hardware module designed to support the implementation of earth-excavating processes.' Why call a disk a mere disk when you can label it a 'peripheral' or, even better, a 'mass-storage subsystem'..."[23] Barry shows how marketing departments in Silicon Valley thrive on writing documents of all kinds, especially press releases, in an affected style.

No one denies the affected style is everywhere. The more complex issue concerns when it's an appropriate style. See Chapter 11 for a discussion of this point.

A Plain Style

A *plain style* is a style that is straightforward and easy to understand. This definition seems simple enough, but there is no universal agreement about the matter. There will probably never be a consensus on the elements of a plain style. Veda Charrow, for example, asks "What is plain English, anyway?" and answers that "although most of us would probably not agree on what plain English is, we could probably agree on what it is not."[24]

This style is the dominant style in technical prose. Readers of technical documents want prose that enables them to read quickly and to apply the information correctly. Most technical writing textbooks urge writers to write in this simple, no-nonsense style, a style also called a *reader-friendly* style.

In spite of the abundance of ineffective writing in both academe and industry, many documents are written in an appropriate plain style. Harley Hahn and Rick Stout's *The Internet Complete Reference* is a good example of a book written in this style. The opening paragraph is typical of the clear, simple, and straightforward style throughout:

> The Internet is, by far, the greatest and most significant achievement in the history of mankind. What? Am I saying that the Internet is more impressive than the pyramids? More beautiful than Michelangelo's David? More important to mankind than the wondrous inventions of the industrial revolution?
>
> Yes, yes and yes.
>
> Do I expect you to believe this? Of course not—not right now anyway. After all, the Internet is just a computer network and—let's face it—most of what we use computers for is pretty dull.[25]

In a plain style simpler words are more common than long words, nontechnical words are more common than technical words, simple and compound sentences are more common than compound-complex sentences, first or second person is more common than third person, and a personable tone is more common than a distant tone. The strategies for a plain style are almost endless.

Qualities commonly associated with the plain style are simplicity, clarity, accuracy, sincerity, directness, and objectivity. The word *simplicity* has many meanings: innocence or naivete, folly or silliness, freedom from pretense. For our purposes, simplicity means restraint in ornamentation. Simplicity is closely tied to clarity. One definition of simplicity is directness of expression or clarity. *Clarity* is best defined as free from obscurity, ambiguity, or undue complexity. *Accuracy* means free of mistakes or errors. Writing that has *sincerity* is honest writing, writing free of hypocrisy. *Directness* is accuracy of a different kind, accuracy in course or aim. Direct writing is straightforward, but, of course, a plain style may also be indirect. For example, if you must write about a delicate matter, and if you want to avoid offending your peers, you might write a memo which states the purpose indirectly rather than directly or straightforwardly. The politics of the workplace require all kinds of indirect strategies conveyed in all kinds of styles.

Finally, the plain style is often characterized as an *objective* style. To be objective in writing is to express or deal with facts or conditions without distortion by personal feelings, prejudices, or interpretations. Of course, total objectivity in technical writing (or any other kind of writing) is not possible (see Chapter 4 for a more detailed discussion of the issue of objectivity), but the belief that the plain style is an objective style persists.

The plain style is not a new style. In fact, it has a long tradition. The plain style originated in the 17th century as a reaction to the ornate style which dominated at the time. Writers such as Sir Thomas Browne were noted for their heavy use of rhetorical devices, long sentences, Latin and Greek quotations, and exotic diction. In Merrill Whitburn's study of this style, he tells us, "The stylistic ideal was copiousness and ornamentation. A writer was often so caught up in verbal exuberance, so delighted with word play, that he seemed more intrigued with rhetorical devices than the search for truth." [26]

The most famous attack on this style came from Thomas Sprat, in his *History of the Royal Society*. Sprat appealed to members of the Society "to reject all the amplifications, disgressions, and swellings of style: to return back to the primitive purity, and shortness, when men delivered so many things, almost in an equal number of words." [27] The goal of plainness became part of the statutes of the Royal Society when they were published in 1728.

A Complex Style

What makes complex prose complex? Is complex prose a legitimate style? Why would anyone want to write in a complex style instead of a simple style? Are all complex styles affected styles? These are difficult questions to answer briefly, and they are the concern of this entire textbook more than this one section of this chapter.

As with the plain style, there is no consensus on the qualities of a complex style. Complexity in writing is a matter of content and all of the elements of style. Many subjects are more technical, abstract, theoretical or difficult than others. A subject unfamiliar to a reader will be inherently more difficult than a subject that's familiar. Many elements of style also contribute to complexity. Sometimes it is necessary to use a big word instead of a simple word, a technical term instead of a nontechnical term, a complex or compound-complex sentence rather than a simple sentence, a long paragraph rather than a short paragraph, an impersonal tone instead of a personal tone.

The plain style is so strongly emphasized we tend to forget a complex style is also a legitimate style. For a style to be complex is not necessarily a fault. According to *Merriam-Webster's Collegiate Dictionary,* the word *complex* "suggests the unavoidable result of a necessary combining and does not imply a fault or failure."[28] Many technologies, theories, issues, and problems cannot easily be discussed in simple language. The word *complex* also means "hard to separate, analyze, or solve."[29] Many topics are complex in this sense of the term.

Why would anyone want to write in a complex style instead of a simple style? Complexity in prose is a matter of content and style, but it also a matter of the discourse community in which the prose is written (see Chapter 2). Different discourse communities have different values, assumptions, expectations, standards, and so on. The discourse community of the scientist is more exacting, more demanding of precision, than the discourse community of the science journalist. The discourse community of defense contractors requires a formal and complex response to a request for proposals from the federal government. An informal and simple response would not be competitive.

Are all complex styles affected styles? Of course not. The answer seems obvious, but there are so many prescriptions about keeping the style simple that the issue must be discussed. An affected style is *unnecessarily* complex, a complex style is not, as the following example illustrates:

> It is probably true that symbiotic relationships between bacteria and their metazoan hosts are much more common in nature than infectious disease, although I cannot prove this. If you count up all the indispensable microbes that live in various intestinal tracts, supplying essential nutrients or providing enzymes for the breakdown of otherwise indigestible food, and add all the peculiar bacterial aggregates that live like necessary organs in the tissues of many insects plus all the bacterial symbionts engaged in nitrogen fixation in collaboration with legumes, the total mass of symbiotic microbial life is overwhelming. For sheer numbers, nothing can match the vast population of bacteria—or, at any rate, the lineal descendants of bacteria—which, taking the form of mitochondria...became essential symbionts for the production of oxidative energy in the cytoplasm of all nucleated cells, including our own, and, in even greater numbers, the photosynthetic bacteria which became the chloroplasts of plants, without which neither plant nor animal life could ever have existed on the earth. Alongside, the list of important bacterial infections of human beings seems a relative handful.[30]

Lewis Thomas's style in this paragraph is complex for a number of reasons. In this case, the paragraph is from an article published in a medical journal, so the subject matter and technical terms are not necessarily troublesome for many in the audience. Still, the technical terms do contribute to the complexity of the style. More important, most of the sentences are quite long. They are grammatically correct and punctuated correctly. The sentence length, however, makes the reading a little more difficult for any reader. The complexity in this paragraph is completely appropriate considering the purpose, audience, subject, and context of the article.

A Literary Style

Earlier we mentioned the high, middle, and low styles of literature, and we discussed the fact that many authors use all of these styles. Of course, there are many other styles within these broad ranges of style. Literary style is a vast and complex subject, and is, as we discussed in Chapter 1 the chief focus of stylistics.

Why discuss literary style in a textbook on technical writing style? Look at the following example of a literary style in Thoreau's *Walden*. It's difficult to imagine that a scientist or a technical writer would describe a lake in the same way:

> In such a day, in September or October, Walden is a perfect forest mirror, set round with stones as precious to my eye as if fewer or rarer. Nothing so fair, so pure, and at the same time so large, as a lake, perchance, lies on the surface of the earth. Sky water. It needs no fence. Nations come and go without defiling it. It is a mirror which no stone can crack, whose quicksilver will never wear off, whose gilding Nature continually repairs; no storms, no dust, can dim its surface ever fresh;—a mirror in which all impurity presented to it sinks, swept and dusted by the sun's hazy brush,—this the light dust-cloth,—which retains no breath that is breathed on it, but sends its own to float as clouds high above its surface, and be reflected in its bosom still.[31]

Thoreau's style is ornate, even poetic. There are many literary elements in this one paragraph (as there are in the entire text of *Walden*). Thoreau uses metaphors ("forest mirror," "a mirror which no stone can crack," "quicksilver," "sun's hazy brush," "the light dust-cloth," and "bosom"), imagery, sentence rhythm and balance, and other elements to heighten this strong visual image.

Some define a literary style narrowly as an ornate or highly learned style. If we limit ourselves to this definition, then we have to wonder how often such a literary style should be used in technical documents. But a definition of a literary style need not be so narrow. A literary style may also be defined as one that relies heavily on any number of literary devices, whether it's a high, middle, or low style. Elements common to a literary style include anecdote, figurative language (for example, metaphors and similes), imagery, hyperbole, understatement, irony, alliteration, crisis, plot, structure, dialogue, subtlety, suspense, tension, theme, voice, and so on.[32] Many science writers—Lewis Thomas, Oliver Sacks, and Stephen Jay Gould, to mention only a few—show that literary devices are essential tools for making science clear to a lay audience.

Oliver Sacks, for example, uses all kinds of literary devices in all of his popular science books. The following passage is the opening paragraph of a clinical case called "The Man Who Mistook His Wife for a Hat":

> Dr P. was a musician of distinction, well-known for many years as a singer, and then, at the local School of Music, as a teacher. It was here, in relation to his students, that certain strange problems were first observed. Sometimes a student would present himself, and Dr P. would not recognize him; or, specifically, would not recognize his face. The moment the student spoke, he would be recognized by his voice. Such incidents multiplied, causing embarrassment, perplexity, fear—and, sometimes, comedy. For not only did Dr P. increasingly fail to see faces, but he saw faces when there were no faces to see: genially, Magoo-like, when in the street he might pat the heads of water hydrants and parking meters, taking these to be the heads of children; he would amiably address carved knobs on the furniture and be astounded when they did not reply. At first these odd mistakes were laughed off as jokes, not least by Dr P. himself. Had he not always had a quirky sense of humour and been given to Zen-like paradoxes and

jests? His musical powers were as dazzling as ever; he did not feel ill—he had never felt better; and the mistakes were so ludicrous—and so ingenious—that they could hardly be serious or betoken anything serious. The notion of there being "something the matter" did not emerge until some three years later, when diabetes developed. Well aware that diabetes could affect his eyes, Dr P. consulted an ophthalmologist, who took a careful history and examined his eyes closely. "There's nothing the matter with your eyes," the doctor concluded. "But there is trouble with the visual parts of your brain. You don't need my help, you must see a neurologist." And so, as a result of this referral, Dr P. came to me.[33]

Sacks provides a very effective opening with many literary elements: imagery, suspense, dialogue, anecdote, irony, plot, and so on.

Are there any examples of a literary style or literary elements in technical writing? Of course, there are. John Harris has shown the role of metaphor and shape imagery in all kinds of technical writing.[34] In writing about an automobile, for instance, you might mention brake drums, brake shoes, a carburetor that may be throttled or choked, an engine that might die, or a fuel pump diaphragm. It's difficult, if not impossible, to talk or write about any technology without using all kinds of metaphors. As for shape imagery, there are all kinds of shape terms drawn from human anatomy (*jaws* of pliers), nature (*bubble, wave*), animals (*Sea Horse, hog-nosed bat*), and the alphabet (*A-frame, c-clamp*). Writers of all kinds of technical documents rely on all kinds of literary devices to make the unfamiliar familiar, the complex simple, the difficult easy.

Finally, is it true that scientific writing has nothing to do with literary style? Not according to the research.[35] Scientists employ all kinds of literary devices in their scientific journal publications: word games, metaphor and simile, allusion, understatement, and so on. For one brief but famous example, look at how James Watson and Francis Crick announced their discovery of the structure of DNA: "It has not escaped our notice that the specific pairing we have postulated immediately suggests a possible copying mechanism for the genetic material."[36] The two scientists agreed to downplay the importance of their discovery and, in doing so, created one of the best known understatements in scientific literature.

Chapter Summary

Why are there differences of opinion about grammatical and stylistic correctness? Why can't writers agree about what good writing is? Are there only appropriate and inappropriate styles, but no bad styles? What kinds of styles are available to technical writers for writing technical documents—memos, letters, reports, proposals, manuals, and so on?

Writers disagree about correctness because language is complex, ever changing, alive, and rebellious to many seemingly logical rules. Language will resist many of these rules no matter how much some people want to impose them. Writers do agree on many of the basic elements of good writing, but agreement on the fine points of writing is another matter. Because writing is a complex art and craft, experienced writers will naturally disagree about many qualities that makes one piece of writing more effective than another. Concerning stylistic appropriateness, it's more difficult to observe firm rules about style than it is to observe rules about grammar. Put simply, *any* stylistic choice that helps writers achieve their purpose contributes to an appropriate style.

Finally, like all other writers, writers of technical prose have a wide range of styles to choose from. To suit their purposes, writers of technical prose may choose a primer style, a telegraphic style, a passive style, a verb style, a nominal style, an affected style, a plain style, a complex style, a literary style. A keen awareness of audience helps you determine which style is most appropriate for a particular document.

Questions/Topics for Discussion

1. How is style different from grammar? Can you give an example of a stylistic choice which is ungrammatical? How is an ungrammatical stylistic choice justifiable?
2. Why do prescriptivists and descriptivists disagree about "correct" English? Identify some issues they disagree about that were not already mentioned in this chapter.
3. Is it necessary to understand the prescriptive and descriptive approaches to language to understand approaches to style? Why or why not?
4. Identify situations where a writer might prefer to use an indirect approach instead of a direct approach in a document.
5. Why can't writers of technical prose just write all of their documents in a plain style?
6. What are some of the reasons you have found a paragraph in a technical document to be complex? Overly complex?
7. Review the qualities mentioned for good writing early in this chapter. What other qualities do you think define or describe good writing?
8. Review the qualities mentioned for bad writing early in this chapter. What other qualities do you think define or describe bad writing?
9. What's the difference between bad writing and an inappropriate style?
10. Why is it a better idea to talk about appropriate and inappropriate styles rather than good and bad styles?
11. Do you agree with Ian Pringle that Joseph Williams's claims about good style are essentially an ideological position? Why or why not?
12. William Lutz suggests that doublespeak "is the product of clear thinking and is carefully designed and constructed to appear to communicate when in fact it doesn't." What are some other possible causes for doublespeak?
13. Given what Richard Lanham says about The Official Style, why does this style have a place in bureaucracies? Does The Official Style have a place in technical instructions?

Exercises

1. Find a memo, letter, or short document you consider to be inappropriate or poor writing. Discuss what characteristics of the writing make it ineffective.
2. One way to revise an overly complex or affected style is to reduce the sentence or paragraph to a telegram. [37] Then add the appropriate words to make more sense of the text.

    For example, look at the following difficult passage:

> As specifically regards the operator Interface equipment, it will be necessary to breach the MDB protection of the unit to perform maintenance, but the integrity of the rest of the terminal equipment protection will be uncompromised if interlocks disconnect the unit from the UPS power when the protective enclosure is opened.

This passage is 52 words and is unnecessarily complex. Of course, a lot of people think such a passage is good technical writing. It's impersonal, difficult, and, well, technical.

To reduce this sentence to a telegram, underline the principal nouns and verbs, and the most indispensable modifiers:

> As specifically regards the operator Interface equipment, it will be <u>necessary</u> to <u>breach</u> the <u>MDB protection</u> of the <u>unit to perform maintenance,</u> but the integrity of the rest of the terminal <u>equipment protection</u> will be <u>uncompromised</u> if interlocks <u>disconnect</u> the unit from the UPS power when the protective <u>enclosure</u> is <u>opened</u>.

A telegram of this passage would read:

> NECESSARY BREAK MDB PROTECTION OF UNIT TO PERFORM MAINTENANCE. INTEGRITY REST OF EQUIPMENT PROTECTION UNCOMPROMISED IF DISCONNECT UNIT WHEN ENCLOSURE OPENED.

This passage is only 21 words, less than half the length of the original.

Now, it's necessary to add a few words to avoid the problems sometimes created by a telegraphic style. Many ways of rewording the telegram are possible. One of the simplest revisions would be to state the text into two simple sentences:

> To perform maintenance, the unit's MDB protection must be breached. However, if the unit is disconnected when the enclosure is opened, the rest of the equipment will still be protected.

Note that this revision is only twenty four words. Contrary to what some might think, this revision doesn't change the meaning of the original. Instead, it improves upon the original by simplifying it.

Now try the same technique with the following paragraph. Revise the paragraph by reducing it to a telegram and then adding any necessary words.

> It is very important to note that the Gamma setting listed in Photoshop's Monitor Setup under the Preference menu has a different function from the Gamma Control Panel software that ships with the Macintosh version of Photoshop. The Monitor preference is not used to set the gamma of your display. This setting tells the program how you have adjusted the actual gamma of your display using either a hardware calibrator or the Gamma Control Panel. The Monitor Setup preference compensates for a variety of displays and viewing conditions when you convert an image from RGB to CMYK Mode.[38]

3. The following example is written in a plain style. Rewrite the document in an affected style. Make a list of the choices you had to make in revising this document into an affected style.

> You're driving down the highway when suddenly you have car trouble. The National Safety Council suggests the following measures when your car breaks down or has a flat tire.

At the first sign of car trouble, gently take your foot off the accelerator. Do not brake hard or suddenly. Carefully work your vehicle toward the side of the road. If you're on an interstate, try to reach an exit. Signal your intentions to drivers behind you. If you have to change lanes, watch your mirrors and traffic around you closely. Once off the road, make your car visible. Use your emergency flashers. If it's dark, turn on the interior dome light. Put flares, warning flags or reflective triangles behind your vehicle to alert other drivers.

When you have a flat tire, make sure you can change it safely without being close to traffic. If that's possible, change the tire as you normally would. Remember, safety must take precedence over your schedule or whatever other concerns you may have. When your car is beyond repair, it's best to get professional help. Don't try to flag down other vehicles. Raise your hood and tie something white to the radio antenna or hang it out a window so police officers or tow truck drivers will know you need help. Don't stand behind or next to your vehicle. If your car is in the roadway, stand away from the vehicle and wait for help to arrive.

If your car is safely out of traffic, wait inside the vehicle with the doors locked. If someone stops and offers to help, open the window slightly and ask them to call police.

4. Rewrite the literary style of Thoreau's description of Walden on page 55 into a less literary, scientific style.
5. Find an example of a complex paragraph on a technical subject. Discuss why the paragraph is complex, but not overly complex.
6. Choose a subject and write a short paragraph on that subject in three of the following styles: primer, telegraphic, passive, verb, nominal, affected, plain, complex, or literary.
7. Choose a page from one of your favorite novel. Discuss what, if anything, makes the author's style literary.
8. Choose a page from one of your favorite essay. Identify some typical features of the writer's style.
9. Find an example of short document written in an affected style. Discuss what characteristics make the style affected.
10. Find an example of short document written in a plain style. Discuss what characteristics make the style plain.

Notes

1. Steven Pinker, *The Language Instinct* (New York: Morrow, 1994) 372.
2. Sidney Greenbaum, Dennis Baron, and Tom McArthur, "Grammar," in *The Oxford Companion to the English Language,* ed. Tom McArthur (New York: Oxford UP, 1992) 446.
3. David Crystal, *The Cambridge Encyclopedia of the English Language* (New York: Cambridge UP, 1995) 215.
4. Crystal, p. 214.
5. See the following texts: John Simon's *Paradigms Lost* (New York: Clarkson Potter, 1980), Edwin Newman's *Strictly Speaking* (New York: Warner, 1974), and William Safire's *Coming to Terms* (New York: Henry Holt, 1991).
6. Geoffrey Nunberg, "The Decline of Grammar," *The Atlantic Monthly* December 1983: 32.
7. Nunberg, p. 32.
8. Pinker, p. 373.

9. Dwight Bolinger, *Language, The Loaded Weapon: The Use and Abuse of Language Today* (New York: Longman, 1980) 2.

10. Ian Pringle, Review of *Style: Ten Lessons in Clarity and Grace,* by Joseph Williams, *College Composition and Communication,* 34.1 (February 1983): 95.

11. Pringle, pp. 95–96.

12. Jo Allen, "The Case Against Defining Technical Writing," *Journal of Business and Technical Communication,* 4.2 (1990): 73.

13. See Daniel Kies, "Some Stylistic Features of Business and Technical Writing: The Functions of Passive Voice, Nominalization, and Agency," *Journal of Technical Writing and Communication,* 16.4 (1986): 299–310; Lilita Rodman, "Anticipatory *It* in Scientific Discourse," *Journal of Technical Writing and Communication,* 21.1 (1991): 17–27; and George Schindler, "Why Engineers and Scientists Write As They Do—Twelve Characteristics of Their Prose," *IEEE Transactions on Professional Communication,* 18.1 (March 1975): 5–10.

14. See **Style** in *The Oxford Companion to the English Language* (New York: Oxford UP, 1992) 992–995, edited by Tom McArthur, for a good overview of important issues concerning style.

15. John S. Fielden, "'What do you mean you don't like my style?'" *Harvard Business Review* 60.3 (May–June 1982): 128–38.

16. Fielden, p. 138.

17. Richard Lanham, *Revising Prose,* 3rd ed. (New York: Macmillan, 1992) 60.

18. Lanham, p. 72.

19. Lanham, *Analyzing Prose* (New York: Scribner's, 1983) 29.

20. Lanham, p. 23.

21. Stephen T. Tyman, "Ricoeur and the Problem of Evil," *The Philosophy of Paul Ricoeur,* ed. Lewis Edwin Hahn (Open Court, 1995).

22. Russell Baker, "Little Red Riding Hood Revisited," *The New York Times,* 13 January 1980: 10.

23. John A. Barry, *Technobabble* (Cambridge, MA: MIT P, 1991) 174–75.

24. Veda Charrow, "What is plain English anyway?" Document Design Center, Technical Report (Washington, DC: American Institutes for Research, December, 1979) 2.

25. Harley Hann and Rick Stout, *The Internet Complete Reference* (New York: Osborne McGraw-Hill, 1994) xix.

26. Merrill Whitburn, "The Plain Style in Scientific and Technical Writing," *Journal of Technical Writing and Communication* 8.4 (1978): 350.

27. Whitburn, p. 350.

28. *Merriam-Webster's Collegiate Dictionary and Thesaurus,* CD-ROM, 1995.

29. *Merriam-Webster's Collegiate Dictionary and Thesaurus*, CD-ROM, 1995.

30. Lewis Thomas, "A Meliorist View of Disease and Dying," *The Journal of Medicine and Philosophy,* 1.3 (1976): 216.

31. Henry David Thoreau, *Walden and Civil Disobedience,* ed. Owen Thomas (New York: Norton, 1966) 126.

32. See Jerry Stern's *Making Shapely Fiction* (New York: Norton, 1991) for a good discussion of these and many other elements of fiction.

33. Oliver Sacks, *The Man Who Mistook His Wife for a Hat and Other Clinical Tales* (New York: Summit, 1985) 7.

34. John Harris, "Metaphor in Technical Writing," *The Technical Writing Teacher,* 2.2 (1975): 9–13, and John Harris, "Shape Imagery in Technical Terminology," *Journal of Technical Writing and Communication,* 16:1/2 (1986): 55–61.

35. Joseph Harmon, "Perturbations in the Scientific Literature," *Journal of Technical Writing and Communication,* 16.4 (1986): 311–317; Debra Journet, "Parallels in Scientific and Literary Discourse: Stephen Jay Gould and The Science of Form," *Journal of Technical Writing and Communication,* 16.4 (1986): 299–310.

36. James D. Watson, *The Double Helix,* ed. Gunther S. Stent, (New York: Norton, 1980) 240.

37. This example is borrowed from Judith Jack's article, "Teaching Analytical Editing," *Technical Communication* 31.1 (1984): 9–11.

38. Rob Day. *Designer Photoshop: From Monitor to Printed Page* (New York: Random House, 1995) 30.

Chapter 4

Persuading through Style

If technical communicators actively create versions of reality instead of serving merely as windows through which reality in all of its pre-existent configurations may be seen, then technical communication must be fundamentally rhetorical: it builds a case that reality is one way and not some other.[1]
—RUSSELL RUTTER

Science understood as argument asks for assent, for an act of will on the part of the audience. Good technical writing becomes rather than the revelation of absolute reality, a persuasive version of experience.... If we pretend for a minute that technical writing is objective, we have passed off a particular political ideology as privileged truth.[2]
—CAROLYN MILLER

The rhetoric of the modern communicator demands a variety of skills, the most important of which is a knowledge of the audience and what motivates it, what it wants, and what it will approve of. The audience becomes the determinant for the form the final communication takes. This is the challenge of the rhetoric of technical communication.[3]
—HARLEY SACHS

Chapter Overview

Technical Prose as Persuasion
The Rhetoric of Science
Some Practical Consequences of the New Rhetoric of Science

Persuasion and Technical Prose Style
Rhetorical Choices and Style
Persuasive Strategies and Stylistic Choices
Case Study 3: **The Exxon Valdez** *and Persuasion*
Chapter Summary
Questions/Topics for Discussion
Exercises
Notes

Technical Prose as Persuasion

Some of you may not think of your technical documents as persuasive documents. Perhaps you think you are merely making the subject as clear as possible to the intended audience. You see yourself as being fair, objective, unbiased, and knowledgeable about the subject for your audience, and you hope the audience will recognize your objectivity and understand or act on the information you provide accordingly. You don't think you're being persuasive; you're being objective. So, you wonder, where does persuasion fit in?

Of course, the reality is that writers of technical prose write all kinds of persuasive documents: proposals for grants, employment letters, feasibility reports, general reports, user guides, reference guides, site assessments. All of these documents and many others may be either directly or indirectly persuasive. As a writer of technical prose you are often doing more than informing or explaining. You are trying to persuade the reader to agree with your point of view or to perform the instructions in the sequence you think they should be performed. A writer of a grant proposal most often tries to convince a group of readers—a committee or organization, for example—that a project is worthwhile and worthy of funding and that the applicant or team is qualified to do the work. A writer of a feasibility report tries to convince readers that one course of action is preferable to another. An engineer's letter of application tries to convince readers that he or she is the best applicant for the job opening. Technical prose can be and often is overtly persuasive.

The first concern in this chapter is to show that all fields and all kinds of technical prose are fundamentally persuasive or rhetorical. What's important to recognize here is that no matter how objective you think you are in your treatment of your subject, you are still taking a position or advocating a point of view. Also, viewing technical prose as fundamentally rhetorical, fundamentally persuasive, frees you to think more creatively and broadly about the kinds of stylistic choices available. Persuasion is not the same as style. But style is so closely tied to persuasion, especially in terms of the ethos or character of the writer, that it's almost impossible to separate persuasive strategies from stylistic strategies.

In Chapter 1 we briefly discussed the old and new paradigms concerning technical prose. In the old paradigm, writers of technical prose were seen as merely passing on information, much like a conveyor belt passes on packages from point A to point B. Writers were somehow seen as mere observers passing information on to readers. Some have even seen this objectivity as an important distinguishing characteristic of technical writing. The following comment is typical of this point of view: "If a writer is objective, he is free from any bias toward his subject. If he writes objectively, he presents the facts as they are, unaffected by his thoughts and feelings about them."[4] The new paradigm challenges this view of writing technical prose and suggests that writers—whether they are scientists, engineers, programmers, or professional technical communicators—are interpreters of information. And whether these writers are writing grant proposals, scientific journal articles, lab reports, or user manuals, they are representing a point of view, advocating a position, arguing for one way of understanding information or doing something rather than another way.

This paradigm shift in the nature of technical prose derives from a much larger paradigm shift concerning the nature of science. As you can imagine, the subject is complex. But it's necessary to summarize the issue here to have a better understanding of why and how technical prose is persuasive and how many persuasive strategies are stylistic strategies.

The Rhetoric of Science

The Traditional Rhetoric of Science

As you may recall from Chapter 1 the term *rhetoric* has many meanings. One broad meaning is the way information is known and expressed in a discourse community. Every discourse community has its own rhetoric—the way its members explain and debate ideas. The traditional rhetoric of science is the reliance on objectivity, the scientific method, and a supposedly nonexpressive writing style.

In *Science As Writing,* David Locke captures this traditional rhetoric well:

> It is a hallmark of the official rhetoric of science that it denies its own existence, that it claims to be not a rhetoric but a neutral voice, a transparent medium for the recording of scientific facts without distortion. If the prose of science is largely agentless—if no one in science "does" what "is done"—so are the writings of science presumed to be readerless; they are written solely for the record, to become part of the inscribed knowledge of science. The message is codified in an "objective" text. It is not the goal of official scientists to influence the reader; they aim not to persuade but only to demonstrate. And just as they seek to expunge themselves as writers from the scene of discourse, they aim to eliminate the reader from that same scene.[5]

It's difficult to assess the tremendous impact of this traditional rhetoric of science. These values of objectivity, impartiality, and disinterestedness are highly regarded in the scientific community. And books advising young scientists on how to write in their profession are full of advice on how to maintain the appearance of neutrality. For example, in *Scientific English: A Guide for Scientists and Other Professionals,* Robert Day comments: "there is a world of difference between creative writing and scientific writing. The one deals primarily with feeling, emotion, opinion, and persuasion. The other emphasizes the dispassionate, factual recording of the results of scientific investigations."[6]

In addition to Locke, many others have commented on this traditional rhetoric of science. Scientists in particular offer some helpful comments on this rhetoric. Gunther Stent, a molecular biologist, in his Preface to *The Double Helix,* refers to the "traditional view of the scientific enterprise as an autonomous exercise of pure reason by disembodied, selfless spirits, inexorably moving toward a true knowledge of nature."[7] Stephen Jay Gould, an evolutionary biologist, in *The Mismeasure of Man* comments that for many "science itself is an objective enterprise, done properly only when scientists can shuck the constraints of their culture and view the world as it really is."[8]

Many rhetoricians have also commented on this traditional discourse community. Carolyn Miller discusses the "window pane theory of language."[9] The "window pane theory of language" is "the notion that language provides a view out onto the real world, a view which may be clear or obfuscated. If language is clear, then we see reality accurately; if language is highly decorative or opaque, then we see what is not really there or we see it with difficulty."[10] Miller labels this traditional view of technical prose "the positivist view of science."[11] Language in this view is less important than content. Rhetoric as persuasion becomes "just irrelevant."[12] This view of science sees science as privileged. Science is accepted as offering absolute scientific truth.

The New Rhetoric of Science

> *There is no such thing as an impersonal person—even a scientist—and, consequently, no such thing as purely objective discourse.*[13]
>
> —DAVID LOCKE

What is the new rhetoric of science? Basically, this field suggests that science itself is a rhetorical activity. Because scientists use language, they invite an analysis of the rhetorical dimensions of that language. Instead of representing privileged truth, science and scientists advocate points of view or arguments. The hypotheses, theories, and laws of science are interpretations, however exact or accurate, not final truth. As interpretations these hypotheses, theories, and laws are open to rhetorical analysis. In other words, these hypotheses, theories, and laws are persuasive statements suggesting other scientists accept one view over another. In "Technical Writing and the Rhetoric of Science," Michael Halloran suggests, "To know anything at all is to entertain an arrangement of symbols, which is to command a system of symbols that can be arranged and rearranged, which is to participate in the community that defines and is defined by that symbol system, or language. Science, then, necessarily involves rhetoric."[14]

The new rhetorical view of science denies that both science and scientists are as objective, disinterested, and impartial as some believe them to be. Philosophers of science, rhetoricians, and many scientists argue, as does Miller, that

> Reality cannot be separated from our knowledge of it; knowledge cannot be separated from the knower; the knower cannot be separated from a community. Facts do not exist independently, waiting to be found and collected and systematized; facts are human

constructions which presuppose theories. We bring to the world a set of innate and learned concepts which help us select, organize, and understand what we encounter.[15]

Both Gunther Stent and Stephen Jay Gould are well aware of this new rhetoric of science. Stent sees *The Double Helix* as "probably the book that contributed most to the demise of the traditional view of the scientific enterprise."[16] Gould wrote *The Mismeasure of Man* to expose the myth of objective science so prevalent in the traditional rhetoric of science. Gould summarizes the new rhetoric of science quite effectively:

> Science, since people must do it, is a socially embedded activity. It progresses by hunch, vision, and intuition. Much of its change through time does not record a closer approach to absolute truth, but the alteration of cultural contexts that influence it so strongly. Facts are not pure and unsullied bits of information; culture also influences what we see and how we see it. Theories, moreover, are not inexorable inductions from facts. The most creative theories are often imaginative visions imposed upon the facts; the source of imagination is also strongly cultural.[17]

On Induction and Deduction

A simple way to be more specific about this new rhetorical view of science is to examine more closely what's involved in induction and deduction. *Induction,* simply defined, is reaching a generalization from particulars. You observe, for example, that water samples taken from lakes A, B, and C over a period of three days indicate a high level of certain kinds of bacteria in the water of these lakes. *Deduction* affirms a particular from a generalization. You suggest that lakes in the area probably have a high level of certain bacteria in them at this time of the year. Tests of lakes A and B support your hypothesis. You deduce that nearby lake C is also probably similarly infected. In these examples induction and deduction seem quite different from each other.

But the differences are not that clear-cut. You are led to believe that induction begins with raw data from which you make observations and draw conclusions. The truth of the matter, however, is that even the data are selective. Karl Popper weakens the distinction between induction and deduction through the following example:

> But in fact the belief that we can start with pure observations alone, without anything in the nature of a theory, is absurd; as may be illustrated by the story of the man who dedicated his life to natural science, wrote down everything he could observe, and bequeathed his priceless collection of observations to the Royal Society to be used as inductive evidence. This story should show us that though beetles may profitably be collected, observations may not.[18]

So the problem with assuming that induction begins with raw data and leads to conclusions ignores that you begin with prior conclusions. The human mind does not accumulate raw data unfiltered. Information is always sensed, filtered, interpreted, assimilated, or shaped in some way.

The Role of Models in Science

Another way to explain this new rhetorical view of science in practical terms is to examine the role of models in science. Scientists work with models in most if not all of the scientific disciplines. Jim Watson and Francis Crick, for example, relied extensively on model building to discover the double helical structure of DNA. Physicists rely on models of atoms to discuss the nature of the atom and the subatomic world. But what's the correspondence between these models and the so called "original" out there, an original somehow independent of our perceptions of it? The answer, of course, is not simple. David Locke summarizes the issue effectively:

> There is no a priori original "out there," known independently of the conceptual framework utilized in modeling it. The very notions of atomic dimensions and atomic bonds are comprehensible only in terms of the way scientists have come to study molecular structures. They don't "know" any "real" atoms, independently of how they study atoms and how they talk about them. The notion of an ultimate original and a contrived model, a true object and a fabricated image, must give way to a view in which both the original and the model are seen to have elements of the construct about them. The original is a concept; what is being modeled is itself a model, an image of reality—indeed, the scientist's world is all model.[19]

Most practicing scientists recognize and accept this view of the representational nature of science, but some scientists forget this view. The traditional rhetoric of science encourages a "science-tells-it-like-it-is" view. When scientists say "A is B," we accept the statement as true. Locke suggests, "We do not remind ourselves that what they are really saying is that what scientists choose to call A appears to them to be that which they designate as B."[20]

A Defense of Science

Significantly, challenging the traditional rhetoric of science and accepting the new rhetoric of science is not to attack or belittle science. Rather, the new rhetoric of science reminds us that science is not done by automatons, but by people. As Gould puts it, "science must be understood as a social phenomenon, a gutsy, human enterprise, not the work of robots programmed to collect pure information."[21] Science is a human activity and therefore a social activity. Its language—however specialized, diverse, and complex—and its methodology are discourses, and as discourses they are rhetorics.

Some Practical Consequences of the New Rhetoric of Science

So what's the significance of the new rhetoric of science for writers of technical prose? Maybe you just want some practical tips on how to write more effectively. "So why all of this theory?" you may wonder. Accepting this new rhetorical model of scientific and technical prose affects how technical prose style is both taught and practiced. In "Personality in

Scientific and Technical Writing," Merrill Whitburn argues that "Teachers in the field do their students a disservice by too great an insistence on objectivity."[22] He suggests, as Halloran and others do, that science is rhetorical and that we should be "increasingly uncomfortable with 'impartial,' 'disinterested,' and 'objective' as words associated with science and scientific writing."[23] This new rhetoric of science has many practical consequences for writers of technical prose.

A Different View of Science and Technology

The first major consequence concerns how your view of science and technology may change. The scientific and technological discourse communities become more human to those within these communities and to outsiders. Scientists and other technical people have their strengths and weaknesses, their beliefs, their prejudices, and their predispositions. And these elements are part of the work, part of the methodologies, in these communities. More broadly, the differences between the so-called two cultures of the arts and religion, and technology and the sciences become less clear. Both the scientist and the poet, in spite of legitimate differences in their approaches, become interpreters.

This theme of the underlying unity of the arts and sciences is a major theme of many works of the 1970s, 80s, and 90s. This theme is pervasive, for example, in the popular essays of Lewis Thomas. In Thomas's essay "Humanities and Science," he suggests that the two cultures have more in common than not:

> Science will, in its own time, produce the data and some of the meaning of the data, but never the full meaning. For getting a full grasp when significance is at hand, we shall need minds at work from all sorts of brains outside the fields of science, most of all the brains of poets, of course, but also those of artists, musicians, philosophers, historians, writers in general.[24]

For Thomas, the humanities and the sciences may not share the same methods and standards, but they do share a common ignorance of the complex mysteries of both nature and human nature. Practitioners in both the humanities and the sciences share a common bewilderment concerning why things and people are the way they are. Robert Pirsig's *Zen and the Art of Motorcycle Maintenance* is another enormously popular work on the same theme.[25]

A Different View of Invention in Technical Prose

A second major consequence of the new rhetoric of science concerns how we may see the role of invention in technical prose. Invention is one of the five canons of classical rhetoric. The others are style, arrangement, memory, and delivery. With the new rhetoric of science, invention in technical prose is restored to a place of importance. If we accept, as some people believe, that "science does not invent, it discovers,"[26] then the rhetorical canon of invention has no place in shaping technical prose. In this view, writers of technical prose are merely transmitting a reality—their subject. Form and style become "as self-evident as content."[27] Invention is the process of discovery and finding arguments to support your position. When you write technical prose, you invent every time you interpret a technical idea to an audience less familiar with your idea. You find your own position in the proposal, report, or manual and find arguments to support that position.

For example, while writing a technical proposal to obtain funding for a new computer laboratory, you may decide to describe the current state of the lab in part of your proposal. You may also decide to argue how outdated the current lab hardware and software have become. You may also decide to include a section on current industry expectations that college graduates be familiar with various computer applications. And you will, of course, provide a detailed listing of the new hardware and software you hope to purchase if the proposal is funded, and you will try to convince your audience that only this hardware and software will meet your needs. As you write, you are discovering or inventing these topics for your proposal. You lab needs are not self-evident. Nor should they merely be listed. You must invent the best topics to cover, and, in addition to invention, you must decide upon the most appropriate style, a clear organization, an effective page layout and design, helpful illustrations, a suitable binding, and many other essential elements.

A Different View of Stylistic Freedom in Technical Prose

The third major consequence of the new rhetoric of science concerns having more stylistic freedom in creating technical prose. Instead of meeting the requirements of an ideal plain style, writers of technical prose may discuss technical content using many different rhetorical strategies. In "The Plain Style in Scientific and Technical Writing," Whitburn mentions several ways this new rhetorical view can help us write more effective technical prose. He suggests we should challenge the stylistic ideal of plainness in scientific and technical writing by using a variety of rhetorical devices. Writers of technical prose can be taught to express the same content in many different ways. They can be taught to amplify their writing through comparison, example, description, digression, and other techniques. As Whitburn suggests, "no one can better achieve brevity in his style than he who knows what words and figures to choose from among a great variety."[28]

Writers of technical prose now have good reason to experiment with tone. Traditionally, writers of technical prose have been told "be objective, be unemotional, be impersonal."[29] Now writers of scientific and technical prose have good reasons to be more expressive in their writing. Additionally, the new rhetoric of science also helps to counter the notion that first-person point of view must be avoided in scientific and technical writing.

A Different View of What Technical Prose Is

Finally, a fourth important consequence of the new rhetoric of science concerns how we define technical writing. The traditional rhetoric of science makes many writers, as Carolyn Miller suggests, "look upon writing as a superfluous, bothersome, and usually irrelevant aspect of their technical work."[30] Too many definitions focus upon content or the subject and appeal for absolute clarity. A view of science based on the new rhetoric allows us a much broader possible definition of technical prose. Technical prose, like all other writing, defines, describes, narrates, informs, argues, and persuades. Technical prose, like all other writing, interprets and advocates a point of view. Technical prose, like all other writing, is fundamentally rhetorical.

Persuasion and Technical Prose Style

> *To argue that the appropriate, well-turned phrase can be as admirable in the scientific paper as in the poem can hardly be said to denigrate the paper or the poem.*[31]
>
> —DAVID LOCKE

Scientific and technical prose are, contrary to what many people believe, expressive forms of writing. In *Science as Writing,* David Locke argues: "I vigorously deny that traditional assessment in literary and scientific camps that one, if not *the,* discriminating feature between literary language and scientific language is that the former is expressive and the latter is not."[32] Locke believes "that scientific language can be expressive…expressive of the individual self and expressive of ideas and of powerful feelings."[33]

Locke comments on the many aspects of feeling scientists have for their work. First, scientists enjoy what they do: "But people who have not associated personally with them may not recognize the degree to which their work gives scientists of many emotional stripes a special feeling, a kind of joy, not daily, perhaps, but when things go especially right, when results begin to appear. Scientists often remark at how lucky they are to be able to spend their lives doing what gives them so much pleasure."[34] Second, scientists "develop this intense feeling for, this intimate awareness of, the objects they study…."[35] Third, "Not only are scientists drawn to their scientific object; they seek to gain mastery over it. It is the aim of the exploration of the object to understand and ultimately to control or dominate it."[36] Fourth, scientists are compelled by their discourse community "to suppress such feelings when they write or at least to avoid their overt expression in their writing,"[37] but peers in their discourse community are able to pick up on the slightest clues about the writer's feelings. Fifth, the excitement of scientific discovery—the belief "that they are uncovering the fundamental laws of nature"[38]—further shows the tremendous feeling scientists have in their work. Finally, what scientists do and their writing style cannot be separated. As Locke comments, "the verbal formulation of scientists' ideas is part and parcel of what they do from the beginning, that they have no work—hence can have no style—apart from the language in which the work and the style are enunciated."[39]

Rhetorical Choices and Style

> *What is involved here surely* is *rhetoric. The scientist who decides to write "20 ml of sulfuric acid was added to the solution," rather than "I added 20 ml of sulfuric acid to the solution," is making a rhetorical decision.*[40]
>
> —DAVID LOCKE

Scientists, engineers, programmers, and other technical professionals choose from a large number of rhetorical devices to communicate to others about their work. Every word, phrase, clause, sentence, paragraph or other segment is a rhetorical decision. What the focus of the content will be is a rhetorical decision. How the conclusions will be emphasized is a rhetorical decision. What will be amplified and what will not be is a rhetorical decision. No matter what people in the technical professions may think of rhetoric, they cannot escape it.

Scientists, like other technical professionals, draw upon a variety of well recognized rhetorical devices. Joseph Harmon has shown how some scientists use word games, metaphors, similes, understatements, and neologisms in their professional scientific journal articles. One of the most famous examples, of course, is the use of understatement in James Watson's and Francis Crick's article for *Nature* announcing their discovery of the double-helical structure of DNA. Toward the end of their article the authors write: "It has not escaped our notice that the specific pairing we have postulated immediately suggests a possible copying mechanism for the genetic material."[41] Harmon notes that this particular kind of understatement is a *litotes,* "an affirmation expressed by the negative of the contrary ('It has not escaped our notice...')."[42] This device "draws the reader's attention, without fanfare, to the important message that follows; namely, the proposed structure...offers a simple mechanism for replication."[43]

Persuasive Strategies and Stylistic Choices

The method of argument you choose directly influences your technical writing style. Your general diction, technical terminology, tone, voice, sentence kind and variety, paragraph coherence, and many other stylistic preferences in large part derive from the method or methods of argument you use in your technical documents.

Technical prose is unquestionably persuasive in the sense that it advocates a point of view regardless of how objective or neutral it attempts to be. Sometimes, however, technical documents must employ more direct means of persuasion.

Rogerian Argument

Technical prose can be made more persuasive and therefore more effective in many ways. For example, you can employ some of the rhetorical strategies of Rogerian argument.[44] American psychologist Carl Rogers emphasized communicating from the other person's point of view. His goal in therapy was to gain the trust of his clients. His principles of client-centered therapy have been applied to rhetorical theory and named *Rogerian argument.*

Rogerian argument is particularly persuasive because it depends not only on your ability to show your reader that you understand his or her position, but that your own position is, in important ways, informed by that understanding. Using Rogerian argument, you show a constant respect for the reader's point of view. You are persuasive because you emphasize cooperation over conflict and you show the need to build or discover links to encourage trust.

Language is used in different ways in Rogerian argument than in traditional argument. According to Richard Young, Alton Becker, and Kenneth Pike, "Traditional argument often exploits language's capacity for arousing emotion in order to strengthen a position; Rogerian argument emphasizes the descriptive, dispassionate use of language."[45] The language of traditional argument is confrontational; the goal is to make your position prevail. The language of Rogerian argument is mutually supportive and cooperative: "The goal of Rogerian argument is to create a situation conducive to cooperation; this may well involve changes in *both* your opponent's image and your own."[46] Rogerian argument shows clearly the close connection between persuasion and style. All of your language choices are aimed at creating a friendly and cooperative tone rather than a confrontational one.

Look at the following example of technical prose to see how it can be revised using Rogerian argument:

> Oilco cannot afford to ignore the accompanying abstract. It should prove to be of great interest to our company because of its application to both mineral processing and oil production technology. Indeed, it is probable that the proposed program could only be undertaken by a company of our size and diversity. Thus the competition is minimized. The program involves research and development of the largest Red Sea geothermal deposit, the Atlantis II Deep. This deposit consists of unsolidated sediments in several mineralogical of varying metallic content. These sediments would be recovered by fluidization of the facies. The material would then be pumped to a surface platform. Oilco now has the great opportunity to develop and patent fluidization technology. This will enable us to become the sole purveyor of this technology which must be used to profitably extract ores from various marine deposits. Our position astride the oil and mineral extraction fields puts us in the best position to exploit such marine deposits. I feel that this abstract deserves the most serious consideration by you and your staff.[47]

Rogerian argument can be reduced to a three-step progression.[48]

1. State your understanding of the facts of the situation as neutrally as you can. Avoid offering any evaluations. Establish an impersonal style by, for example, using third-person point of view and passive verb constructions so the reader will focus on the situation or action. Note the following revised opening:

> The accompanying abstract proposes research and development of the largest geothermal mineral deposit in the Red Sea, the Atlantis II Deep. This deposit consists of unconsolidated sediments in several mineralogical facies of varying metallic content. These sediments would be recovered by fluidization of the facies. The material would then be pumped to a surface platform.

2. Next show you understand the reader's point of view by discussing the subject from the reader's interests. Use an active style (active voice, second- or third-person verb constructions) to show the reader as the major actor. Then change to a first-person plural point of view to combine your actions and understanding of the situation with the reader's. In this manner, you establish a common ground, as in this revision:

> Oilco's position astride both the oil and the mineral extraction fields puts us in a good position to develop and apply the necessary mineral processing and oil production technology. Because the proposed program probably can be undertaken only by a com-

pany of Oilco's size and diversity, our competition would be minimal. We could possibly become the sole purveyor of a new extraction technology and world leader in marine mining.

3. Finally, use a bold and straightforward style (as well as first-person singular and plural verb constructions) to state your aims or goals as clearly as possible. Tie everything to the mutual understanding and concern you established in steps 1 and 2. Now is the time to offer value judgments and ask for what you want.[49] See the following closing:

> Oilco cannot afford to ignore the accompanying abstract. I feel that it deserves the most serious consideration by you and your staff. May I begin researching the project as soon as I have your approval?

The persuasive strategy of the first step helps the reader to think of you as also being an fair or objective observer. (Of course, as the discussion earlier in the chapter shows, you cannot be completely objective. Your strategy here is to try to appear to be as objective as possible. Your objectivity is, in effect, a rhetorical stance.) You've stated the facts as unemotionally as you can. You've established your credibility with your reader by your balanced presentation of the facts.

The second step is a bigger challenge. Here you have to develop your understanding of the reader's point of view. You carefully determine the reader's needs and the purpose of your document. Everything you choose to say about your subject in your document must be said with the reader's interests in mind. You try to establish that the reader's interest in the subject is the same as your interest. If you are successful, you have built a trust that will now allow you to be more direct in informing the reader of what you want.[50]

Aristotle's Three Modes of Persuasion

The three steps of Rogerian persuasive rhetoric may be seen as a helpful modification of Aristotle's means of persuasion: ethos, logos, and pathos.[51] *Ethos* is the character of the writer, the writer's integrity, competency, credibility. *Logos* is the nature of the message or the quality of the data. *Pathos* is the emotion of the audience, the possible benefits for the audience from understanding and applying the data.[52]

For example, in the proposal for a new computer laboratory mentioned earlier, the *ethos* of the writer (or writers) may be detected in the knowledge of the subject, the command of detail, the credibility of the writer(s), the seriousness of the tone, and other elements. The *logos* of the proposal is evident in the arrangement of the supporting arguments for a new lab and the accuracy of the specifications for the software and the hardware. The *pathos* of the proposal is revealed in how emotionally receptive the audience is to the proposal and how the proposal makes the audience aware of various benefits in having a new computer lab.

Rogerian argument redefines the importance of these three means of persuasion. Ethos, the writer's character or integrity, becomes the most important mode. According to Scott

Sanders, "The technical writer employing Rogerian persuasion wants readers to understand appeals to pathos (assessments of the readers' use of the data derived from audience analysis) and appeals to logos (the quality of the data itself) in the established context of the writer's ethos, the writer's competent, scientifically objective character."[53] Again, the writer cannot *be* scientifically objective. Instead, the writer *adopts* a sincere persona of open-mindedness, objectivity, and fairness, and the writer adopts a sympathetic and friendly tone toward the reader. Of course, this persona and tone may be insincere, but Rogerian argument depends upon sincerity to be successful. This sincerity is a major quality of the ethos of the writer. As Chapter 9 will show, the ethos of the writer is one of the most important elements of style.

Aristotelian and Rogerian argument both have their advantages. Aristotelian or traditional argument was described earlier as argument that exploits language's capacity for arousing emotion in order to strengthen a position whereas Rogerian argument emphasizes the descriptive, dispassionate use of language. And the language of traditional argument was described as confrontational; the goal is to make your position prevail. In contrast, language in Rogerian argument is mutually supportive and cooperative. Of course, both kinds of argument are using language and logic for their own purposes: Aristotelian to heighten emotional appeal and to arrange the supporting arguments in the most effective way possible to the advantage of the writer; Rogerian to minimize emotional appeal and to arrange the supporting arguments in a manner which appears to be mutually advantageous for both the writer and the reader. Simply stated, if the ethos is highly credible, the logos overwhelmingly in your favor, and pathos easy to establish, Aristotelian argument often works effectively. If the audience is highly resistant to your ideas from the outset, then the ethos, logos, and pathos you may establish through Rogerian argument may be a more successful approach.

Of course, there are other forms of argument. For example, Toulmin logic, named after Stephen Toulmin, is another important persuasive strategy which helps shape the content, style and organization of a document. Toulmin logic helps you anticipate questions readers may have, so it helps you provide appropriate and adequately developed details throughout your argument.

The five parts of Toulmin logic are:

1. *Claim*—The major proposition or conclusion of the argument
2. *Grounds*—The evidence upon which the claim rests (facts, experimental research data, statements from authorities, and so forth)
3. *Warrant*—That which justifies the grounds and makes them relevant to the claim
4. *Backing*—Further evidence for accepting the warrant
5. *Rebuttal*—Counter arguments, exceptions to the claim, warrant, or backing, or reasons for not accepting them

By applying Toulmin logic you may become more aware of any potential flaws in your argument, you will make a better effort to develop your paragraphs and larger sections more fully (see Chapter 8), and you will identify any weaknesses in the organization or arrangement of your document.

Persuading through Style **75**

Chapter Summary

Writers of technical prose must recognize that they write persuasive documents and they need to be familiar with a variety of rhetorical strategies to make these documents more successful. An understanding of the traditional rhetoric of science offers many insights into why so many scientists communicate the way they do with each other and with laypeople, but an understanding of a new rhetoric of science is also essential for writing effective technical prose. The new rhetoric of science reveals more clearly the persuasive nature of the scientific and technological community. This new rhetoric of science also has important consequences for understanding the place of the sciences alongside the arts, for understanding the role of invention in technical prose, and for accepting and using helpful stylistic elements in scientific and technical prose.

The persuasive nature of scientific and technical prose directly shapes and influences the stylistic elements of this prose. This prose is more expressive, more creative, and more dynamic than some have previously thought. Acquiring a knowledge of the many basic rhetorical devices helps writers of technical prose become even more persuasive in their documents.

Finally, by looking at the nature of Aristotelian and Rogerian argument, you can see how persuasive strategies influence your stylistic choices. Aristotelian and Rogerian argument, as well as Toulmin logic, depend on many stylistic choices to succeed. Issues concerning the ethos of the writer, point of view, and choice of person, tone, and diction are at the center of effective persuasion.

Case Study 3: *The* Exxon Valdez *and Persuasion*

On 24 March 1989, the 987-foot-long supertanker *Exxon Valdez* ran aground on Bligh Reef, spilling 10.8 million gallons of crude oil into the waters of Prince William Sound. The spill became the environmental disaster of the decade. The Alaska Oil Spill Commission was created by the Alaska legislature to find some answers to what went wrong.[54] Members of this commission are described below:

Members of the Alaska Oil Commission

Walter B. Parker, Chairman—Anchorage, a former technical staff director of Alaska's Office of Pipeline Coordinator, currently is president of his own transportation and resource consulting firm and president of the Alaska Academy of Engineering and Sciences. Parker served on the Federal Field Committee for Planning in Alaska and co-chaired the Joint Federal–State Land Use Planning Commission for Alaska 1976–79. He was Alaska Commissioner of Highways and an Anchorage municipal assemblyman during the 1970s. He was chairman of the Alaska Oil Tanker Standards Task Force 1975–1977 and served twenty-four years with the Federal Aviation Administration.

Fisher Wunnicke, Vice Chair—Anchorage, is an attorney who served as commissioner of the Alaska Department of Natural Resources in the early and mid-1980s. She managed the U.S. Department of the Interior's Alaska Outer Continental Shelf Office, co-chaired the Joint Federal–State Land Use Planning Commission for Alaska in the mid- and late 1970s, and served on staff of the Federal Field Committee for Development Planning in Alaska.

Margaret Hayes—Anchorage, is a geologist and former director of the Alaska Department of Natural Resources Division of Land and Water Management.

She was employed by the department in various capacities from 1975 through 1988.

Tim Wallis—Fairbanks, is president of Tim Wallis and Associates, a consulting firm. The firm is currently representing a municipality and other interests as a lobbyist in Juneau. Wallis is a former state legislator, past president of Doyon, Ltd., an Interior Native corporation, as well as the past president of Alaska Federation of Natives and the Fairbanks Native Association.

John Sund—Ketchikan, is a former state legislator and commercial fisherman who now practices law and operates a fish-processing firm. Sund served on the Resources Committee as a state house member from 1984 to 1988, and from 1981 to 1985 was president and chief executive officer of the Waterfall Group Ltd., a resort operation.

Edward Wenk, Jr.—Seattle, professor emeritus of engineering, public affairs, and social management of technology at the University of Washington, is a former advisor to three presidents and Congress. An expert on the strength of ships, Wenk was a test pilot on the initial deep dive of America's first nuclear submarines and developed a world-class lab on the structural mechanics of submarine pressure hulls. The author of more than 150 papers and books, many on the interaction of technology with people and politics, he holds a master of science from Harvard University and a doctorate of engineering from Johns Hopkins University.

Michael Herz—Berkeley, California, has studied previous oil spills and tanker accidents and is currently baykeeper and executive director of the San Francisco Bay-Delta Preservation Association, a nonprofit corporation that monitors oil and chemical spills. An advisor on oil spill dispersants, waste disposal, and the impact of oil spills on fisheries, Herz studied and produced a major report on the 1984 *Puerto Rican* tanker spill and has co-written three books and more than 80 technical reports and papers. He holds a doctorate from the University of Southern California, was a postdoctoral fellow at UCLA's Brain Research Center, and has been involved in marine research and policy since 1973.

The following selection is the Conclusion from the final report, *Spill: The Wreck of the Exxon Valdez,* from this commission.[55]

Conclusion from Final Report of Alaska Oil Spill Commission

Eight fundamental points emerged from the work of the Alaska Oil Spill Commission:

I. Moving oil by sea involves a complex, high-risk megasystem whose breakdown can threaten the welfare of entire coastlines.
II. Risk is unavoidable in modern oil transportation. It can be reduced but not eliminated.
III. Prevention of major oil spills must be a fundamental goal in the oil trade, for cleanup and response methods remain primitive and inadequate.
IV. Enforcement zeal in government and industry has declined over the last decade. Rigor flagged, complacency took root. Prevention was neglected, with disastrous results.
V. Without continuing focus on the safety of the entire system by government and industry leaders, the oil transportation system poses an increasing risk to the environment and people of Alaska.
VI. The State of Alaska has primary responsibility for protecting the resources of the state and the welfare of the its people, who bear the risk of unsafe conditions in oil transportation.
VII. Privatization and self-regulation in oil transportation contributed to the complacency and neglect that helped cause the wreck of the *Exxon Valdez*.
VIII. The safety of oil transportation demands review and overhaul. Not just new technology, but new institutions and new attitudes in old institutions are required. These are the basic premises we believe policy makers should understand in designing remedies for a flawed system of oil transportation. They are the foundation for this report.

Risk is an unavoidable part of any complex technological system. The magnitude of risk facing the Valdez tanker trade became powerfully apparent in the wake of the *Exxon Valdez* spill. That should have been no surprise. The losses suffered along Alaska's coasts had been anticipated for twenty years, and safeguards had been installed to prevent such a disaster, or at least mitigate its impact. Those safeguards had eroded dangerously by the time the *Exxon Valdez* set sail last 23 March. Shortsighted decisions or simple neglect at the highest levels of the oil industry, the state, and the federal government brought on serious lapses in the oversight and preparedness promised for the trans-Alaska pipeline system when it was approved in 1973.

But neglecting such a serious risk eventually brings a heavy cost. The bill came due at Bligh Reef.

Where it may come due next has been a matter of considerable discussion in recent months and properly so: Corrosion problems in major portions of the trans-Alaska pipeline threaten the integrity of the land system. The Valdez tanker fleet is aging—and weakening—in the grueling conditions of the Gulf of Alaska. The risk of further disaster remains high. Alaskans, who are both stewards of a wondrous natural environment and partners (through their royalty share) in the production of North Slope oil, must confront that risk honestly and prudently—or they will be lulled again into complacency and neglect, to their continuing peril.

Experienced mariners express astonishment that a modern, well-equipped supertanker ran aground at Bligh Reef. The *Exxon Valdez* was traveling through well-charted waters in conditions of moderate weather and visibility. Bligh Reef was a well-known hazard, and all mechanical and navigational systems on the ship were working properly. Coast Guard Commandant Paul Yost engaged in only slight hyperbole when he said after inspecting the accident scene that his ten-year-old son could have steered the tanker safely through the area.

Yet even the events leading to the grounding, and the institutions and procedures reflected in them, revealed a situation where the risk of disaster had increased steadily through years of relatively incident-free tanker trade. Success bred complacency; complacency bred neglect; neglect increased the risk—until the right combination of errors finally led to an accident of disastrous proportions. All parties—the shippers, Alaska, the Coast Guard, and the State of Alaska—shared in the complacency that produced this result.

At one level it is obvious that a combination of human actions and errors led to the *Exxon Valdez* disaster. Many have been scrutinized in the public record, particularly the proceedings of the National Transportation Safety Board. Students of maritime disaster will not be surprised: human error is involved in 85 percent of all marine casualties. The root of this disaster—departing from traffic lanes—was not unique: The 1967 *Torrey Canyon* grounding off England took place when the captain left traffic lanes to save time.

Yet behind all human actions in the Valdez tanker trade, supporting the men and women who load and operate the tankers, is a system—one whose design and function clearly failed that night in Prince William Sound.

The system includes hardware in the form of pipelines, terminals, storage tanks, loading facilities, tankers, and all the associated gauges, meters, and machinery that operate them. It also involves operating instructions in the form of technical and design standards, international protocols, capacity ratings, terminal procedures, loading instructions, contingency plans, pilotage rules, maritime rules of the road, local navigation regulations, vessel traffic monitoring, and economic and career pressures on all participants. Finally, the system involves institutional oversight in the form of corporate management, private insurance systems, state inspection and enforcement, local port management, and Coast Guard regulation.

The objective is to move oil safely across the seas regardless of inevitable human error. System design must provide for redundancy—backup systems to prevent error from becoming disaster, and overbuilding to provide for wider margins of error. Proper functioning of the whole system requires constant testing, inspection vigilance, cooperation, discipline, expertise, and commitment of organizations at every level of government and industry.

Yet for reasons of maritime tradition, economics, politics, public policy, and modern practice, the maritime oil transport system is relatively more error-prone than safety-inducing. Industry tends to measure success as operating the biggest vessel with the thinnest hull and the smallest crew at the highest speed with the quickest port turnaround consistent with meeting minimum government requirements. Efficiency in a competitive world dominated by profit is all-important in the oil transportation business, even in the Alaska trade where transportation competition is muted.

A comparison between the nation's passenger air transport system and the maritime transport system is instructive, if not exact. Air transport safety is better reinforced, backed up, and institutionally safeguarded than maritime transport.

- Mistakes in the cockpit are more easily challenged than on the bridge. Air pilots share responsibility with co-pilots and foster teamwork in the cockpit.

Marine masters hold absolute authority, sharing little command responsibility with other ship officers.
- Air traffic control is mandatory, and ground controllers share responsibility with air pilots for safety of takeoffs, landings, and approaches. There is no equivalent to ground control in marine transport, and vessel traffic systems are typically only advisory.
- The federal government imposes strict standards and enforcement carried out by the Federal Aviation Administration in air transport. Federal presence in the marine environment falls to the Coast Guard, already stretched thin.
- Strong international cooperation governs air transport practices. Competition reigns in the maritime field, and cooperation and safety suffer.
- Air transport crew working conditions reflect strictly enforced limits on numbers of hours. Overwork and long hours are routine aboard ship and resulting fatigue considered part of the job.
- Airline accidents get extensive media coverage, partly because most of us travel by plane from time to time and can identify with the victims and their families. Victims of marine accidents—crew, fishers, villagers, wildlife—are more likely to be anonymous.

The analogy to air transport is not perfect. The issues described here reflect institutional settings, demands, and traditions that go beyond considerations of safety. But two points illustrate the relevance of the comparison.

First: Every day there are approximately 17,000 airliner departures in the United States. Ordinarily, every single one arrives safely at its destination. The *Exxon Valdez* was a catastrophic failure—the oil transport equivalent of a major airliner crash. Studies performed for the commission indicate that a catastrophic failure such as the *Exxon Valdez* disaster can be expected to occur in the Valdez tanker trade approximately every thirteen years, or about once every 11,600 transits. At a similar rate of catastrophic failure, the air transport system would produce 1.5 airliner disasters every single day, or 550 per year. If an average of 150 people died in each airline crash, such an accident rate would result in the loss of about 82,500 human lives per year—an unthinkable carnage that is prevented by a tight, safety-reinforcing system of regulation and oversight.

Technological and human systems aren't perfect: Airliners occasionally do crash. But we have built a system that does not tolerate in air traffic anything like the catastrophic failure rate we expect in the Valdez tanker trade. Because of that system, air travel can be considered safe and reliable. Risk cannot be eliminated, but it can be reduced—if we accept the costs involved.

Second: As vessels carrying oil and other hazardous materials impose higher and higher risks upon the world's oceans and coastlines, the environmental and social costs of marine transport accidents increase. The growth of a massive international system of transportation of oil by sea since World War II has not been accompanied by the development of organizations and active constituencies of those affected by the environmental hazards inherent in the trade. Those stakeholders, however, deserve increasing attention, for the risks they suffer are growing as the world's oil transportation system grows. And the marine transport system must become tighter and more safety inducing as the costs of failure grow more serious and more pervasive.

Prince William Sound, like most of Alaska, is a gift to us all—"God's finest creation," in the words of one commission witness, "next to human beings." As stewards of Alaska's resource wealth, natural beauty, and environmental integrity, Alaskans (indeed all Americans) have an obligation to account for both risks and benefits in the development of the state and its resources. For a time, the *Exxon Valdez* disaster hocked sensibilities, numbing confidence that oil can be transported with a decent respect for both environment and economic opportunity. As the shock fades, however, what matters is our ability to face present risks squarely.

In the realm of oil transportation, the social tradeoff that must be faced is this: How much risk to the environment are we willing to tolerate in order to gain the benefits of inexpensive, efficient delivery of crude oil to market? What is the cost of that risk? Who pays? For more than a decade before the wreck of the *Exxon Valdez* the managers and overseers of the Valdez tanker trade looked away while tanker safeguards decayed. They behaved as though the risk had been overcome, as though tunnel vision and luck somehow could protect us from disaster. The oil industry and the Coast Guard established policies that pursued this myopic vision; Alaskans and their leaders tolerated them. But the fantasy of a risk-free, high-tech world is just that: a fantasy we cannot afford. The risk is real and serious; the *Exxon Valdez* disaster is a powerful demonstration that as a people we must carefully review that risk and choose a balance between remedies and benefits.

The grounding of the *Exxon Valdez* was not an isolated, freak occurrence, but simply one possible result of policies, habits and practices that for nearly two decades have infused the nation's maritime oil transportation system with increasing levels of risk. The *Exxon Valdez* was an accident waiting to happen, the link that broke first in a chain with many unreliable couplings. The specific lapses that permitted the *Exxon Valdez* to run aground on Bligh Reef are being remedied, but similar circumstances easily could be repeated in some other combination to allow some other disaster. What is required now is comprehensive action to reduce overall risk in the system.

The recommendations in this report—safety inspections, crew levels, double hulls, traffic control systems, responses depots, training policies, citizens oversight, and all the rest—are intended to accomplish just that. Alaskans, indeed all Americans, must insist that these safeguards be implemented to protect an increasingly threatened natural environment.

Questions

1. What elements of this Conclusion are persuasive writing? Are the views expressed here convincing? Why or why not?
2. What kind of ethos does the Alaska Oil Commission establish in this Conclusion? How is this ethos established?
3. How are Aristotle's three modes of persuasion relevant to this Conclusion? (Remember, *ethos* is the character of the writer, the writer's integrity, competency, objectivity. *Logos* is the nature of the message or the quality of the data. *Pathos* is the emotion of the audience, the possible benefits for the audience from understanding and applying the data.)
4. What is the tone of this Conclusion? How is this tone achieved?
5. In what ways is the airlines analogy effective? In what ways does this analogy fall short?
6. What other major conclusions are possible but not mentioned here?
7. How would you apply the principles of Rogerian argument to this conclusion? What kinds of style decisions and organizational changes would have to be made?
8. What possible different views of the *Exxon Valdez* disaster might Exxon have that are not reflected in this report?

Questions/Topics for Discussion

1. What is persuasion and why is it important for writers of technical prose to understand persuasive strategies? Is it important to recognize technical prose as rhetorical and persuasive rather than as objective?
2. What do you think is the meaning of "scientific objectivity"? Do you think the phrase is misunderstood?
3. What is the "window pane theory" of language? Do you think this theory is still pervasive?
4. Do you agree with the four practical consequences for writers of technical prose of the new rhetoric of science? Why or why not? Can you suggest any other consequences?
5. What are some of the key differences between the traditional rhetoric of science and the new rhetoric of science?
6. In what ways is *ethos,* or the character of the writer, important in a technical document? How does a document which is collaboratively written have an *ethos*?
7. What kinds of stylistic choices do you make when using the Rogerian method of argument?
8. How can Rogerian argument be effectively used to write a proposal for a grant?
9. In what ways is a letter of application for a job a persuasive document? Is a resume a completely objective document?
10. In *Zen and the Art of Motorcycle Maintenance* Robert Pirsig comments on a set of instructions for a barbeque rotisserie. "What's really angering about instructions of this sort is that they imply

there's only one way to put this rotisserie together—*their* way. And that presumption wipes out all the creativity. Actually there are hundreds of ways to put the rotisserie together..." (p. 166). What does Pirsig's comment have to do with the rhetorical view of technical prose discussed in this chapter?

Exercises

1. Find a technical document—a memo, letter, report, proposal, or manual—which you think is very persuasive. What are the persuasive strategies used? Why are they successful?
2. Read the following passage. Do you think it is persuasive? If so, what strategies contribute to its persuasiveness? Why are they successful? If you think it is not persuasive, in what ways does it fail to persuade?

   ### Important Notice

   It has been brought to our attention by several policyholders that an outside insurance agent has contacted them proclaiming they had a better insurance program offering more cash value and higher death benefits using approximately the same premium. They encourage you to cancel your current plan and sign up with them.

   ### Beware not to Fall Victim to this Approach

   The insurance market today is very complicated. It is very difficult to decide which company or product is the right one for your individual needs. Every insurance agent is provided sales material stating in one form or another that their insurance company is "the best" and their products out perform their competitors. Most sales people with the help of personal computers have the ability to run ledgers and change the products to make their program stand out.

   Remember, there are a lot of good insurance companies offering some good products. If there was one company that was so much better than all the others, we would certainly make it available to you.

   If you are presented a plan that you like and feel comfortable with and want to purchase it as an additional program, go ahead. If you see a plan you are interested in and want to send us the information, we will be glad to look at it for you. But NEVER let anyone tell you to cancel your present program to start a new one. Usually the only one to benefit is the salesperson.

3. Using the methods of Rogerian argument, write a letter of complaint requesting an adjustment for a product or service. Identify the ways using these methods helped make the letter more persuasive. Then rewrite this letter of complaint relying, as much as possible, only on stating the facts as neutrally as possible. Try not to employ any appeals to the point of view of your reader and avoid stating your view directly. How do these decisions affect the potential success of the second letter?
4. Select a few paragraphs from a scientific journal article. Do you think these paragraphs are expressive writing? Why or why not?
5. Consider the following passage. How does the writer express his enthusiasm for nature? How does this passage differ from an article on the same subject you might read in a scientific journal?

It was then that I saw the flight coming on. It was moving like a little close-knit body of black specks that danced and darted and closed again. It was pouring from the north and heading toward me with the undeviating relentlessness of a compass needle. It streamed through the shadows rising out of monstrous gorges. It rushed over towering pinnacles in the red light of the sun or momentarily sank from sight within their shade. Across that desert of eroding clay and windworn stone they came with a faint wild twittering that filled the air about me as those tiny living bullets hurtled past into the night.

It may not strike you as a marvel. It would not, perhaps, unless you stood in the middle of a dead world at sunset, but that was where I stood. Fifty million years lay under my feet, fifty million years of bellowing monsters moving in a green world now gone so utterly that its very light was traveling on the farther edge of space. The chemicals of all that vanished age lay about me in the ground. Around me still lay the shearing molars of dead titan-otheres, the delicate sabers of soft-stepping cats, the hollow sockets that had held the eyes of many a strange, outmoded beast. Those eyes had looked out upon a world as real as ours: dark, savage brains had roamed and roared their challenges into the steaming night.[56]

6. In the following example taken from a computer operation and installation guide, the writer conveys important information to the new computer user. Discuss ways the writer attempts to be objective or expressive.

Drive Names

DOS assigns a drive designation letter to each diskette drive and fixed disk drive that is connected to the computer when DOS loads. The operating system starts assigning letters with the diskette drives, and assigns the letter A to the first diskette drive, B to the second diskette drive (if installed), C to the optional fixed disk drive's first partition, and D to the next partition (if more than one partition exists).

NOTE: Drive letter B is reserved for diskette drive use even when a system has only one diskette drive. In that case, DOS allows you to specify B as a source or destination in a command, and prompts you to change diskettes at the proper times.

If a fixed disk drive contains multiple partitions, they are assigned the next letters available.

If a RAM disk is installed (by a command in your CONFIG.SYS or AUTOEXEC.BAT file), it is assigned the next letter available after all others have been assigned. See your MS-DOS User's Guide for more information about RAMDRIVE. SYS (virtual or "RAM" disks).[57]

7. After the reading the following passage, do you think the writer has established a strong credibility? If so, how has he done that; if not, what prevented his writing from being credible to you?

Every time he enters a hospital, the visitor enters with two simultaneous thoughts: He hates hospitals, and only people working in them lead serious lives. Everything else is selfish. Entering a hospital he always thinks, I should work for a year as a nurse, an aide, a volunteer helping people, coming to terms with disease and death. This feeling will pass the moment he leaves the hospital. In reality the visitor hopes his fear and depression are not evident on his face as he walks down the gleaming, silent hall from the elevator to his friend's room. He is trying hard to stay calm.

The door of the room the receptionist downstairs has told the visitor his friend is in is closed—and on it are taped four signs that are not on any of the other doors and are headlined, WARNING. The visitor stops as much to read them as to allow his heartbeat to subside before going in. He knows—from the accounts of friends who have already visited—he must don a robe, gloves, mask, and even a plastic cap. He is not sure if the door is closed because his friend is asleep inside or because the door to this room is always kept closed. So he pushes it open a crack and peers in. His friend is turned on his side, a white mound of bed linen, apparently sleeping. The visitor is immensely relieved....[58]

8. Read the following passage. How does the author convey a friendly tone to the reader?

Welcome to the Electronic Universe!

It's an Electronic Universe in which messages and information streak across the continent or around the world at the speed of light. It's a place where you can find a fact or find a job, play a game, publish a poem, meet a friend, pay your bills, or do thousands of other things without ever leaving your home or your office.

The Electronic Universe, in short, is a realm of myriad opportunities destined to forever alter the way each of us lives, works, and plays. And it is here—waiting for you—today!

This book will show you how to tap in. It will show you where the goodies are and how to avoid the junk, regardless of whether you go online for work or for pleasure. It will build a foundation of understanding—a matrix that will show you where everything fits. And it will do so as quickly and easily as is humanly possible.[59]

9. Find an impersonal technical document. What elements make the document impersonal?

Notes

1. Russell Rutter, "History, Rhetoric, and Humanism: Toward a More Comprehensive Definition of Technical Communication," *Journal of Technical Writing and Communication* 21.2 (1991): 144.
2. Carolyn R. Miller, "A Humanistic Rationale for Technical Writing," *College English* 40 (1979): 616.
3. Harley Sachs, "Rhetoric, Persuasion, and the Technical Communicator," *Technical Communication* 25.4 (1978): 15.
4. Patrick Kelley and Roger Masse, "A Definition of Technical Writing," *Technical Writing Teacher* 4 (1977): 95.
5. David Locke, *Science as Writing* (New Haven: Yale UP, 1992) 112.
6. Rob Day, *Scientific English: A Guide for Scientists and Other Professionals* (Phoenix, AZ: Oryx, 1992) 1.
7. Gunther S. Stent, preface, *The Double Helix* (New York: Norton, 1980) ix.
8. Stephen Jay Gould, *The Mismeasure of Man* (New York: Norton, 1981) 21.
9. Miller, p. 611.
10. Miller, pp. 611–612.
11. Miller, p. 612.
12. Miller, p. 612.
13. Locke, p. 59.
14. Michael Halloran, "Technical Writing and the Rhetoric of Science," *Journal of Technical Writing and Communication* 8.2 (1978): 82.

15. Miller, p. 615.
16. Stent, p. ix.
17. Gould, pp. 21–22.
18. Karl Popper, *Conjectures and Refutations: The Growth of Scientific Knowledge* (New York: Harper & Row, 1963) 46.
19. Locke, p. 44.
20. Locke, p. 210.
21. Gould, p. 21.
22. Merrill Whitburn, "Personality in Scientific and Technical Writing," *Journal of Technical Writing and Communication* 6.4 (1976): 305.
23. Whitburn, p. 299.
24. Lewis Thomas, *Late Night Thoughts on Listening to Mahler's Ninth Symphony* (New York: Bantam, 1984) 150.
25. Robert Pirsig, *Zen and the Art of Motorcycle Maintenance* (New York: William Morrow, 1974).
26. Miller, p. 614.
27. Miller, p. 614.
28. Merrill Whitburn, "The Plain Style in Style in Scientific and Technical Writing," *Journal of Technical Writing and Communication* 8.4 (1978): 353.
29. Miller, p. 614.
30. Miller, p. 615.
31. Locke, p. 208.
32. Locke, p. 86.
33. Locke, p. 85.
34. Locke, p. 66–67.
35. Locke, p. 67.
36. Locke, p. 67.
37. Locke, p. 67.
38. Locke, p. 68.
39. Locke, p. 64.
40. Locke, p. 91.
41. Quoted by Joseph Harmon in "Perturbations in the Scientific Literature," *Journal of Technical Writing and Communication* 16.4 (1986): 315.
42. Harmon, p. 315.
43. Harmon, p. 315.
44. See Scott Sanders' "How Can Technical Writing Be Persuasive?" *Solving Problems in Technical Writing,* ed. Lynne Beene and Peter White (New York: Oxford UP, 1988) 65–67.
45. Richard E. Young, Alton L. Becker, and Kenneth L. Pike, *Rhetoric: Discovery and Change* (New York: Harcourt Brace Jovanovich, 1970) 282.
46. Young, Becker, and Pike, p. 282.
47. Sanders, pp. 71–72.
48. Sanders, p. 67.
49. Sanders, p. 67.
50. Sanders, pp. 67–68.
51. I am indebted to Sanders (pp. 68–69) for this connection.
52. Sanders, p. 68.
53. Sanders, p. 69.
54. Alaska Oil Commission Final Report, *Spill: The Wreck of the Exxon Valdez.* (State of Alaska. February 1990) ix.

55. *Spill: The Wreck of the Exxon Valdez*, pp. 205–210.

56. Loren Eiseley, "More Thoughts on Wilderness," *A Writer's Reader*, 7th ed. (New York: Harper-Collins, 1994).

57. *Personal Computer Operation and Installation Guide*. Rev. D. Text Block, 1992.

58. Andrew Holleran, "Bedside Manners," *A Writer's Reader*. 7th ed. (New York: HarperCollins, 1994) 227.

59. Alfred Glossbrenner, *The Little Online Book: A Gentle Introduction to Modems, Online Services, Electronic Bulletin Boards, and the Internet* (Berkeley: Peachpit, 1994).

Chapter 5

Choosing Appropriate Words—Diction

> Polonius: What do you read, my lord?
> Hamlet: Words, words, words.[1]
> (Hamlet, II, ii, 191–192)
> —WILLIAM SHAKESPEARE

> Writing is easy. All you have to do is cross out the wrong words.[2]
> —MARK TWAIN

> Technical writing shares much of the "common" English vocabulary. The verbs heard on trolleys and at tennis matches are also used in managerial conferences and over drawing boards; and the verbs are conjugated the same in technical writing as in other prose styles. The workhorse articles a, an, and the appear in reports just as they appear on the front page of the daily paper.[3]
> —ROBERT HAYS

Chapter Overview

On Diction

Levels of Diction

Denotation and Connotation

Challenges Presented by Specialized Diction

Major Diction Strategies in Technical Prose

Common Diction Faults

Technical Prose and a Literary Style

On Euphemisms and Dysphemisms

Chapter Summary
Case Study 4: The Ruby Ridge Incident and Diction
Questions/Topics for Discussion
Exercises
Notes

On Diction

Diction is your choice of words. A writer's style is characterized at this level, the level of the word, just as it is at the larger levels of the phrase, clause, sentence, paragraph, or other segment. Whether the style is informal or formal, plain or complex, unpretentious or affected, technical or literary, writers have a tendency to prefer consistently some kinds of words to others to achieve these distinct styles. Some writers prefer to write as plainly as possible, regardless of the subject. They choose the simpler, concrete, nontechnical word instead of the complex, abstract, and scientific one. Some writers favor a more complex style and select bigger words over shorter words, some abstract over some specific words, and some technical terms over some nontechnical terms. Some writers habitually use emotionally charged words while others prefer nonemotive words.

Writers of technical prose face special challenges with diction. They have, for example, the ever present demands of handling scientific and technical terms. Other writers also use many kinds of specialized terms, but for writers of technical prose a technical vocabulary is a key part of their style. How to deal with the technical terminology of a profession or trade (and how to avoid the temptation of using pretentious jargon) will be covered in detail in Chapter 6.

In addition to handling technical terms, writers of technical prose are faced with the challenges of achieving accuracy, precision, appropriateness, clarity, conciseness, concreteness, consistency, and sincerity in their word choices. And many writers of technical prose make the common diction mistakes of using cliches or creating redundancies, to mention only a few of the many examples covered later in this chapter. In all of these areas, writers of technical prose face the same challenges and often make the same kinds of mistakes as do any other kind of writers. After all, knowing how to choose the best or most appropriate words to suit the subject, purpose, audience, and context in large part determines the success of any kind of writing.

So the focus of this chapter is not only on common diction strategies and common errors for writers of technical prose, but also diction problems that all writers encounter. All writers have problems with words. It's not easy to know in a particular communication situation what are the best, most effective, most persuasive, most accurate, or most appropriate words.

It's also impossible to address all of the diction concerns that are important to writers of technical prose because there are so many technologies, industries, specialties, and professions. Some writers in some industries have special diction challenges. For example, many industries employ specifications writers, people who write about exactly what kinds and qualities of materials will be used to build everything from a house to a missile. These writers have diction problems that may not occur for many writers of software documentation or training manuals. Specifications writers must be continually alert to the legal ramifications of every word chosen so that a house or a missile is built the way it's supposed to be built. Lots of words are meaningless in specifications writing. Words such as *about, highest grade,* and *substantial* should not be used in specifications unless they are accompanied by words that convey additional information. (See **Meaningless Words** later in this chapter.)

Levels of Diction

Just as there are many levels of formality in style—from highly formal to highly informal—there are many levels of diction. The four generally accepted levels of diction are *formal, informal, colloquial,* and *slang.* Any of these four levels may be correct in a particular context but incorrect in another. And sometimes the levels are mixed unintentionally with poor results.

Formal Diction

Formal means following an established form, custom, or rule. A formal diction is most often used in a formal context. In speech, for example, a formal diction is usually expected at a swearing-in ceremony for a government official, keynote addresses at conferences or conventions, and other kinds of public functions. In writing, a formal diction is found, for instance, in legal documents such as software licensing agreements or most contracts, many types of government documents such as requests for proposals or specifications writing, numerous kinds of abstracts, and most academic journal articles.

You often choose a formal style if the seriousness of the occasion demands it. As with the other levels of diction, whether you must use a formal diction depends on your purpose, the genre of the technical document, the audience, the context, and other factors. But sometimes genres may be more informal than some writers imagine. The following prose example illustrates a typical formal diction from an academic collection of articles on software documentation:

> The emergence of professional writers who document computer systems is a phenomenon of the last two decades. During this period, the development of software applications has experienced extremely rapid growth. There is a continuing need for well-trained, competent writers to document computer software applications for users of these products.
> In the late 1960s and early 1970s, many people became software documenters quite by accident. They were either technicians who could (or were forced to) write, or they were people trained in the humanities who were hired by software development organizations because they could write and were able to assimilate the required technical information. With the development of more software applications for nontechnicians, software companies gave increased attention to their documentation as an important component for marketing their products and for establishing a satisfied customer base. The need for better trained "professional" software documenters has not only rapidly escalated, but academic degrees, certificates, and company training programs for training them have proliferated.[4]

Informal Diction

Informal refers to ordinary, casual, or familiar use. Many people believe most technical documents must be formal because, after all, the formality of the style, they believe, is part of what makes the document technical. Actually, there are probably as many occasions to write informal technical documents as there are to write formal ones. Many types of manuals, for example, use an informal diction to set the user at ease and to help the user understand the steps more readily. The following is an example of typical informal diction in a guide to a software program:

> What is a field guide?
>
> Sometime during grade school, my parents gave me a field guide to North American birds. With its visual approach, its maps, and its numerous illustrations, that guide delivered hours of enjoyment. The book also helped me better understand and more fully appreciate the birds in my neighborhood. And the small book fit neatly in a child's rucksack. But I'm getting off the track.
>
> This book works in the same way as that field guide to North American birds. It organizes information visually with numerous illustrations. And it does this in a way that helps you more easily understand and, yes, even enjoy working with Microsoft Windows 95. For new users, the Field Guide provides a visual path to the essential information necessary to start using Windows 95. But this Field Guide isn't only for beginners. For experienced users, it provides concise, easy-to-find descriptions of Windows 95 tasks, terms, and techniques.[5]

Colloquial Diction

Colloquial refers to conversation or diction used to achieve conversational prose. The term *colloquial* also may be used pejoratively for informal, everyday speech, including slang. Colloquial language is used in or characteristic of familiar and informal conversation. *Colloquial* may also mean unacceptably informal. The following is an example of colloquial diction in a technical manual, a level of diction that rarely occurs in technical writing:

> June 4
>
> Major brainstorm last night! Woke up in a babbling frenzy. Images of gears, motors, fan belts, generators, all sorts of gizmos and gadgets buzzing through my brain. I'm jotting down what I can remember in this journal. These notes will record my progress as I undertake construction. The basic concept is this: the world is in danger of becoming boring and brain-dead and what we all need is some kind of major outrageous mind-blowing challenge! I'm talking about BIG TIME challenge. Something that will make you really wrack your brain and THINK. But it's gotta be fun! It's gotta be wild and intense and full of weird twists and tricks. And that's when it hit me: a machine! An incredible machine! A machine that changes into a hundred different puzzles, games, and gadgets! And so...I'm taking on this outrageous task of inventing this machine. What will it do? How will it work? The possibilities are endless!

July 16

Another major eureka today! Found a totally wild way of producing mechanical energy via critter power: the hamster and the monkey bike. Low overhead, too: a few hunks of moldy cheese and a crate of bananas, and I'm in business!

It won't be long now! The even more Incredible Machine is almost ready....[6]

Slang

Some people make a distinction between colloquial language and slang, and some people do not. The two types of diction certainly overlap. Yet slang has its own distinctive qualities. It is the most informal, and often it is the most subversive of the levels of diction.

Slang may be a language peculiar to a particular group or an informal nonstandard vocabulary composed typically of coinages, arbitrarily changed words, and extravagant, forced, or facetious figures of speech. In *American Slang* Robert Chapman offers a helpful distinction: "Slang is language that has little to do with the main aim of language, the connection of sounds with ideas in order to communicate ideas, but is rather an attitude, a feeling, and an act. To pose another paradox: Slang is the most nonlinguistic sort of language."[7]

Chapman distinguishes two kinds of slang: primary and secondary. According to Chapman, "Primary slang is the pristine speech of subculture members, so very natural to its speakers that it seems they might be mute without it.... Much teenage talk, and the speech of urban street gangs, would be examples of primary slang."[8] For Chapman, secondary slang has a different function:

> Secondary slang is chosen not so much to fix one in a group as to express one's attitudes and resourcefulness by *pretending,* momentarily, in a little shtick of personal guerilla theater, to be a member of a street gang, or a criminal, or a gambler, or a drug user, or a professional football player.... To express one's contempt, superiority, and cleverness by borrowing someone else's verbal dress. Secondary slang is a matter of stylistic choice rather than true identification.[9]

Slang comes very close to jargon. But, according to Katie Wales in *A Dictionary of Stylistics,* "jargon is best reserved for technical vocabulary arising from rather specialized needs."[10] Slang is closer to a secret language whereas professional jargon is generally more formal. Wales tells us "slang is less socially 'acceptable' than jargon, and more socially subversive, to be defined almost as a DEVIATION from standard usage."[11] The following example illustrates some of the slang typically used in a pc newsgroup (a discussion group devoted to personal computer issues) on the Internet:

> Apple computer is taking the pipe and there is nothing, no force in the universe, that can stop it. Their products are being dropped like hot potatoes by nearly every big corporation and being relegated to a few—very minor—roles in the departments where the weenies and air-heads work. Every day, fewer and fewer software vendors support this platform and increasing numbers of THEM announce their first or only releases for Windows/PC. Sales fall, Apples's prices and, as a result, margins slip downward—this MUST affect R&D spending. And, virtually nobody else is spending R&D money on this platform, at least as far as hardware is concerned.[12]

At its best, slang is often fresh, colorful, and vigorous language. Inventing or using slang is a way of filling in language gaps, finding a more precise word or phrase to articulate, and reinventing and building up an already existing language.

People use slang for a variety of reasons: just for fun; for humor; to be different; to be startling or surprising; to avoid clichés; to be concise; to enrich the language; to make the abstract, idealistic, or remote more concrete, earthy, or immediate; to lessen or enhance a refusal or rejection; to reduce the solemnness of a conversation or document; to address an inferior or amuse a superior; to facilitate conversation; to enhance friendliness; to show membership in a certain discourse community; and to be secretive.[13]

Denotation and Connotation

Words are like bullets. Fire one into a mattress and you get a predictable reaction. That's denotation. Fire one into a crowded auditorium and you can't be sure what reaction you will get beyond the noise. Will people freeze or panic? Will the sound echo or go dead? Will there be backfiring, ringing, reverberation? Whatever the result, that's connotation.[14]

—THOMAS WHISSEN

Denotation

Denotation, also known as the *cognitive* meaning, concerns the direct and specific relationship between a word and an object, action, or idea. Colors, for example, can be described in terms either of wavelengths of reflected light, or simple associations to everyday life: *red* with blood, *white* with snow, *green* with grass, *blue* with sea and sky.[15] These are all denotative meanings.

Connotation

Connotation, also known as the *affective* meaning, concerns the suggesting of a meaning by a word apart from the thing it explicitly names or describes, or something suggested by a word or thing. For colors, for example, more implicit associations can be made—*red* with anger or irritation, *white* with purity and innocence, *green* with inexperience or envy, *blue* with sadness and depression.[16] These meanings are connotations. Or consider the example of an emotionally charged word such as *feminist.* One person may associate the word *feminist* with individual freedom, the right to choose, sophistication, intelligence. Another person may associate *feminist* with overbearing women, forgotten rights of the unborn, social naiveté, and stubbornness. One person may associate the word *computers* with what's wrong with technology while someone else may associate *computers* with medical miracles.

Challenges Presented by Specialized Diction

As we mentioned earlier, writers of technical prose are more challenged by the demands of a specialized vocabulary than other kinds of writers. This topic is so complex and important that an entire chapter is devoted to it (See Chapter 6). Whether you are a long-standing member of a discipline, profession, trade, or hobbyist group, or a novice, the specialized vocabulary of your discourse community presents many challenges.

For example, one of the biggest problems faced by novices in any profession or discipline is the challenge of dealing with overwhelming information about a technical subject during their first year on the job. Whether the position is in aerospace, computers, engineering, or telecommunications, entry-level employees are faced with the enormous challenge of learning the specialized language of their particular industry. What they learned in college is helpful, but it is only the beginning. This information often consists of far more than just the terminology specific to an industry, but it is this terminology that most often presents the biggest challenge.

Of course, the more experienced members of any technical profession, trade, discipline, or hobbyist group also require a great deal of time to stay current with the language changes in their discourse community. These experienced members are still challenged with all kinds of do's and don'ts concerning the terminology within the community and even specialized terminology outside the field. For example, in *Science and Technical Writing: A Manual of Style,* Philip Rubens devotes an entire chapter to specialized terminology, offering guidelines for handling personal names, titles, honors and awards, foreign names, forms of address, geographic names, military terminology, health and medical terms, scientific terms and symbols, and technology symbols, to mention only some of the most important topics.[17] Experienced writers of technical prose and novices alike have difficulty keeping up with all of the rules concerning these topics and others.

Unfortunately, there are no shortcuts to learning the technical vocabulary you need to know to write well in your subject area. A strong command of English grammar, a basic knowledge of Latin, an affinity for science and technology, and the ability to do good research all help, but the knowledge of a technical subject is never acquired easily or casually. However, the strategies covered in the next section are a good beginning for gaining control over the particular challenges of a specialized diction.

Major Diction Strategies in Technical Prose

In addition to becoming familiar with the strategies for handling technical terms in the previous section (and in Chapter 6), there are some larger strategies you can use to help you control your technical prose—especially the diction of this technical prose. In every document you need to cultivate the following: accuracy and precision, active verbs and active voice (if applicable), appropriateness, clarity, conciseness, concreteness, consistency, simplicity, and sincerity.

Accuracy and Precision

To be accurate means to achieve freedom from mistake or error. To achieve a technical accuracy is to achieve an exactness in your technical prose. The information must conform to truth or to a standard or a model. Writers of technical prose must take great care to use accurate technical terms. *Accuracy* is close in meaning to the term *precision,* but there are some important differences. *Precision* refers to the degree of refinement with which an operation is performed or a measurement stated.

Consider the following terms, for example: *absorb, adsorb; cement, concrete;* and *mass, weight.* To a layperson there may not be much of a difference between each word in these pairs. To a specialist, however, there are essential differences. A blotter will *absorb,* but a glass beaker will *adsorb.* People in the construction business know that *cement* comes in bags, and *concrete* is something made by mixing *cement* with sand, crushed stone, and water. To a specialist *weight* varies depending on the location where the measurement is made; *mass* will be the same (for example, a gram) regardless of where the measurement is done. It's essential that these technical terms, and countless others, be used accurately and not interchangeably.

The following are some examples of inaccuracy in meteorology. A meteorologist recently said, "A severe storm has dumped a lot of moisture over Central and South Georgia." *Moisture* is "water or other liquid causing a slight wetness or dampness." Is it possible to dump *moisture*? Another meteorologist, standing in front of a Doppler radar screen, points to an incomplete eye in Hurricane Erin and tells us the eye is incomplete because there is "not enough weather" for it to close up. *Webster's* defines *weather* as "the state of the air or atmosphere, as dry weather, or wet weather." The meteorologist should have said that there were not enough winds converging on the low-pressure center to complete the eye. A third meteorologist refers to *pop-up showers. Showers* are minor episodes of rainfall, without the more-threatening aspects of thunderstorms. Also, rain falls down; it does not *pop up.*

As Chapter 6 shows, writers of technical prose have problems not only with technical terms but also terms in the general vocabulary. Writers of technical prose are just as likely as other writers to confuse *that* and *which, affect* and *effect, lay* and *lie,* and numerous other commonly confused pairs. See **Confusibles** later in this chapter.

Accuracy, of course, is not only a matter of style. It's also a matter of content. One of the biggest challenges facing writers of technical prose is the challenge of assuring that statements about the subject are correct, that is to say, are technically accurate. Writers of technical prose are faced with the same problem of any writer—you have to know your subject well. Technical subjects, however, can be and often are extremely complex. As we discussed in Chapter 2, there are levels of technicality—low, medium, and high—and achieving all three levels in terms of both style and content requires a thorough knowledge of the subject. As Chapter 12 discusses, one important part of the editing process is substantive editing, and an important part of this kind of editing is checking for technical accuracy.

Active Verbs and Active Voice

Two important strategies for adding more energy or power to your writing are to express action in your verbs and to use the active voice. Both strategies help to focus your writing on the actors (or doers) and the action.

Often writers bury the action of a sentence in nouns, adjectives, or other parts of speech. Writers create sentences such as the following: *The computer technician accomplished the diagnosis of the computer problem and recommended the addition of more memory.* In this sentence the nouns—*diagnosis, addition*—are trying to carry the sentence. With a little revision, the verbs could do most of the work: *The computer technician diagnosed the computer problem and recommended adding more memory.* This revision is shorter and more direct.

Good writers make abundant use of action verbs. Action verbs are specific verbs which descriptively tell the audience what was or will be done. They make your writing more direct and forceful. Action verbs in an employment letter may include: *performed, evaluated, supervised, achieved, reduced, simplified, streamlined, managed, deposited, distributed, maintained, increased,* and *won.* The list of possible action verbs for an employment letter is almost endless.

In instructions, simple imperative statements use action verbs to tell readers exactly what needs to be done: "tighten by hand only," "turn one quarter turn clockwise," "open the Control Panel and double-click the system icon," or "remove the cover and locate the internal modem card."

Lewis Thomas offers us an effective example of an action verb when he writes: "We explode clouds of aerosol, mixed for good luck with deodorants, into our noses, mouths, underarms, privileged crannies—even into the intimate insides of our telephones."[18]

To place the action in the verbs of your sentences, ask yourself what the main action of the sentence is, identify the actor, and try to begin your sentences with the actor followed by the main action. For example, consider the following sentence: "The new computer was set up by the technician so the home user could begin working immediately." The main action concerns setting up the computer. The actor is the technician. A possible revision would be: "The technician set up the new computer so the home user could begin working immediately."

Active voice is another strategy for helping your reader recognize the agent or doer of an action, and, like action verbs, active voice helps to make your writing more direct and forceful. Your readers are told quite clearly who is doing what to whom. To write in the active voice, place the actor in the subject position of the sentence. For example, write "The systems administrator set up the Novell Netware network in the lab so twenty workstations could share programs and printers." Or write "This program installs itself after you double-click the executable file." In the first sentence, you know who set up the network, and in the second sentence you know the program installs itself. Active voice helps readers recognize more quickly who is doing what action, active voice helps to make sentences shorter, and active voice also helps avoid the ambiguity created by using passive voice.

Passive voice makes the agent of the action less clear. Sometimes passive voice weakens a sentence because it increases the ambiguity. The following is an example of passive voice: "Every evening the lab must be cleaned and the computers checked for viruses." In this sentence it is not clear who must perform the actions. In the following revision, the actor is clearer: "Every evening you must clean the lab and check the computers for viruses."

Passive voice makes sentences more ambiguous and often longer, but sometimes passive voice is necessary. For example, sometimes you do not want to identify the actor, as in the following sentence: "The lab door has been left unlocked overnight for three days. In the future check this door before you leave the building." If you do not know who is leaving the door unlocked, you may want to use passive voice to avoid pinning blame on any one individual. As you learned in Chapter 3, a passive style is a legitimate style. There are many reasons for using passive voice, and you shouldn't hesitate to do so when it suits your purposes.

Appropriateness

Appropriateness means suitability or compatibility. Of course, the topic of appropriateness refers to far more than just your word choices. For example, the appropriateness of your tone is also an important issue in style. (Chapter 9 discusses this issue in far more detail.) But the concept of appropriateness is so important that a definition is necessary. David Crystal comments that appropriateness tries to capture a notion of *naturalness* in language use: an appropriate use of language is one which does not draw attention to itself, does not motivate criticism. Informal language on a formal occasion is inappropriate because it stands out, as does formal language on an informal occasion.... The best uses of everyday language...are those which do not draw attention to themselves, and where the structures do not get in the way of the meaning they are trying to convey. To say that a usage is "appropriate" in a situation is only to say that it is performing this function satisfactorily.[19]

You must be careful to choose words that have the appropriate associations, and the most effective way to achieve this goal is to consider carefully the connotations of the words you use. (See the discussion of Denotation and Connotation earlier in this chapter.) Every word you choose helps to shape your readers' perceptions of your subject matter. Consider, for example, the word *scrutinize*. The word means to look at or over critically and searchingly. Synonyms are *eye, eyeball,* or *watch*. Telling a client you are going to *scrutinize* a contract has a different connotation than suggesting you are going to *eyeball* the contract. *Scrutinize* is a more appropriate diction choice and suggests you are going to examine the contract very carefully. *Eyeball* is a less formal word choice and suggests a more relaxed approach to examining the contract. Every word you choose can sway a reader one way or another concerning your subject and your seriousness, credibility, and persuasiveness.

Clarity

Clarity, simply defined, is the state of being clear. It depends on using the right words in the right way. Achieving clarity in technical prose is one of the biggest challenges for a writer because it is so easy to choose words, phrases, clauses, sentences, and larger segments that confuse or mislead the reader. Often you may think your choices are clear, but after a thorough edit of your work by someone else, you discover that a word may be misconstrued or a sentence interpreted in a different way from what you intended. Few strategies are more helpful (and often humbling) to writers of technical prose (or any kind of prose, for that matter) than the strategy of having someone point out to you the ambiguities in prose you think is perfectly clear. (See Chapter 12 for some helpful strategies for editing your work or the work of others.)

One way to gain a better understanding of clarity is to discuss its opposite: obscurity or ambiguity. *Ambiguity,* the quality or state of being unclear, is one of the biggest problems in technical prose. This ambiguity may occur at any or all levels of a technical document: the word, phrase, clause, sentence, paragraph, or other segment level. Words may be ambiguous for a variety of reasons.

For example, **double meanings** are one problem. If you tell someone to *check* a setting, do you mean to look at it, record it, or remember it? Or consider the following sentence: "Now that you've made your shell impregnate to the advances of even the most nasty overtures of the logout and exit commands, it's time to take advantage of the unique state they put your terminal into."[20]

Another problem is **relative words.** Words such as *top* and *bottom, front* and *back,* and *left* and *right* are relative words and depend on the location or position of the reader.

Another cause for ambiguity is **unusual words.** The word *protocol,* for example is an unusual word that has many meanings. In the example below, the term applies to a type of Internet connection. To establish a full Internet connection over a telephone line

> you must first arrange for some other Internet host to act as your connection point. You then install a set of programs on your computer called *PPP* (Point to Point Protocol). Once a telephone line connection is made between the two computers (using modems, of course), PPP will endow your computer with TCP/IP capabilities. This enables your computer to be a real Internet host with its own official electronic address.[21]

A frequent cause for ambiguity is **familiar words** used in an unusual sense, for example, "Check *okay, cancel,* or *apply* in the dialogue box." Is *apply* the same as *okay*?

Ambiguity is also caused by confusing syntax in a sentence. Commenting on the Chernobyl containment building, a consultant said: "It would be one of the biggest civil engineering projects ever." Does the consultant mean *expensive, architecturally challenging,* or *physically large*? Or consider this confusing dictionary definition: "Martingale: 2a: a lower stay of rope or chain for the jibboom used to sustain the strain of the forestays and fastened to or rove through the dolphin striker."[22]

Finally, ambiguity may also be intentional. In *The Careful Writer* Theodore Bernstein comments on **equivocal ambiguity:** "When a statement is *equivocal,* however, the ambiguity is usually considered to be intentional. *Equivocal,* therefore, carries with it nasty overtones that are not normally associated with *ambiguous.*"[23] See Chapter 11 for a discussion of intentional ambiguity, pretentiousness, and deception.

Conciseness

Conciseness refers to the brevity or leanness of your writing. It's a big challenge for writers to write sentences, paragraphs, and larger segments that express the intended meaning in the fewest possible words. Every writer tends to include many unnecessary words, especially in the early drafts of a document. Cutting out wordiness, the opposite of conciseness, requires a strong command of the language, a keen editorial eye, and lots of rewriting. Two weaknesses to look out for are unnecessary words and long words.

Consider, for example, the unnecessary words in the following sentence: "Boot up the computer, place disk 1 of the program disk in drive A, and, after placing disk 1 in drive A, run setup.exe in drive A." Instead try: "After starting the computer, place disk 1 of the program in drive A and run setup.exe."

And consider the long words in the following examples. Instead of "This work has involved the acquisition of skills and techniques required when working with...," write "In this work we have learned new ways of working with..." Instead of "The precise mechanism responsible for this antagonism cannot be elucidated...," write "We do not know what causes this antagonism...."[24]

Concreteness

Concrete, or *specific,* words identify specific realities, or anything that can be seen, heard, tasted, felt, or smelled. They refer to specific persons, places, objects, and acts. *Lewis Thomas* is more concrete than *homo sapiens, Disney World* is more concrete than *amusement park, Lockheed Martin* is more specific than *the defense industry,* a *Compaq Presario CDS 972* is more specific than *multimedia computer,* and *using a graphical browser to see links on the World Wide Web* is more specific than *accessing the Internet.*

In contrast to concrete words, *abstract* words refer to general ideas, acts, conditions, relationships, qualities, or anything that cannot be perceived by the five senses: *truth, generosity, tyranny, indifference, cowardice,* or *infinity.*

Of course, there are ranges or hierarchies of abstraction. Words may be highly abstract and general, or more concrete and specific. A *construction* is more abstract than a *building,* a *building* is more abstract than a *dwelling,* and a *dwelling,* in turn, is more abstract than a *house.* [25] Choose words suitable for your audience. You will use both types of words in your writing, but an overabundance of abstract words will mask your meaning and bore readers. Use concrete details in conjunction with abstract words as well as examples to explain the concepts.

Many abstract words present readers too many possibilities for interpretation. In technical prose abstract words are often necessary, but there are times when a more concrete word is a better choice.

Consistency

Another common problem for writers of technical prose is assuring consistency of diction throughout a technical document. Synonyms, of course, may be used in a technical document but should not be used excessively. Readers of technical documents are easily confused if you call one item several different names in the same document. For example, if a writer switches back and forth from *you, readers, users, end-users, customers,* and *audience,* the inconsistency may be confusing to anyone using the document.

And you must be careful about spelling or punctuating certain words consistently. You need to decide, for example, whether you are going to use *online, on-line,* or *on line,* or *discs* or *disks.* Will you use *World-Wide Web* or *World Wide Web*? Of course, decisions must be made about capitalization and usage as well. As we discussed in Chapter 1, many companies decide to use a style guide and standards manual largely to assure that documents produced by employees are consistent not only in their format or design but also in their style.

Simplicity

In Chapter 3 we discussed a wide range of styles available to writers of technical prose or to writers of any other kind of prose. As that chapter illustrates, sometimes an affected style is preferable to a plain style, or a noun style is preferable to a verb style. Most often in technical prose a simple or plain style should be your objective in the technical documents you write. The strategy of simplicity does not mean to patronize your audience or simplify your prose unnecessarily. After all, you are often writing about a complex subject. However, it does mean that as much as possible and as often as it suits your purpose, you should prefer the familiar word to the unfamiliar word, the simpler word to the more complex word, and a plain style to an affected or pretentious style. Many writers of technical prose create prose that is more complex than it needs to be. And many needlessly create jargon, in the sense of pretentious language, which is totally unsuited to their audience. (See the discussion of Affected Style in Chapter 3 and refer to the various meanings of *jargon* in Chapter 6.)

Sincerity

You may not think of sincerity as a problem in the diction of your technical prose. After all, why would someone want to lie or be misleading about how to install a software program in a user guide or how to create columns in an online help screen? In these two examples most writers would not want to deceive their readers. Of course, most software programs have undocumented features, features which are not covered in the printed manuals or on-line help. Users are often frustrated by not knowing how to perform a particular task because the task is not covered. Still, this problem is usually one more of omission than commission. In other words, the feature may have been undocumented due to lack of space, an oversight, or lack of time to do so before the user guide was shipped with the product rather than willful intent to defraud or deceive the customer. Chapter 11 discusses various communication situations in which writers are deliberately deceptive to hide information from an audience. Sometimes writers even have a good reason to withhold information, but, as you will see in this later chapter, often writers want to be ambiguous, deceptive, or pretentious to keep you from knowing what you need to know.

The advice here is to write truthfully and honestly, and, unless using humor suits your purpose, you should also treat your subject and your readers seriously. You should make every effort to choose words that help you to convey your tone, establish your credibility, and build trust with your audience (see Chapter 9). You need to be alert to those words, phrases, clauses, sentences, and larger segments that undermine the relationship you intend to build with your reader.

Common Diction Faults

Now that you have focused on the larger strategies necessary to improve your diction, you also must be careful about many other diction concerns. The faults covered here are common in the craft of writing, any kind of writing. For the sake of brevity, the faults are grouped in categories.

Archaic Words/Dialect/Colloquialisms

Archaic words are words having the characteristics of the language of the past and surviving chiefly in specialized uses. *Whereas* has a special meaning in law, for example. Other examples include: *albeit* (although), *aught* (anything), *methinks* (I think), *nigh* (near), and *zany* (a clown; a fool).[26] If using an archaic word will provide you with the intended effect, then you may consider using such words. But, in most technical prose, archaisms are rare and should be used only for a very good reason.

A *dialect* is a regional variety of language distinguished by features of vocabulary, grammar, and pronunciation from other regional varieties and constituting together with them a single language. *Dialect* may also be a variety of a language used by the members of a group. Another broad meaning of *dialect* is a manner or means of expressing yourself.

As we discussed at the beginning of this chapter, colloquial language is one of the four major levels of diction. A *colloquialism* is a local or regional dialect expression. It has more status than slang, and, at its best, overlaps significantly with familiar or informal English. Colloquialism comes from the root word *colloquy*. This word in turn comes from *colloquium*, which means to converse with or have a conversation. Colloquialisms are abundant in all kinds of writing but are not often used in technical writing. Some examples are the following:

> My father, Haven Peck, killed pigs for a living. Today, when reviewers write about my past, they describe Papa as a man who worked at *pork processing*. Even though Haven Peck was a Vermonter who didn't smile every year, I believe this term would have bent him a grin.[27]

> Sound like the online services market? It was called Videotex. It got skunked. Collectively, the losses were over $1 billion in 1982 dollars.[28]

> At this point, you're probably hankering for some examples. I know I am.[29]

Bloopers/Vulgarisms/Taboo Words/Barbarisms

A *blooper* is an embarrassing public blunder. The term also refers to embarrassments in writing. Church bulletins, accident reports, and signs in English in foreign countries are often a good source of bloopers. Consider for example, *"Women giving milk and cookies host church social,"* or some of the following signs: a sign in an Athens hotel—*"Visitors are expected to complain at the office between the hours of 9 and 11 a.m. daily,"* or a sign in a Rome laundry—*"Ladies, please leave your clothes here and spend the afternoon having a good time."*

Bloopers, of course, may appear in any kind of writing. You have to edit your work carefully to be sure you wrote what you intended to write.

According to Eric Partridge, "Vulgarisms are words that belong to idiomatic English or denote such objects or processes of functions or tendencies or acts as are not usually mentioned by the polite in company and are never, under *those* names, mentioned in respectable circles."[30] Slang words denoting the genitals of either sex, for example, are considered vulgarisms.

Related to vulgarisms are *taboo words,* "items which people avoid using in polite society, either because they believe them harmful or feel them embarrassing or offensive...." Words associated with certain other topics may also be called taboo, from time to time, because society is sensitive to them. During the *recession* of the early 1990s, newspapers would talk about "the R word."[31]

A *barbarism* is an idea, act, or expression that in form or use offends against contemporary standards of good taste or acceptability. Geoffrey Nunberg defines barbarisms as "expressions that could not legitimately be used in serious writing."[32] Barbarisms cover a wide range of categories from using foreign expressions in polite discourse to archaisms, cant, certain idioms, and jargon. One particular barbarism, borrowings from technical usage, concerns many modern critics of language usage. Many object, for example, to *prioritize, source, implement,* and *input* used as verbs. But, as Nunberg points out, "while critics are heated in their condemnation of such usages, they are rarely explicit about the reasons for their objections."[33]

Catchword/Catchphrase/Vogue Words

A *catchword,* or *catchphrase,* is a memorable word or phrase, repeated by many people. The word *"Not!"* as in *"I'm looking forward to the exam. Not!"* The word *awesome.* Typical catchphrases include *"Where's the beef?"* from a Wendy's commercial; *"Read my lips, no new taxes"* from George Bush's campaign; *"Siskell and Ebert give it two thumbs up"* from almost any preview of any movie today. Catchwords or catchphrases may be common on the news in the form of sound bites, and they may be especially common during an election year, but they are rare in technical prose, especially in any context requiring formal technical prose.

Vogue words are words that are in fashion or popular. See the vogue words *techies, tripping, surfers,* and *cyberhero* in the example below:

> But for many techies and even unsophisticated surfers, Macs and PCs look to be hot tickets for tripping on archetypal tales. The computer as looking glass tempts the isolated cyberhero to dive bravely into terra incognita via windows, portals, arcane codes, laser-beams, through webs and nets to super-highways, domains, and homesites, with Gophers, Archies, and Veronicas for squires in an endless—and sometimes aimless—quest for knowledge, adventure, a great yarn.[34]

Cliches

A *cliche* is a trite phrase or expression, something that has become overly familiar or commonplace. The term comes from a French word for stereotype plate, a block for printing. The reader pays little attention to the words because they have been so overused. Examples include: *a crying shame, slept like a log, stick to your guns,* and *in this day and age.*

Clipped Words/Contractions

Clipped words are created when part of a word, usually the beginning or end, is dropped to create a shorter word: examples include *info, exam, rep, lab,* and *exec.* This clipping of a word is also referred to as a *clipped form.* Such words are commonplace in conversation, but should not be used in formal writing unless they are well known to a particular occupational group. Additional examples of clipped words include: *ad* from *advertisement, chimp* from *chimpanzee, deli* from *delicatessen, lab* from *laboratory,* and *reg* from *regulations.*[35]

A *contraction* is a shortening of a word, syllable, or word group by omission of a sound or letter. An apostrophe is used to substitute for the missing letter or letters: *can't* for *cannot; shouldn't* for *should not.* You may use contractions in technical prose if you're trying to achieve an informal style. Avoid them if you are trying to achieve a more formal style.

Commonly Confused Words/Confusibles/Suggestibles

All writers confuse some words with others. Sometimes words are confused because they are **homonyms** (words with the same pronunciation but different spellings or meanings) such as *bare* or *bear* or *ware* and *wear;* **homographs** (words with the same spelling but with different pronunciation or meanings) such as *lead* and *lead;* **homophones** (a word pronounced the same as but different in meaning from another, whether spelled the same way or not) such as *heir* and *air* or *to, too,* and *two;* **synonyms** (words with the same or similar meanings) such as *advice, advise,* and *inform;* and words that often have multiple meanings such as *wear.* [36]

In *Room's Dictionary of Confusibles,* Adrian Room suggests there are two main categories of words that get mixed up: "one that is fairly specialised, erudite or bookish" and one that is "a rather more familiar word."[37] Examples from the first category include *apogee/perigee,* or *atmosphere/stratosphere/ionosphere/troposphere/tropopause.* Examples from the second category include: *affect/effect, it's/its,* and *lie/lay.*

Definitions of *confusibles* are confusing. According to Tom McArthur, a confusible is "one of two or more words that are commonly or easily confused with one another: *luxuriant* with *luxurious; they're* with *there* and *their.*"[38] Adrian Room provides a more exacting definition: "*a word that not only resembles another in spelling and pronunciation, but one that additionally has a similar or associated meaning.* Put rather more formally, it is a word having a lexical and semantic (but not necessarily etymological) affinity with another."[39] Some common confusibles include the following:

acid/acrid/acerbic	hovercraft/hydrofoil
assure/ensure/insure	motive/motif/leitmotiv
carbon monoxide/carbon dioxide	perimeter/parameter/periphery
constitute/compose	quote/cite
discrepancy/disparity	sated/satiated/saturated
e.g./i.e.	temerity/timidity
glance/glimpse	zeal/zest[40]

A *suggestible* is "a word which is involuntarily evoked by another, although unrelated to it in origin or even in meaning."[41] Examples include *bombast/bomb, whet/wet, absinthe/absent, greyhound/grey,* and *walnut/wall.*

Double Negatives

A double negative is using two negatives within a sentence for emphasis. Often double negatives are used colloquially: *I can't hardly get any peace and quiet* or *I can't get no satisfaction.* In technical prose, you may occasionally come across a sentence such as the following: "Do not unplug the computer during an electrical storm if you don't want to risk being shocked." Readers must reread such a sentence to be sure of its meaning. It would be better to revise such a sentence to something such as the following: "To avoid the risk of being shocked, do not unplug the computer during an electrical storm." You should avoid double negatives in your writing because they create unnecessary ambiguities.

Foreign Words and Phrases/Latin Abbreviations

English has borrowed many foreign words and phrases over the centuries. Often these words or phrases more aptly capture the precise meaning required than does an English word or expression. The following sentences reflect an appropriate use of foreign words: "But when they ask the technical writer what she does, the answer she gives becomes, *de facto,* a powerful statement about her professional identity as well as about the field as a whole."[42] Or "The grep command is UNIX's numero uno search command."[43] Or "Unlike emacs, it is available in nearly identical form on nearly every UNIX system, thus providing a kind of text-editing lingua franca."[44]

Latin abbreviations are shortened versions of words or phrases derived from ancient Latium and ancient Rome. Numerous Latin abbreviations are available for all kinds of expressions. Sometimes authors use too many Latin abbreviations in their writing, contributing to a choppy style. Common Latin abbreviations include the following:

i.e.	id est	that is
e.g.	exempli gratia	for example
etc.	et cetera	and so forth
ibid	ibidem	in the same place or cited just before
A.D.	Anno Domini	in the year of the Lord
cf.	confer	compare
NB	Nota Bene	take notice, note well
viz.	videlicet	namely
A.M.	ante meridiem	before noon
P.M.	post meridiem	afternoon[45]

Idioms

Idioms are the language peculiar to a people or to a district, community, or class. An idiom is also an expression in the usage of a language that is peculiar to itself either grammatically (*no, it wasn't me*) or in having a meaning that cannot be derived from the conjoined meanings of its elements (as *Monday week* for *the Monday a week after next Monday; painting the town red; putting on the dog*).[46]

According to David Crystal, "Two central features identify an idiom. The meaning of the idiomatic expression cannot be deduced by examining the meanings of the constituent lexemes. And the expression is fixed, both grammatically and lexically. Thus *put a sock in it!* means 'stop talking,' and it is not possible to replace any of the lexemes and retain the idiomatic meaning."[47]

"Izing"/Nominalizations/Verbizing/-Wise Words

"*Izing*" is adding the suffix *-ize* to all or part of a noun or adjective to form a verb: *prioritizing, maximizing, energizing, optimizing,* or *customizing.* Some people frown on such usage, but, at the same time, use such words as *itemizing, patronizing,* or *generalizing* with no reservation. Some recent *-ize* words include: *ruggedize*—to make more durable for military use; *definitize*—to iron out all the details in a contract.

A *nominalization* occurs when you turn a verb into a noun or when nouns do the work of other parts of speech. Nominalizations may be created by turning a verb into a noun—*nominalize* into *nominalization,* by nominalizing adjectives—*careless* into *carelessness,* or by adding *-ing* to a verb—*The naming of the boat.* [48] Nominalizing verbs or adjectives is a common practice, but overuse of nominalization often creates a passive style. And nominalizing often requires using more words to say the same thing, writing, for example, "He was guilty of carelessness" instead of "He was careless."

Verbizing is the practice of turning nouns into verbs. The process of nouns becoming verbs is a normal process in language. We have many commonly accepted examples: "He is a good contact if you want someone to take you on a tour of the city"; "Contact him if you want to tour the city." Verbizing becomes objectionable to some as in the following sentences: "We will have to gorilla him if we want to get our way"; "Maybe my friend will gift me this holiday season"; "The police did not Mirandize him before taking him to jail."

-Wise words are created by adding the suffix *-wise.* Examples include *health-wise, teeth-wise,* and *weather-wise.* Such words are often poor substitutes for more exact terms.

Logic Trap Words/Meaningless Words/Totality Words/Vagueness/Wooliness

Logic trap words are often found in specifications, and their logic is not immediately obvious:

accept or reject	This allows no option for partial rejection.
approve	It is not wise to say that we hold something in esteem. Try *authorize.* It makes less of a commitment.
designed to	It may be "designed" as such, but will it be built and installed that way?
either	Be sure you don't really mean "both."
extent possible	Often found after "greatest" or "maximum," this may require an unbounded effort.
maximum, minimum	These two words may denote an unbounded or impossible task. Is the requirement practically attainable?
as a minimum	You will probably not get more. Contractors nearly always do the minimum required.[49]

Many words are meaningless unless they are accompanied by other words that provide additional necessary information. In specifications writing, for example, it is unlikely that the meaning of "a long bolt" could ever be agreed upon, but "a seven-inch long bolt" is precise. Writing that building materials must be of the *highest grade* or *highest quality* is not helpful either. What do these terms mean, and *highest* according to whom? To require that information be provided in a *timely* fashion does not specify when or how often the information must be provided. To demand that the appearance of a building facade should be *pleasing* faces the same problem: *pleasing* how and *pleasing* whom?

Representative meaningless words for specification writers include:

about	*et al.*	*neat*	*simple*
acceptable	*etc.*	*neat*	*smooth*
accordance	*excessive*	*normal*	*stable*
accordingly	*general*	*not necessarily*	*substantial*
accurate	*good*	*optimum*	*sufficient*
additional	*greater*	*other*	*suitable*
adequate	*high*	*periodically*	*temporarily*
adjustable	*higher*	*pleasing*	*temporary*
affordable	*highest grade*	*possible*	*the like*
applicable	*highest quality*	*practicable*	*timely*
appropriate	*immediately*	*practical*	*typical*
average	*improper*	*proper*	*uneconomical*
better	*instant*	*quick*	*unsafe*
careful	*insufficient*	*reasonable*	*variable*
convenient	*irrelevant*	*recognized*	*various*
deep	*knowledgeable*	*relevant*	*wide*
desirable	*less*	*reputable*	*workmanlike*
easily	*long*	*safe*	*worse*
easy	*low*	*satisfactory*	*worst*[50]
economical	*lower*	*secure*	
efficient	*major*	*significant*	
essential	*multiple*	*similar*	

Totality words do not allow for exceptions or for qualifying. They are comparable to blanket statements. Totality words can create problems if they are taken literally. In specifications writing, for example, " Contractors will take totality statements literally when it is profitable to do so. The following words are a few that are characteristic of totality statements: *all, always, continually, entire, identical, never, none,* and *simultaneous.*"[51]

Vagueness concerns language not clearly expressed. For example, a specification may call for *stainless steel nuts and bolts.* Does the specification require that both the nuts and bolts be stainless steel? If so, this specification must be revised to be less vague.[52]

According to Eric Partridge, "Wooliness is that fault of style which consists in writing around a subject instead of on it; of making approximations serve as exactitudes; of resting content with intention as opposed to performance; of forgetting that whereas a haziness may mean something to the perpetrator, it usually means nothing (or an ambiguity) to the reader or the listener."[53] The word *wooliness* is a good example of an unusual term. It may occur at the word, phrase, clause, sentence, paragraph, or larger segment level, and is a fault not only in style but in content and organization.

Legal Terms/Overuse of Prepositional Phrases

Legalese is the specialized language of the legal profession, and it is appropriate in this community. Writers in other communities may create problems for readers when they use legal terms in their documents. Some typical legal terms are *negligence, falsification, aggravated damages, without prejudice, perjury,* and *circumstantial evidence.*

Sometimes writers use too many prepositional phrases in a sentence or paragraph. Example: "This report is intended to show Intel Pentium processor (610\75) performance on standard benchmarks that can be used as one of the predictors of performance on real applications that you can expect to see and use on your mobile system."[54]

Modifiers/Noun Phrases/Noun Strings

A *modifier* is any word or group of words that restricts, limits, or makes more exact the meaning of other words. The modifiers of nouns and pronouns are usually adjectives, participles, adjective phrases, and adjective clauses. The modifiers of verbs, adjectives, and adverbs are adverbs, adverb phrases, and adverb clauses: *a nervous, high pitched* giggle, *the* boy *in the first row, a royal* purple, *sickeningly* sweet, *scorching* hot.

Meaningless Modifiers

Meaningless modifiers are modifiers that neither restrict nor make the meaning of other words clear. These modifiers are often used to make the prose sound more important or impressive. "She was *quite* happy to see her boyfriend again." "He was *very* excited to be at the concert." "The audience talked about how *extremely* hot the theater was." These modifiers or intensifiers are not always meaningless. The biggest problem many writers have with them is a tendency to overuse them in their writing.

Modifier Strings

Modifier strings are groups of modifiers used together without a noun or phrase in between. "He was arrested on *conspiracy and concealing stolen property* charges." "A 1-cent *per $100 assessed valuation road tax increase* was proposed."

A noun phrase is simply a noun with all of its attached modifiers. Each noun phrase has one main word, the head noun, that all the other words in the phrase modify: *enhanced mouse buttons, multimedia pentium computer.* Noun phrases may be used anywhere simple nouns are used, as subjects, objects of verbs, or prepositions, or in other noun functions.

Noun strings are formed by stringing together two or more noun phrases: *multimedia pentium computer with a 1.5 gigabyte hard drive.* Noun strings are often a problem in technical prose if there are too many modifiers for the nouns: "The 4019NYS and 4019NYO Single Monitor Units upgrade a Western Electric 551A Channel Service Unit to provide a self contained, in line, signal monitor, loopback and performance diagnostic system."[55]

Negative Words and Phrases/Negative Instructions

Negative words and phrases are words or expressions that often create a defensive or hostile attitude in the audience. In a letter of complaint, for example, expressions such as *your fault, your mistake, your stupidity, your employee's blatant negligence,* or *your shoddy workmanship* may cause a hostile reaction.

Negative instructions may be confusing to a reader. Instead of writing, "The computer should not be shut off until Windows is no longer running," write "Shut off the computer after exiting from Windows."

Neologisms/Nonsense Words

Neologisms may be new words, usages, or expressions, or meaningless words coined by a psychotic. (A **nonce-word** is a word coined for one occasion.) Neologisms are often introduced into the language to enhance style. Note, for example, the many *gate* suffix scandals after *Watergate,* or the term *blandiose* to describe both *bland* and *grandiose.* Writers and speakers employ neologisms for a variety of reasons: novelty, cleverness, laziness, to avoid the obvious word, a preoccupation with precision, or a desire for brevity that the current language does not permit.[56] The following sentences illustrate neologisms: "For a true commandlineeditophile, it's not enough."[57] Or a longer example:

> This could be the Year of Rejuberation. The teams we love to love, the Dallas Cowboys and Chicago Bulls, are back on track; the sport we love to hate, boxing, may be coming back again, and the Olympics has discovered money. Baseball, the sport of the 20th Century, may have found its wedge into the 21st, interleague play.
>
> Rejuberation is a great word, not yet in the dictionary although far juicier than mere rejuvenation, which contains fresh vitality but not such joyous celebration. The word's author is Terrible Tim Witherspoon, one of the many former but still active heavyweight champions. Last week, at a Madison Square Garden news press conference before the Roy Jones, Jr. show, on which he was featured, Witherspoon described himself as 'rejuberated.' Freed from the clutches of Don King, he said, he felt young again and ready to reclaim another title.[58]

Nonsense words are words or language having no meaning or conveying no intelligible ideas. Nonsense words can also be defined as language, conduct, or an idea that is absurd or contrary to good sense. Lewis Carroll's poem "Jabberwocky" (1908) is full of nonsense words, and the word *jabberwocky* itself has come to mean meaningless speech or writing.

Redundancy/Repetition/Tautology/Expletives

Redundancy is one of the biggest problems all writers face. Often writers use more words than necessary to express an idea. In an early draft a writer may write: *The **reason** for this price increase in this software program was **probably due** to the fact that the company's competitors also increased their prices.* A shorter version may read: *This software program costs more because competing programs also cost more.*

Repetition is the act or instance of restating or repeating an action. Repetition may help a writer make transitions between sentences and paragraphs. Repeating key words or phrases may help emphasize a feeling or idea. Eric Partridge tells us, "Words or phrases should be repeated only if the repetition is effective or if it is essential to clarity."[59] Of course, writers may have useless repetition that obscures the meaning of a sentence or paragraph, for example: "For the most part, the differences between different UNIXes don't make much difference."[60]

Or consider the following example that uses the word *create* in one form or another five different times: "When the Menu widget menu bar is created, it will read this XmtlN items resource and automatically create the three cascade buttons it describes. Since these cascade buttons have attached panes, however, it will also create new Menu widgets for each of these panes. Thus, creating the single top-level menu bar widget will cause the entire menu system to be created."[61] One possible revision of the sentence might be: "When the Menu widget menu bar is created, it will read this XmtlN items resource and automatically generate the three cascade buttons it describes. Because these cascade buttons have attached panes, however, it will also set up new Menu widgets for each of these panes. Installing the single top-level menu bar widget creates the entire menu system."

A *tautology* is "a repetition of the same statement" or "the repetition (especially in the immediate context) of the same word or phrase, or the same idea or statement in other words, usually as a fault of style."[62] Examples of tautological expressions include: *adequate enough; attach together; collaborate together; final completion; finish up; link together; meet together; new innovation; recall back; refer back; rise up; seldom ever; still remain; unite together; where at; whether or not; widow woman; young infant.* [63]

An *expletive* is a syllable, word, or phrase inserted to fill a vacancy (as in a sentence or a metrical line) without adding to the sense. Examples of expletives in sentences include writing *There was a short which burned out the circuit* instead of *A short burned out the circuit,* or writing *It is absolutely essential to ground the device before charging it* instead of *You must ground the device before charging it.* The expletive *there was* in the first sentence adds nothing to the sentence. The *it is* in the second sentence likewise adds nothing.

An expletive is also an exclamatory word or phrase, especially one that is obscene or profane.

Sexist and Discriminatory Language

Although the issue is still controversial to some people, especially many of the pop grammarians discussed in Chapter 3, the days when you could use the generic "he" with impunity are gone. To be taken seriously by most publishing houses and in most public speaking situations, you must no longer write or say "the user...he," "the consumer...he," "man," or "mankind," to mention only a few examples of what is now considered sexist language. See Chapter 10 for a more thorough discussion of this point.

Solecisms/Improprieties

According to Geoffrey Nunberg, *solecisms* "are violations of the rules of verb agreement, pronoun case, and so on, rules that are supposed to ensure that prose will be constructed logically."[64] For example, injunctions against dangling modifiers—*Walking on the sidewalk, the parking meter was in violation* —are necessary to avoid confusing constructions.

An *impropriety* is an improper or indecorous act or remark, especially an unacceptable use of a word or of language. Improprieties are good words used inappropriately: *set* for *sit, accept* for *except, lay* for *lie, belabor* for *labor.*

Technical Prose and a Literary Style

In Chapter 3, we discussed a variety of styles, including several literary styles. We discussed how some styles are considered literary because they are ornate and how other styles are simpler but literary because they use a variety of stylistic elements more commonly associated with literary prose. We discussed, for example, the literary style of Oliver Sacks in a selection from his highly successful book *The Man Who Mistook His Wife for a Hat.* The purpose of this section is to show that writers of technical prose may select from a wide variety of diction choices. They may make diction choices more commonly associated with literary prose, but choices that are also effective in technical prose, especially when they are writing for a lay audience.

Alliteration

Alliteration is the repetition of usually initial consonant sounds in two or more neighboring words or syllables (as *wild and woolly, threatening throngs*). Writers of technical prose tend to use alliteration much less often than other writers, but technical writers, science writers, and many scientists do make use of alliteration in their prose. The following sentences illustrate the effect: "It is, basically, a response to propaganda, something like the panic-producing pheromone that slave-taking ants release to disorganize the colonies of their prey."[65] And "UNIX's much-maligned man pages have a certain charm."[66]

Anthropomorphizing

Anthropomorphizing is to attribute human form or personality or other human attributes to things not human. John Barry suggests that "the computer industry out paces all others in the area of anthropomorphic language."[67] Barry lists some of the more common computer terms which are "rooted in mind and body."[68] Some examples of anthropomormism are:

> *dumb:* the opposite of intelligent. A dumb terminal, for example, is a mere conduit to the intelligence of, say, a computer
>
> *intuitive:* essentially means "easy to figure out"—as in an "intuitive interface"
>
> *memory:* location for data storage, generally on a chip made for the purpose. Can be "dynamic," in which case the data vanish when the computer is turned off, or "static," in which case the data remain in the storage location.[69]

Barry also calls our attention to the opposite of anthropomorphism: *mechanism* or "describing people in terms of technological phenomena."[70] Examples include *bandwidth* ("He doesn't have enough bandwidth to deal with all those concepts") and *hardwired* ("People are not hardwired for data; computers are not good at pattern recognition").[71]

Assonance/Consonance

Assonance occurs when identical vowel sounds are preceded and followed by different consonant sounds. Examples of assonance include: *hide-mine, throat-soak, blot-clot, drain-blade, farm-hard, curl-heard, broil-moist, loud-clout, grand-black, crude-tune, mud-run, grist-ship, throttle-bottom, sick-prince.* Some familiar assonant expressions are: *hat rack, shop talk, jail bait, bake sale, lug nut, home phone.* [72]

Consonance is a recurrence or repetition of consonants especially at the end of stressed syllables without the similar correspondence of vowels (as in the final sounds of "stroke" and "luck." Examples include: *fail-feel, rough-roof, ruin-rain, jog-jig, groan-grin,* and *simple-sample.* Some common expressions are also good examples: *pitter-patter, wishy-washy, mish-mash, flim-flam,* and *tip-top.* [73]

Elegancies and Literarisms

Elegancies are words or phrases that writers believe to be more elegant than the terms they replace. Elegancies are also referred to as *genteelisms.* Examples include: *abode* (home), *anon, beverage, consume* (to eat), *denizen, espousal, expectorate, hither, mentor, modicum, perchance, petite, retire* (go to bed), *thither, visage, whither,* and *wont* (custom, habit).[74]

Literarisms are the journalese or the unusual words used only by the literary or learned.[75] Examples of literarisms include: *aura, avid, avocation, bucolic, catharsis, crux, denigrate, dichotomy, effete, ethos, homo sapiens, ineluctable, intrinsic, ivory tower, literati, neophyte, paramount, quietude, superimpose, tantamount, untoward, viable,* and *writ* (written).[76]

Figures of Speech

A *figure of speech* is a form of expression used to convey meaning or heighten effect. A figure of speech often compares or identifies one thing with another that has a meaning or connotation familiar to the reader or listener. Typical figures of speech include **similes, metaphors, personification, hyperbole,** and **understatement.**

A **simile** is a figure of speech comparing two unlike things that is often introduced by *like* or *as* (as in *cheeks like roses*): "Seeing Linus jumping up and down on the demonstration table and moving his arms like a magician about to pull a rabbit out of his shoe made them feel inadequate."[77] "Using ed is like trying to pick up one grain of rice with chopsticks: possible, but difficult for the uninitiated."[78]

A **metaphor** is a figure of speech in which a word or phrase literally denoting one kind of object or idea is used in place of another to suggest a likeness or analogy between them (as in *drowning in money*). A metaphor implies a comparison whereas a simile makes the comparison explicit by the use of *as* or *like.*

A **dead metaphor** is a word or phrase (as *time is running out*) that has lost its metaphoric force through common usage. A **mixed metaphor,** or what Theodore Bernstein calls a **mixaphor,** is a figure of speech combining inconsistent or incongruous metaphors (*Yet the President has backed him to the hilt every time the chips were down*).[79]

Technical prose is filled with all kinds of examples of metaphors: *c-clamp, A-frame, orange-peel finish, host computer, the Web, the Net, dialogue box, windows, network, buttons, menus.*

Personification is a figure of speech that attributes human characteristics to nonhuman things or abstract ideas: *The sailboat unfolded her sails.* An important distinction can be made between personification and anthropomorphism: "Whereas *anthropomorphism* is the conceptual presentation of some nonhuman entity in human form, *personification* is the much more limited rhetorical presentation of some nonhuman entity in figuratively human form or with figuratively human qualities."[80]

The following sentences are examples of personification: "Cairo is being strangled, Laocoon-like by a mess of ill-thought-out concrete traffic arteries."[81] And "What works great for a VT100 will lock up an HP terminal and cause a Data General terminal to rush to the kitchen and consume the contents of your freezer."[82]

A **hyperbole** is an extravagant exaggeration. Examples include *mile-high ice-cream cones; Our softball team was killed by the opposition; The sun was murder today; The world's most famous beach*.

An **understatement** is an expression that states or presents an idea with restraint especially for effect. A comment made during a blizzard: *I imagine it's a little cold and windy outside.*

On Euphemisms and Dysphemisms

Finally, no chapter on diction would be complete without some attention to euphemisms and dysphemisms.

Euphemisms

The word *euphemism* comes from Greek meaning "words of good omen." Euphemisms are words or expressions that aim at politeness. They attempt to avoid the truth by substituting pleasant or inoffensive words for ones that might be considered coarse or rude. Using euphemistic expressions makes delicate or unpleasant subjects less offensive. Euphemisms are used for many reasons: to adapt to a particular audience, time, or place; to enhance the value of what you possess or wish to give; to show respect or to impress or please the person you address; to soften painful news; to avoid discussing intimate details in good company; and to handle superstitious taboos and religious interdictions.[83]

Hugh Rawson informs us that

> euphemisms have very serious reasons for being. They conceal the things people fear the most—death, the dead, the supernatural. They cover up the facts of life—of sex, and reproduction and excretion.... They are beloved by individuals and institutions... who are anxious to present only the handsomest possible images of themselves to the world. And they are embedded so deeply in our language that few of us, even those who pride themselves on being plainspoken, ever get through a day without using them.[84]

In *The Writer's Art* James Kilpatrick suggests that most euphemisms can be classified under one of three headings: Euphemisms of Inflation, Euphemisms of Modesty or Taste, and Euphemisms of Deception. Examples of Euphemisms of Inflation include *sanitation engineer* for *garbage man* and *custodian* for *janitor*. Examples of Euphemisms of Modesty or Taste include *put some powder on our nose* for *go to the bathroom* or *occasional irregularity* for *constipation*. Examples of Euphemisms of Deception range from mild to seriously unethical. Kilpatrick objects to using *protective reaction strike* for *bombing*, for example.[85]

Dysphemisms

Dysphemisms are expressions with connotations that are meant to be offensive to the reader or listener. They are used for the opposite reason a euphemism is used. In *Euphemism & Dysphemism,* Keith Allan and Kate Burridge comment: "Dysphemisms...are used in talking about one's opponents, things one wishes to show disapproval of, and things one wishes to be seen to downgrade."[86] Examples include *The nasty little creep!* or, said sarcastically, *If you could spare me a FEW moments of your time.* [87]

Chapter Summary

Writers of technical prose have the same levels of diction—formal, informal, colloquial, and slang—available to them as do writers of nontechnical prose. And writers of technical prose have numerous occasions for writing both informally and formally. Writers of technical prose must also be alert to the denotative and connotative meanings of the words they choose. Technical writers may offend others with their word choices as easily as any other kind of writer.

Writers of technical prose are faced with special challenges presented by writing for certain kinds of industries or professions (specifications or legal writing, for example) and by writing about a variety of scientific and technological subjects. Specialized vocabulary is a major concern. By focusing on specific strategies—accuracy and precision, active verbs and active voice (if applicable), appropriateness, clarity, conciseness, concreteness, consistency, simplicity, and sincerity, you can make major improvements in your diction.

Writers of technical prose are faced with many of the same diction problems that all other writers must make decisions about. By understanding common diction faults—from archaic words to wooliness, technical writers can strengthen their prose.

And technical writers may also strengthen their prose by understanding the many ties between technical prose and the many elements—from anthropomorphizing to many figures of speech—that characterize a literary style.

Finally, technical writers must understand the distinctions between euphemisms and dysphemisms. Like all writers, technical writers have to be vigilant about every word they choose and the effects these words will have on readers and listeners.

Case Study 4: The Ruby Ridge Incident and Diction[88]

Introduction

Fast, precise, and explicit communication is an indispensable element in cases of emergency, war, or any other situations in which human life may be endangered. The necessity for such narrow definitions is illustrated every day in hospital emergency rooms, fire and police stations, army headquarters, and airport control towers.

In recent years, one of the most striking examples of the fatal effects of communication failure is the incident at Ruby Ridge. The ten-day standoff between Randy Weaver, his family, and over 400 federal agents in August 1992 at Ruby Ridge, Idaho, has been used to demonstrate various theories on the problems with law enforcement, gun-control, the Second Amendment, and human rights. Although all of these issues were raised in the Ruby Ridge case, a key element that turned the operation into a fatal incident resulting in the deaths of Weaver's wife, his son, and a decorated deputy U.S. Marshal was ambiguous, insufficient communication. The episode proved how, in such extreme situations, every little word, a certain word choice, and even a minimal ambiguity in wording can alter the meaning of certain commands and produce deadly confusion.

How It All Began

The story of Ruby Ridge began when the U.S. Marshal Service decided to initiate an eighteen-month-long investigation of Weaver, known for his extremist views. The ultimate purpose of the mission, called Operation Northern Exposure, was to recruit Weaver as an informant against the illegal activities of Aryan Nations members.

On 17 January 1991, Weaver and his wife, Vicki, stopped to help a couple stranded on the roadside. The two Bureau of Alcohol, Tobacco and Firearms (BATF) agents who posed as the motorists arrested Weaver on weapons charges. Weaver was soon arraigned and released on bond, but was required to appear in court on 19 February 1991. Two weeks before the set court date, U.S. district court clerk Karl Richins sent a letter to Weaver noting erroneously that the date for the trial had been postponed until March 10, when in fact it had not. When Weaver failed to appear in court, Chief U.S. District Court Judge Harold Ryan issued a bench warrant for him.

On 14 March 1991, a federal grand jury indicted Weaver for failing to appear in court. A few days later, Chief Deputy Marshal Ron Evans asked for assistance from the U.S. Marshal's Special Service Group. After a week of jet reconnaissance photography of the Weavers' estate, surveillance with night vision equipment, and the recording of 160 hours of tape from high-resolution video equipment, the team developed a plan to arrest Randy Weaver, describing him as "extremely dangerous and suicidal." The facts on which the description was based were never disclosed.

When a Special Service team was dispatched in September to proceed with the arrest, the team discovered inaccuracies with the gathered information and canceled the plan. A second plan to survey and arrest Weaver was scrapped after reports (later proven inaccurate) surfaced that a helicopter from Geraldo Rivera's television program *Now It Can Be Told* had been shot at while flying over the property and that a surveillance camera from a ridge over the Weaver's land had been stolen. A third plan, to purchase land next to Weaver's to be used as a base for a surveillance mission, was underway on 21 August when its performance went badly awry.

Deputy marshals, engaged in the surveillance mission of Randy Weaver's cabin, were surprised by Weaver's son Sammy, his dog, and Kevin Harris, a guest of the family. A shootout ensued, ending in the deaths of the dog, the boy, and William Degan, a federal marshal.

These are the initial tragic events that began the ten-day standoff between the self-proclaimed white separatist Randy Weaver and three government agencies: the U.S. Marshal's Service, the BATF, and the Federal Bureau of Investigation (FBI).

The Involvement of the FBI

21 August 1992

1:30 P.M. After the Marshal's Service Crisis Center is alerted that the marshals are under attack at Ruby Ridge, a Special Operation Group (SOG) is deployed to the scene. In Washington D.C., officials of the Marshal's Service and the FBI meet to decide on a response strategy. The first decision of that meeting is the additional deployment of the FBI's Hostage Rescue Team (HRT).

6:30 P.M. On the plane to Idaho, HRT Commander Richard Rogers talks extensively on the phone with his boss Larry Potts and Deputy Assistant Director Coulson. They draw up the operation's Rules of Engagement—the backbone of any emergency mission where human lives are concerned. These rules state that the agents "can and should" fire at any armed men on the Weaver's property.

22, August 1992

9:00 A.M. At the National Guard Armory in Bonners Ferry, Idaho, HRT Commander Rogers briefs his men on the Rules of Engagement, which are still in draft form. Later that afternoon, the Rules of Engagement, along with the HRT and SOG joint operations plan, will be forwarded to FBI headquarters in Washington and approved.

According to one source, part of the final Rules of Engagement states:

> If any adult in the compound is observed with a weapon after the surrender announcement is made, deadly force can and should be employed to neutralize the individual. If any adult male is observed with a weapon prior to the announcement, deadly force

can and should be employed if the shot can be taken without endangering the children.[89]

5:58 P.M. The eleven snipers/observers from the rescue teams take their positions around the Weavers' cabin. When Weaver, his daughter Sara, and the armed Kevin Harris leave the cabin to retrieve Sammy's body, a marksman aims at Weaver's neck and fires. His bullet goes through Weaver's arm, and he and the others run back toward the cabin. The marksman aims for Weaver again and this time kills Weaver's wife, badly injuring Harris in the process.

After a ten-day standoff between the Weaver family and federal forces, two friends of the Weavers' persuade the wounded Kevin Harris to surrender. The following day, Randy Weaver and his daughters follow suit.

The Investigation

Even today it is not clear who wrote and who approved the fateful Rules of Engagement stating that agents "can and should" use deadly force against any armed adult men. The FBI apparently blocked access to some of the most important documents pertaining to the incident, impeding the progress of the Department of Justice investigation.

Justice Department frustration at the reluctance of the FBI to cooperate in its investigation was thinly disguised: "During our investigation there were a number of instances in which an interviewee told that he had prepared a document, but no one could produce a copy of it." The report also added: "We never located any original facsimiles or notes of headquarters personnel that were prepared during the crisis at Ruby Ridge. We are troubled by the apparent lack of a system to preserve such critical records."[90]

Although the original draft of the Rules of Engagement was never found, all parties involved in the incident and its investigation agreed that the ill-fated phrase "can and should" was the reason for the fatal shots.

The incident led to one of the most intensive internal reviews of an FBI mission ever. In an unprecedented testimony before the Senate Subcommittee investigating the Ruby Ridge incident, FBI director Louis Freeh admitted that the FBI operation was flawed by multiple mistakes:

We know today that law enforcement overreacted at Ruby Ridge. FBI officials promulgated rules of engagement that were reasonably subject to interpretation that would permit a violation of FBI policy and the Constitution—rules that could have caused even worse consequences than actually occurred. Rules of engagement that I will never allow the FBI to use again.

At Ruby Ridge, the Hostage Rescue Team ("HRT") was operating in accordance with rules of engagement that were reasonably subject to interpretation that would permit a violation of FBI policy and the Constitution. Those rules said that, under certain circumstances, certain persons "can and should" be the subject of deadly force. Those rules of engagement were contrary to law and FBI policy. Moreover, some FBI SWAT personnel on-scene interpreted the rules as a "shoot-on-sight" policy—which they knew was inconsistent with the FBI's deadly force policy. Such confusion is entirely unacceptable.[91]

Freeh was praised by the Subcommittee for his quick response to the internal problems of his agency and for his new policy designed to prevent another Ruby Ridge tragedy. However, Senator Arlen Specter, the Subcommittee Chair, pointed out the remaining ambiguity in the new rules that, by allowing the use of deadly force when an "imminent" rather than an "immediate" threat is perceived, still leaves room for a personal interpretation.[92] Therefore, the question of whether the language used for communication between the agents and their commanders has been made explicit and direct remains open.

Questions

1. What do you think were the ultimate consequences, if any, of the court clerk's letter to Randy Weaver that stated a wrong trial date?
2. Considering the Rules of Engagement, do you believe the agent who killed Randy Weaver's wife acted appropriately under the circumstances? Why or why not?
3. Read the comments of the Justice Department official on the FBI and BATF's reluctance to cooperate in the investigation. Do you think that the last sentence ("We are troubled by...") represents a sincere bewilderment at the agencies' record-keeping methods? What other possibilities could account for the choice of words here?

Questions/Topics for Discussion

1. Why is it helpful to be able to recognize different levels of diction—formal, informal, colloquial, and slang? What are some key differences between colloquial diction and slang?
2. What characteristics make some words emotive and some words nonemotive? To what extent can a writer control a reader's emotional responses?
3. Of the major strategies for improving your diction—accuracy and precision, active verbs and active voice (if applicable), appropriateness, clarity, conciseness, concreteness, consistency, simplicity, and sincerity, which topic or topics present the biggest challenges to your writing? Why?
4. Which topics discussed in the section on common diction faults appear most often in your writing? What are some examples from your prose?
5. What is the difference between accuracy and precision in technical prose? Provide additional examples to illustrate your point.
6. Specifications writers seem to have more diction problems to consider than do other writers of technical prose. Do you agree? Why or why not? Can you name other professions where writers have many of the same concerns?
7. What is the role of metaphors in technical prose? How do metaphors in technical prose differ from metaphors in literary prose?
8. Are there other important elements of literary prose that are often used in technical prose but omitted in this chapter? Provide several examples.
9. What are the essential differences among barbarisms, improprieties, and solecisms?
10. Why is it important to know what euphemisms are and how to use them in your writing?

Exercises

1. Now that you have read five chapters of this textbook, how would you characterize the writing style? As formal, informal, colloquial, or slang? What qualities of the style support your choice?
2. Read the following selection from *The Adventures of Huckleberry Finn* and discuss the level of diction and what word choices help to create that level of diction:

   > Now the way that book winds up, is this: Tom and me found the money that the robbers hid in the cave, and it made us rich. We got six thousand dollars apiece—all gold. It was an awful sight of money when it was piled up. Well, Judge Thatcher, he took it and put it out at interest, and it fetched us a dollar a day apiece, all the year round—more than a body could tell what to do with. The Widow Douglas, she took me for her son, and allowed she would sivilize me; but it was rough living in the house all the time, considering how dismal regular and decent the widow was in all her ways; and so when I couldn't stand it no longer, I lit out. I got into my old rags, and my sugar-hogshead again, and was free and satisfied. But Tom Sawyer, he hunted me up and said he was going to start a band of robbers, and I might join if I would go back to the widow and be respectable. So I went back.[93]

3. Reread the examples of formal, informal, colloquial, and slang levels of diction at the beginning of this chapter. Write a paragraph on each example discussing its specific diction qualities.
4. Review the report concerning the *Exxon Valdez* in Chapter 4. Discuss the level of diction in this report and how this level is achieved.
5. Reread the memo written in a passive style in Chapter 3, and then rewrite the memo using active verbs and active voice throughout.

6. Review the discussion of denotation and connotation at the beginning of this chapter. Provide your own definitions of these two terms, and provide examples of both denotation and connotation that show clear differences between these two terms.
7. Read the following two paragraphs concerning the presentation program PowerPoint. Characterize the appropriateness of the word choices and the conciseness of the prose:

> Whether presentations help you deliver company results to a shareholders' meeting or report sales figures at a hastily scheduled business meeting, they play a major role in how business people communicate. Microsoft PowerPoint for Windows 95, the leader in presentation graphics software, has all the tools you'll need to put together professional, compelling presentations quickly and easily.
>
> *Microsoft PowerPoint for Windows 95 Step by Step* is a comprehensive tutorial that shows you how to use PowerPoint to create professional looking presentations. Working through this book you'll learn about the different presentation materials PowerPoint helps you create: slides, overheads, audience handouts, presentation outlines, speaker's notes, and electronic presentations that show on your computer. You'll discover the basic building blocks that make up dynamic slides, including eye-catching clip art, easy-to-use text tools that let you create and format text, drawing tools the [sic] help you create interesting shapes and effects, and special tools that help you import information from other sources, such as charts, graphs, tables, picture, sounds, and movies. This book also teaches you how to find the help you need from PowerPoint's useful aids such as online Help, tips, and wizards.[94]

8. Reread the example of poorly written technical prose at the beginning of Chapter 1. Revise the example to make it more clear and write a paragraph summarizing the changes you made.
9. Choose a sample report, letter, memo, or manual you have written, and analyze the diction you use. How would you characterize your habitual diction choices? What level of diction did you aim for in the document? Why? Give examples to support your conclusion.
10. Choose a sample letter or memo written by a friend or classmate, and analyze the diction used. How would you characterize the habitual diction choices? What level of diction did the author seem to aim for? Cite examples from the document that show whether this is a formal or informal example of writing.

Notes

1. William Shakespeare, *Hamlet,* ed. Harold Jenkins (New York: Methuen, 1982) 247.
2. Widely attributed to Mark Twain.
3. Robert Hays, "What is Technical Writing?" *Word Study* (April 1961): 4.
4. Henrietta Nickels Shirk, "Prologue to Teaching Software Documentation," *Perspectives on Software Documentation: Inquiries and Innovations,* ed. Thomas T. Barker (Amityville, NY: Baywood, 1991) 25.
5. Stephen L. Nelson, *Field Guide to Microsoft Windows 95* (Redmond, WA: Microsoft, 1995) viii.
6. *The Even More Incredible Machine: Journal Entries June thru August* (Coarsegold, CA: Sierra On-line, n.d.) 1, 26.
7. Robert L. Chapman, ed., *American Slang* (New York: Harper & Row, 1987) xv.
8. Chapman, p. xiii.
9. Chapman, p. xiii.

10. Katie Wales, *A Dictionary of Stylistics* (New York: Longman, 1989) 423.
11. Wales, p. 424.
12. From newsgroup alt.binaries.warez.ibm-pc, May 6, 1996, 21:21:55 GMT.
13. David Crystal, *The Cambridge Encyclopedia of the English Language* (New York: Cambridge UP, 1995) 182.
14. Thomas Whissen, *A Way with Words: A Guide for Writers* (New York: Oxford UP, 1982), 54.
15. Tom McArthur, ed. *The Oxford Companion to the English Language* (New York: Oxford UP) 257.
16. McArthur, p. 257.
17. Philip Rubens, ed., *Science and Technical Writing: A Manual of Style* (New York: Henry Holt, 1992) 119–178.
18. Lewis Thomas, "Germs," in *Lives of a Cell: Notes of a Biology Watcher* (New York: Bantam, 1983) 88.
19. Crystal, p. 367.
20. John Montgomery, *The Underground Guide to Unix* (Addison-Wesley, 1994) 88.
21. Harley Hahn and Rick Stout, *The Internet Complete Reference* (New York: Osborne McGraw-Hill, 1994) 40.
22. *Webster's Tenth Collegiate Dictionary and Thesaurus,* CD-ROM, 1995.
23. Theodore M. Bernstein, *The Careful Writer: A Modern Guide to English Usage* (New York: Atheneum, 1977) 38.
24. Christopher Turk and John Kirkman, *Effective Writing: Improving Scientific, Technical and Business Communication,* 2nd ed. (New York: Spon, 1989) 99.
25. Turk and Kirkman, pp. 104–105.
26. All examples from Eric Partridge's *Usage and Abusage: A Guide to Good English* (New York: Norton, 1995) 31–36.
27. Robert Newton Peck, *Fiction Is Folks.* (Cincinnati, OH: Writers's Digest, 1983) 159.
28. Steve Reynolds, "Can Consumer Online Be Master of Its Own Evolution?" *The Red Herring,* March 1995: 38.
29. Montgomery, p. 101.
30. Partridge, p. 370.
31. Partridge, p. 381.
32. Geoffrey Nunberg, "The Decline of Grammar," *Atlantic Monthly,* December 1983, 41.
33. Nunberg, p. 41.
34. Kathleen Murphy, "The Bard: Hi-Tech Storytelling," *Film Comment* 31.4 (July/August 1995): 37.
35. McArthur, p. 223.
36. See Robert Hays, *Principles of Technical Writing.* (Reading, MA: Addison-Wesley, 1965) 267–83.
37. Adrian Room, *Room's Dictionary of Confusibles* (Boston: Routledge, 1979) 1.
38. McArthur, p. 256.
39. Room, pp. 2–3.
40. All examples from Room.
41. Room, p. 1.
42. Richard VanDeWeghe, "What is Technical Communication? A Rhetorical Analysis," *Technical Communication* 38.3 (1991): 295.
43. Montgomery, p. 11.
44. Linda Lamb, *Learning the vi Editor* (Sebastol, CA.: O'Reilly, 1994) 1.

45. Examples from David B. Gurlanik, ed., *Webster's New World Dictionary,* 2nd ed. (New York: Simon and Schuster, 1982).
46. *Webster's Tenth Collegiate Dictionary,* CD-ROM.
47. Crystal, p. 163.
48. Joseph Williams, *Style: Ten Lessons in Clarity and Grace,* 4th ed. (New York: HarperCollins, 1994) 43.
49. Examples from John T. Oriel, NAWCTSD Technical Report 93–022. *Engineering Specification Editing Tools. Final Report.* December 1993: B-6–B-7.
50. Oriel, p. B-1.
51. Oriel, p. B-2.
52. Oriel, p. B-10.
53. Partridge, p. 381.
54. Intel Home Page, World Wide Web, September 1994.
55. Quoted in John A. Barry's *Technobabble* (Cambridge, MA: MIT P, 1991) 38.
56. Partridge, p. 205.
57. Montgomery, p. 99.
58. Robert Lipsyte, "Boxing, Baseball, and 'Rejuberation.'" *The New York Times Online,* January 19, 1996.
59. Partridge, p. 276.
60. Montgomery, p. 5.
61. David Flanagan, *Motif Tools: Streamlined GUI Design and Programming with the XMT Library* CD-ROM (O'Reilly, 1994).
62. Partridge, p. 342.
63. Partridge, pp. 342–343.
64. Nunberg, p. 42.
65. Lewis Thomas, "Germs" in *The Lives of a Cell* (New York: Bantam, 1983) 93.
66. Montgomery, p. 35.
67. Barry, p. 134.
68. Barry, p. 134.
69. Barry, p. 135
70. Barry, p. 136.
71. Barry, p. 136.
72. Whissen, p. 31.
73. See Whissen, p. 32, for these and other examples.
74. All examples from Partridge, 101–103.
75. Partridge, p. 172.
76. All examples from Eric Partridge, *Usage and Abusage: A Guide to Good English* (New York: W. W. Norton & Company, 1995), pp. 172–174.
77. James D. Watson, *The Double Helix* (New York: Norton, 1980) 25.
78. Montgomery, p. 33.
79. Theodore Bernstein, *The Careful Writer: A Modern Guide to English Usage* (New York: Atheneum, 1977) 275–276.
80. C. Hugh Holman and William Harmon, *A Handbook to Literature,* 6th ed. (New York: Macmillan, 1992) 27.
81. Peter Davey, "An Intemperate Argument," *The Architectural Review* (July 1994): 5.
82. Montgomery, p. 70.
83. Partridge, pp. 108–110.
84. Hugh Rawson, *A Dictionary of Euphemism and Other Doubletalk* (New York: Crown, 1981) 1.

85. James Kilpatrick, *The Writer's Art* (New York: Andrews, 1984) 72–74.

86. Keith Allan and Kate Burridge, *Euphemism and Dysphemism: Language Used as Shield and Weapon* (New York: Oxford UP, 1991) 27.

87. Allan and Burridge, p. 27.

88. The following sources were used for this case study:

Louis J. Freeh, *Opening Statement Before the Subcommittee on Terrorism, Technology, and Government Information, Committee on the Judiciary. United States Senate, Washington, DC Oct. 19, 1995.* 6–9. FBI Online. Available at URL: http://www.fbi.gov/rubystat. htm#top

David Johnston, "Report Shows FBI Officials Blocked Access to Documents." *The New York Times News Service.* Online. Available at URL: http://www.tribnet.com/~stories/news/oo768.htm

George Lardner Jr., "Freeh Says Actions at Ruby Ridge Were 'Flawed.'" *Washington Post.* [20 Oct. 1995] Online. Available at URL: http://the-tech.mit.edu/V115/N50/freeh.50w.html

Jim Oliver, "The Randy Weaver Case," *American Rifleman.* Nov. 1993: 1–8. Online. Available at URL: http://eagle.tamn.edu/~carlp/Liberty/Weaver.Case.AR.html

Arlen Specter and Herb Kohl. *Executive Summary of Ruby Ridge Report of the Subcommittee on Terrorism, Technology, and Government Information of the Senate Judiciary Committee.* Online. Available at URL: http://www.nra.org/pub/general /96–02-02_senate_ report_on_ruby_ridge

89. Oliver, p. 4.

90. Johnston, p. 2.

91. Excerpt from the Opening Statement of Louis J. Freeh, Director of FBI, before the Subcommittee on Terrorism, Technology and Government Information. United States Senate. Washington, D.C., October 19, 1995.

92. Lardner, p. 3.

93. Mark Twain, [Samuel Langhorne Clemens], *Adventures of Huckleberry Finn,* 2nd ed., Sculley Bradley, Richmond Croom Beatty, E. Hudson Long, and Thomas Cooley (New York: Norton, 1977) 7.

94. *Microsoft PowerPoint for Windows 95: Step by Step* (Redmond, WA: Microsoft, 1995) xvii.

Chapter 6

Handling Technical Terms and Jargon

Jargon is like television or junk food: we condemn it, but we all indulge in it.[1]
—TOM FAHEY

Another aspect of the computer industry's Tower of Babel is different lexical strokes for different lexical folks. What you call a server, *I may call an* engine. *My colleagues may refer to a piece of software as an* interface, *while yours may call it an* application. *One user's* tool *may be another's* utility.[2]
—JOHN BARRY

Clarity is often the last thing The Official Style really wants to create and, if you find yourself in a bureaucratic context, often the last thing you want to create.[3]
—RICHARD LANHAM

Chapter Overview

Using Technical Language

Words, Terms, and Terminology

Technical Terms as Jargon

The Challenges of Using Technical Terms

The Challenges of Using Jargon

Other Specialized Languages

Abbreviations, Acronyms, Initialisms

Tips for Using Clear Technical Language

Chapter Summary
Case Study 5: **Apollo 13** *and Jargon*
Questions/Topics for Discussion
Exercises
Notes

Using Technical Language

All of us at one time or another have been overwhelmed and exasperated by technical language we don't understand. Whether it's our doctor explaining a surgical procedure, a lawyer explaining the terms of a trust, an auto mechanic explaining what's wrong with our car engine, a credit union officer explaining the terms of a home equity loan, or a tax accountant explaining our taxes, all of us have been perplexed by the difficulties of language that is complex and unfamiliar to us. "Why can't they just use plain English?" we wonder.

Of course, in these situations these experts or technicians should talk to you in plain English or, at the very least, make sure you understand any technical terminology. But there are many other communication situations—peer to peer, expert to expert—where technical words are a necessary shorthand.

Writers of technical prose must use technical terms. Technical terms are the currency of information in trades, professions, hobbies, and the sciences. Automobile mechanics talk about *ignition modules, brake drums, distributors,* and *fuel injectors.* Medical doctors discuss *lymphomas* and *metastatic cancers.* Motorcyclists discuss *clutch alignment, fork springs,* and *dual brake conversion.* Immunologists discuss *genomes, arterioles,* and *peptide hormones.* Physical anthropologists refer to *substrate, cytoarchitectonic,* and *hippocampal formation.* These terms have precise meanings to mechanics, doctors, motorcyclists, immunologists, and physical anthropologists. Usually, as long as they are talking to peers within their discourse community, there is little need for definition, clarification, or elaboration. These terms and countless others save these people incalculable time and effort.

Technical terms are essential to specialists for communicating efficiently and effectively within their discourse communities. People in these trades and professions typically have spent years learning and acquiring their technical vocabularies, and, understandably, they resent lay people casually misusing these precise terms or pretending they know what these terms mean when they do not.

And, of course, the knowledge technical people have consists of far more than just the terminology. Along with the technical words are a wide array of basic assumptions, beliefs, values, theories, histories, and expectations. In brief, discourse communities consist of far more than just the technical words used by people in the technical trades and professions. As we discussed in earlier chapters, each discourse community has its own rhetoric that constitutes the way things are discussed in that community. Technical terminology is but one of the many important elements of that rhetoric.

Some writers of technical prose are offended by the notion they should simplify their prose for any audience. Their expertise is hard earned, and, in many cases, the audience demands evidence of their expertise. Simplifying, these experts feel, may undermine their authority or their rapport with the audience. There is some truth to these concerns. Yet people seldom complain that something was too easy to understand. In an age when experts are mystifying other experts, such experts especially could benefit from employing many of the strategies discussed here. Technical knowledge is not easily acquired, and it's definitely not easy to communicate it well to those less knowledgeable. Yet, in general, people respect those who know how to communicate technical information clearly, concisely, and accurately.

The issue is not whether writers of technical prose should use technical terms. The issue is when technical terms should be used. Many writers of technical prose use technical terms for the wrong purpose, audience, or context. Audience is the most important factor in determining how technical a document or presentation should be. If you have any doubt about the knowledge level of your audience, then you are obligated to make sure your audience understands all of the technical terms you use. Of course, even if you have no doubts about the knowledge level of your audience, you are still obligated to make sure your audience understands all of the terms you use.

This chapter provides an overview of technical terminology and jargon and a discussion of some of the major problems presented in using both to communicate.

Words, Terms, and Terminology

Words, terms, and *terminology* require some discussion. Some useful distinctions can be made among the three. Tom McArthur informs us that a *word* is "a fundamental term in both the general and technical discussion of language."[4] A *term* is usually more specialized: A *term* "has a particular (often unusual) meaning because of the context in which it is used."[5] Usually, a *term* refers to something in a specialized field. For example, *hypermedia* is a term in computer technology and *lexeme* is a term in linguistics. *Terminology* is, according to McArthur, "the vocabulary of a specialized field as contrasted with the general vocabulary of a language. There is, however, no clear dividing line between general and specialist vocabulary...."[6] An everyday word such as *pig* can have a technical meaning—"a mass of metal such as iron, copper, or lead when cast into a simple shape"—or a technical term may be used in the general vocabulary (*neon,* the name for a rare gas, used in *neon sign*).[7]

Technical Terms as Jargon

Like *technical terms,* the term *jargon* applies to the specialized vocabulary of a trade, profession, hobby, or science. But *jargon* also has several negative or pejorative meanings. First, *jargon* refers to technical terminology used inappropriately for an audience. Whenever you use technical terms unfamiliar to all or some of your audience, you are using jargon. Second, whenever you use a pretentious or unnecessarily complex word because you want to impress someone or because you want to make yourself feel more important, you are using jargon.

Jargon in both of these senses has many names: *computerese, engineerese, bureaucratese, governmentese, academese, officialese, tech speak, techno speak, newspeak, doublespeak, doubletalk,* and *pentagonese.* Some of the more unusual names include *bafflegab, scientese, duck speak, bureau quack,* and *sciench.* (Some of these terms—*newspeak, doublespeak,* and *doubletalk* will be discussed in more detail in Chapter 11.)

Examples of inappropriate technical terminology are everywhere. Here's an example of a writer using technical terms for an audience who is not familiar with all of the terms:

> What about Windows NT? Do products that are 'Designed for Windows 95' run on Windows NT?
>
> Part of the logo requirements are that a product be tested on Windows NT. However, a handful of APIs are supported on one platform that are not available in the other. An example of products and APIS [sic] are: games using Direct X and utilities using TAPI. At this time, these two APIs are not supported in Windows NT. The idea is that if an application uses Windows 95-specific application programming interfaces (APIs), then the Windows 95-only functionality must degrade gracefully [meaning, just 'stub' that functionality, as opposed to allowing the system to crash] on Windows NT 3.5 or later. Conversely, if it uses Windows NT-specific APIs, then the Windows NT-only functionality must degrade gracefully on Windows 95. The bottom line is that the product must run successfully on both Windows 95 and Windows NT, unless architectural differences between the 2 OSs prevents it. Additionally, while products may pass testing on Windows NT, it is up to the vendors themselves whether they want to support their product(s) on Windows NT.
>
> The requirement to test on Windows NT is a testing requirement, designed to identify and address bugs, which has the side benefit of enlarging the potential market for an ISV's product as Windows NT takes hold over the next couple years. About 90% of Win32-based applications run without incident on both Windows 95 and Windows NT.[8]

Subscribers to *WinNews* are, in general, users of Microsoft's Windows 95 operating system. These users range in experience from novices to experts. Many novices would not understand many of the terms mentioned above.

Here's an example of pretentious jargon in a typical press release by a marketing department in the computer industry:

> For Immediate Release
>
> XERCOM today announces a state-of-the-art CMOS-based EPROM module conforming to all MIL-SPEC parameters and interfacing to ROMable/RAMable ASCII/ANSI devices.
>
> The ABC-PDQ 381 EPROM environments into hex and alphageomosaic applications virtualized to PIA DIP switch functionality at 1200 (10) baud. A downloading EBCDIC-oriented uploader enhances the enhancements of the enhanced IDRIS protocols that talk to the PABX store-and-forward interface bread-boarded into the unit.
>
> Available in 16, 32, and 48-bit configurations, the 381 utilizes user-keyboardable soft keys and display functionalities maxing out at 265 colors on an RGB CRT configured for modemized operation that concatenates with a 4MHz, 12 VAC, 155 cps, VAX-alike, DEC-emulating, annualized, prioritized, PROMized, and user-friendlyized operation for easy understanding and utilization in an environmentalized environment.
>
> For knowledge workers everywhere.[9]

The example above is actually an example made up by John Barry and submitted to a contest on the best (worst?) press releases. Surprisingly enough, his entry placed third.

The Challenges of Using Technical Terms

More discussion of pretentious jargon will be provided later in this chapter. But because this kind of jargon presents only one kind of challenge to writers of technical prose, more attention must be given to strategies for handling technical terms. It helps, for a start, to know more about technical terms and their characteristics.

Objectivity and Technical Terminology

The scientific and technological communities are still very much dominated by the paradigm of objectivity. They make every effort to appear impartial, unbiased, and neutral. People writing in many scientific and technological communities are encouraged to write as though they had no involvement with what was done or what is being reported on. Much scientific work, in particular, requires giving instructions to readers to act a specific way or reporting on the results of acting in a certain way. David Crystal identifies the following categories:

Verbs of exposition: ascertain, assume, compare, construct, describe, determine, estimate, examine, explain, label, plot, record, test, verify.

Verbs of warning and advising: avoid, check, ensure, notice, prevent, remember, take care; also several negative items: not drop, not spill.

Verbs of manipulation: adjust, align, assemble, begin, boil, clamp, connect, cover, decrease, dilute, extract, fill, immerse, mix, prepare, release, rotate, switch on, take, weigh.

Adjectival modifiers (and their related adverbs): careful(ly), clockwise, continuous(ly), final(ly), gradual(ly), moderate(ly), periodic(ally), secure(ly), subsequent(ly), vertical(ly).[10]

Of course, all of these categories of verbs are found not only in scientific laboratory reports and procedures but appear prominently in all kinds of technical manuals, reports, and proposals.

Note the objective tone and words or phrases suggesting objectivity (noted in bold for clarification purposes here) in James Watson's and Francis Crick's discussion of their discovery of the structure of DNA.

The novel feature of the structure is the manner in which the two chains are held together by the purine and pyrimidine bases. The planes of the bases are perpendicular to the fibre axis. They are joined together in pairs, a single base from one chain being hydrogen-bonded to a single base from the other chain, so that the two lie side by side with identical z-co-ordinates. One of the pair must be a purine and the other a pyrimidine for bonding to occur. The hydrogen bonds are made as follows: purine position 1 to pyrimidine position 1; purine position 6 to pyrimidine position 6.

If it is assumed that the bases only occur in the structure **in the most plausible** tautomeric forms (that is, with the keto rather than the enol configurations) **it is found that** only specific pairs of bases can bond together. These pairs are: adenine (purine) with thymine (pyrimidine), and guanine (purine) with cytosine (pyrimidine).

In other words, **if** an adenine forms one member of a pair, on either chain, **then on these assumptions** the other member must be thymine; similarly for quanine and cytosine. The sequence of bases on a single chain **does not appear to be** restricted in any way. However, **if** only specific pairs of bases can be formed, **it follows that** if the sequence of bases on one chain is given, then the sequence on the other chain **is automatically determined.**

It has been found experimentally that the ratio of the amounts of adenine to thymine, and the ratio of guanine to cytosine, are always very close to unity for deoxyribose nucleic acid.

It is probably impossible to build this structure with a ribose sugar in place of the deoxyribose, as the extra oxygen atom would make too close a van der Waals contact.

The previously published X-ray data on deoxyribose nucleic acid are **insufficient** for a rigorous test of our structure. **So far as we can tell, it is roughly** compatible with the experimental data, but **it must be regarded as unproved until it has been checked against more exact results.** Some of these are given in the following communications. We were not aware of the details of the results presented there when we devised our structure, which rests mainly though not entirely on published experimental data and stereochemical arguments.

It has not escaped our notice that the specific pairing **we have postulated** immediately **suggests a possible** copying mechanism for the genetic material.[11]

Sources for Technical Terms

Sources for technical terms is an ambiguous expression. On the one hand, it refers to the numerous general and specialized technical dictionaries that aid readers and writers of technical prose, general sources such as *McGraw-Hill's Dictionary of Scientific and Technical Terms* or *Chamber's Dictionary of Science and Technology* and profession-specific dictionaries such as *Taber's Cyclopedic Medical Dictionary* or *Mosby's Medical and Nursing Dictionary.* Almost every trade or profession has at least one major dictionary, the authority for terms used by members of the discourse community.

Many more nonspecialized dictionaries are also helpful in the way they treat technical terminology. *Merriam-Webster's Tenth Collegiate Dictionary,* for example, effectively covers a broad spectrum of scientific and technical terms.

On the other hand, "sources for technical terms" also refers to the origins of technical terms. It's helpful for anyone using or encountering technical terms to understand where most technical terms come from. Knowing the origin of technical terms takes some of the mystery away from them and can make them a little less intimidating.

Of course, as is the case with nonspecialized language, nouns and verbs constitute most of technical terminology. Names in the sciences and technology account for many of the nouns.

According to John Harris, the names in the sciences and technology have six major sources of origin:

- Foreign language sources such as Greek and Latin root words, terms such as *telegraph, orthopter, pyrometer, microscope,* and *Pinus ponderosa.*
- Naming for inventor, discoverer, maker, or place. Eponyms such as *MacPherson strut, Allen wrenches, Crescent wrenches, Ohm's Law.* Place names such as *Morocco leather, Artesian wells,* and *Spanish bowline.*
- Naming for shape: *C clamp, A frame, O ring, V block.*
- Naming for function: *starter, generator, carburetor, buffer, valve lifter, sear release stop lever.*
- Acronyms/Abbreviations: *laser, NASA, radar, scuba.*
- Arbitrary terms and press agentry: the word *quark,* for example, as an arbitrary term apparently chosen from James Joyce's *Finnegans Wake.* Some names get applied by advertisers: *Nylon, Orlon, Dacron, Plexiglas, Rayon, Lucite, Kodak.*[12]

Common Difficulties Presented by Technical Terms

In addition to understanding the origin of most technical terms, writers of technical prose must be aware of the special difficulties they face and share with writers in general.

Readers and writers of technical prose are bothered by at least seven types of words.

- Words demanding a specificity of meaning: *absorb, adsorb; acceleration, velocity; concrete, cement; power, torque; mass, weight; flammable, inflammable.*
- Words in the general vocabulary that are often confused with technical words: *capacity, rating; complimentary, complementary.*
- Pairs or triads of words with similar meanings but differences in spelling: *aid, aide; adapt, adopt; accent, ascent, assent; decent, dissent,* and *descent.*
- Pairs or larger groups of words with similarity in pronunciation but differences in spelling, part of speech, or meaning: *ascent, ascend; advice, advise; material, materiel.*
- Words in the general vocabulary demanding a specificity of meaning—words that vex trolley riders as well as technicians: *principal, principle; affect, effect.*
- Words confused because of metathesis (metathesis refers to a transposition of the letters or syllables of a word): *casual, causal; bare, bear; perform, preform; stake, steak; break, brake; great, grate; discreet, discrete.*
- Sets of words causing special problems, problems often technical in nature: *complete, cylindrical, cubic, hexagonal, perfect, perpendicular, rectangular, saturated, true, unique.*[13]

The Challenges of Using Jargon

> What we now generally think of as "jargon" has other, even more fertile sources, in pretentiousness, in the emulation of fashionable trends, in the wish to be on the inside and the woe of being on the outside. There is a social origin of jargon which has less to do with pursuing a profession than with the masquerade of assuming a role and striking an accredited pose. It produces expressions like the following, noted by me in the course of ten minutes during an academic committee meeting: "operating in the tutorial situation"; "maximizing feedback"; "managerialism"; "corporate image"; "range of proven teaching strengths"; "package of proposals"; "involving students in the decision-making process"; "developing entrepreneurial skills".... You can shuffle the words and cut the phrases to the admiration of all, and your brain need never once be up and running.[14]
>
> —WALTER NASH

The meaning of *jargon* discussed here is not the specialized terms of a trade or profession, but the pretentious words almost everyone uses at one time or another. Underlying reasons for this pretentiousness range from ignorance of the accurate technical terms, laziness concerning finding out what the accurate technical terms are, indifference to the accurate technical terms, cliquishness or a sense of wanting to exclude others from your small group, obscurantism or deliberately wanting to confuse others, and self-importance, simply wanting to sound more knowledgeable about a subject than you actually are.

To understand this pretentious jargon, it helps to know the origin of the term and its relationship to *gobbledygook*.

Origin of Jargon

The term *jargon* comes from late Middle English forms *iargo(u)n, gargoun,* and *girgoun* and from Old French *jargoun, gargon,* and *gergon*. These terms refer to "the twittering and chattering of birds, meaningless talk, gibberish."[15]

George Orwell was no doubt aware of this origin of *jargon* when, early in *1984*, he describes Winston Smith overhearing a man and a young woman talking in the cafeteria of the Ministry of Truth:

> Winston knew the man by sight, though he knew no more about him than that he held some important post in the Fiction Department. He was a man of about thirty, with a muscular throat and a large, mobile mouth. His head was thrown back a little, and because of the angle at which he was sitting, his spectacles caught the light and presented to Winston two blank discs instead of eyes. What was slightly horrible was that from the stream of sound that poured out of his mouth, it was almost impossible to distinguish a single word.... it was just a noise, a quack-quack-quacking.... Whatever it was, you could be certain that every word of it was pure orthodoxy, pure Ingsoc. As he watched the eyeless face with the jaw moving rapidly up and down, Winston had a curious feeling that this was not a real human being but some kind of dummy. It was not

the man's brain that was speaking; it was his larynx. The stuff that was coming out of him consisted of words, but it was not speech in the true sense: it was a noise uttered in unconsciousness, like the quacking of a duck.[16]

1984 is concerned not only with the abuse of language for political purposes, but the increasing trend toward abstraction in every day English. Orwell's *newspeak* has become almost synonymous with deceptive language.

Gobbledygook

When jargon, as pretentious and often unintelligible words, is part of long, convoluted sentences, paragraphs, and other segments, *gobbledygook* occurs. The term *gobbledygook* is attributed to congressman Maury Maverick of Texas, who coined the term some time during World War II. Listening to the long-winded speeches of his colleagues in Congress, he reportedly said, "Perhaps I was thinking of the old bearded turkey gobbler back in Texas who was always gobbledygobbling and strutting with ludicrous pomposity. At the end of this gobble there was a sort of gook."[17]

In *A Browser's Dictionary,* John Ciardi defines *gobbledygook* by example: "Linguistic utilizations intermediate to finalized specification and rhetorically structured to maximize optionalization of alternatives while preserving deniability interim-wise."[18] Examples of gobbledygook are everywhere. Consider the following excerpts.

A sentence from a memo by a high school principal to his teachers:

> Intercom utilization will be used to initiate substitute teacher involvement.

Part of a memo written by the commandant of the marine corps:

> It has been decisioned that some form of unit rotation may be a desirable objective. Detailed planning has been held in abeyance because of structural and manning level imbalances between WestPac and EastPac and other associated areas of concern. Recent CMC decisions have alleviated the major inhibitors allowing a fresh approach and revaluation of alternative methods of unit replacement of WestPac personnel. Preliminary staff analyses has concluded that a month TAD unit replacement appears to be an attractive possibility and should serve as the focal point for a full feasibility determination prior to development of any implementation procedures or recommendations.

From a draft of a proposed executive order defining federal policy:

> It should permit knowledge to be gained at various points in the process such that progress can be measured in advance of the termination of an effort where interim fine-tuning or revision is an alternative to postmortem examinations.

From a readme.txt file found on the World Wide Web:

> As part of its investigation into solutions for the interoperability problem associated with use of dissimilar simulators in a DIS network, it is the intent of the Institute for Simulation and Training to define acceptable levels of interoperability. The approach,

in part will consist of an evaluation of those simulators currently in use and considered "interoperable" by users. This evaluation will consist of a series of tests to measure parameters associated with the computer image generator (CIG) of each system to determine difference (if any) in CIG capability. Such differences can then be construed to be acceptable differences insofar as interoperability is concerned.

To accomplish this investigation an interoperability test procedure is being developed to define a series of tests to be performed on simulator CIGOs. The development of this procedure is an on-going process. Any feedback from the user community concerning this document will be much appreciated.

Advantages of Jargon and Gobbledygook

Here we come to the main problem with The Official Style. There is no point in reproaching it for not being clear. *It does not really want to be clear. It wants to be poetic. At its best, it wants to tell you* how it feels to be an official, *to project the sense of numinous self-importance officialdom confers. It wants to make a prosaic world mysterious.*[19]

—RICHARD LANHAM

In *Revising Prose,* Richard Lanham coins the phrase "The Official Style" to refer to the bureaucratic style which "dominates written discourse in our time."[20] Lanham argues that The Official Style "is often stigmatized as bureaucratese or jargon, and it often is both. But it is a genuine style, and one that reflects the genuine bureaucratization of American life.... The Official Style comes to us in many guises but two main ones: as the language of the learned professions and as the language of bureaucracy, whether in government, business, or the military."[21] For Lanham, The Official Style is the opposite of a plain style. Significantly, however, Lanham does not argue that you should avoid The Official Style and always write in a plain style. He suggests (perhaps with a hint of sarcasm) that The Official Style so dominates communication today that you must learn how to write it—"If you can analyze, write, and translate it, maybe you can find your niche in the system—public sector or private."[22] Lanham does not advocate total capitulation to this style, though. He believes you can write in The Official Style "without losing your soul to it. For you may have to write in The Official Style, but you don't have to think in it."[23] We will discuss the ethical implications of using this style in Chapter 11.

Lanham helps us accept the reality of a style that will not go away and that has its advantages. For Lanham, "The Official Style aims deliberately at saying nothing at all, but saying it in the required way. Or at saying the obvious impressively. The Official Stylist must seem in control of everything but responsible for nothing."[24] The reality of this bureaucratic style is that it helps people survive. First, The Official Style "makes what you've done sound important and, still more important than important, Official."[25] Second, The Official Style also helps us to live in "a euphemistic society": "Society may have its pains and problems, but language can sugar-coat them."[26] Third, The Official Style helps us cope with "a society afraid of taking responsibility."[27] One of the major rules of the bureaucratic world is "Keep your head down. Don't assert anything you'll have to take the blame for."[28]

Lanham acknowledges that much of The Official Style also exists because so many people do not know how to write well. While much of The Official Style is "self-conscious put-on," much is due to "real ineptitude, genuine system sickness."[29]

Of course, Lanham is not alone in recognizing the values of The Official Style. Tom Fahey comments that "inventing jargon is a perennial feature of Americana."[30] He believes we are to be congratulated for "our undeniable contribution to that most marvelous of human gifts, language."[31] For Fahey there are four categories of professionals who "get PAID to talk that way": jargoneers, jargonizers, jargoons, and jargonauts.[32] Jargoneers are the "true enthusiasts," the "marketeers, political spin artists, tellers of tall technical tales, computer evangelists."[33] Jargonizers are the "techies": "your neighborhood do-it-yourselfers, hobbyists, ham radio operators, hot rodders, and hackers."[34] A jargoon "is someone in authority who uses language to confuse us."[35] Jargonauts are "the originators of jargon," people like Red Barber who coined words such as *rhubarb* ("a squabble between opposing players or between coaches and umpires"), and *cat bird seat* ("the announcer's press box vantage point").[36]

The following is an example of some typical sales jargon:

Hi-Tech Computer Sales Jargon

advanced design	the advertising agency doesn't understand it
it's here at last!	rush job; nobody knew it was coming
field tested	manufacturer has no test equipment
high accuracy	unit on which all parts fit
direct sales only	factory had big argument with distributor
years of development	we finally got one that works
revolutionary	it's different from our competitors'
breakthrough	we finally figured out a way to sell it
futuristic	no other reason why it looks the way it does
distinctive	a different shape and color than the others
maintenance-free	impossible to fix
re-designed	previous faults corrected, we hope....
handcrafted	assembly machinists worked without wearing gloves
performance proven	will operate through the warranty period
meets all standards	ours, not yours
all solid-state	heavy as hell
broadcast quality	gives a picture and makes noise
high reliability	we made it work long enough to ship it
SMPTE Bus compatible	will be shipped by Greyhound
new generation	old design failed; maybe this one will work
mil-spec components	we got a good deal at a government auction
customer service across the country	you can return it from most airports

unprecedented performance	none of the ones we had before worked this way!
built to precision tolerances	we finally got it to fit together
satisfaction guaranteed	manufacturer's, upon cashing your check
microprocessor controlled	does things that we can't explain
latest aerospace technology	one of our techs used to work for Boeing[37]

Disadvantages of Jargon and Gobbledygook

The disadvantages of jargon and gobbledygook are many. First, if you are trying to explain how to do something (a manual), offering to do something (a proposal), or reporting on something done (a report), and your purpose is to be clear, then using unfamiliar or pretentious jargon defeats your purpose. If even one member of the audience is confused by the terms you use, then you are not achieving your purpose as effectively as you should.

Second, pretentious jargon often strikes listeners and readers as self-serving. Unless it is your purpose in your prose to be pretentious, pretentiousness usually undermines your purpose. Audiences want to know that you're taking their needs, background, interests, and knowledge-level into account. Most audiences want sincerity, not insincerity.

Third, other practical concerns about jargon include possible significant financial losses. Both the public and private sector have lost countless hours in employee productivity as a result of confusing or misleading terminology. Users of software documentation, many of them confused by the terminology used in the documents shipped with the product, are increasingly calling technical support. Some companies have been forced to provide large staffs for technical support to talk customers through all kinds of problems, and this kind of technical support can be very expensive to provide.

Fourth, when writers of technical prose use terms their audiences do not know or use pretentious jargon, all kinds of ethical issues come to the surface. Is the writer being deliberately pretentious, deceptive, or ambiguous? If so, why and to whose advantage? (These concerns are addressed in Chapter 11.)

Fifth, if we deliberately ignore our audience and overwhelm them with jargon and gobbledygook, what kind of society and self are we promoting? In spite of his recognition of The Official Style as a genuine style, Lanham has some concerns about how this style may dehumanize us. For Lanham you should "bother" to write better prose "to invigorate and enrich your selfhood, to increase, in the most literal sense, your self-consciousness. Writing, properly pursued, does not make you better. It makes you more alive."[38] The Official Style distances you from your real self: "It has no human voice, no face, no personality behind it."[39] Ultimately, The Official Style "is a bad style...because it denatures human relations."[40]

Other Specialized Languages

There are many different approaches to the subject of jargon, and the different attempts to distinguish jargon from similar language are confusing. An important point to remember is that jargon, like all of the other specialized languages it is compared to, is just that: a specialized language.

Jargon and Slang

Jargon has much in common with slang, cant, and argot. All are specialized languages, but there are subtle differences. Whereas jargon is chiefly the technical terminology used to practice a trade or profession, *slang* consists mostly of colloquial words and phrases that are considered socially lower than more widely accepted language. Also, as the technical terminology of a trade or profession, jargon aims primarily at helping members of a discourse community communicate more efficiently and effectively. Jargon also may help to enhance the dignity of a profession, as in the case of law and lawyers, for example. Slang, as we learned in Chapter 5, is more of an attitude and one of its primary aims may be to prevent outsiders from understanding. Slang is also a colloquial shorthand enabling members of a group to communicate.

Cant and Argot

In this respect, slang shares more with cant and argot. In fact, it would be accurate to say that cant and argot are kinds of slang. All three aim at misleading and preventing the rest of society and authorities in particular from understanding what is being said. *Cant* is specifically the language of members of the underworld. McArthur defines *cant* as "the jargon of a class, group, or profession, often used to exclude or mislead others: a teenage gang member in Los Angeles saying that he was in his *hoopty* around *dimday* when some *mud duck* with a *trey-eight* tried to *take him out of the box* (he was in his car around dusk when a woman armed with a .38 caliber pistol tried to shoot him)."[41]

Argot is more specialized than cant. *Argot* is specifically criminal cant. Anthony Burgess's *A Clockwork Orange* is written entirely in argot from the point of view of the gang leader Alex. In the opening of the novel, Alex tells us:

> There was me, that is Alex, and my three droogs, that is Pete, Georgie, and Dim, Dim being really dim, and we sat in the Korova Milkbar making up our rassoodocks what to do with the evening, a flip dark chill winter bastard though dry. The Korova Milkbar was a milk-plus mesto, and you may, O my brothers, have forgotten what these mestos were like, things changing so skorry these days and everybody very quick to forget, newspapers not being read much neither. Well, what they sold there was milk plus something else. They had no license for selling liquor, but there was no law yet against prodding some of the new veshches which they used to put into the old moloko, so you could peet it with vellocet or synthemesc or drencrom or one or two other veshches which would give you a nice quite horrorshow....[42]

Although Alex's specialized language is fictive, it demonstrates well the qualities of argot. Alex and his droogs are not only members of the underworld, but they are criminals who use their language in part to shield their activities.

Of course, as you might expect there is no consensus on these subtle distinctions. For example, for Samuel Johnson and many current language pundits, cant is humbug or hypocritical affectation. Johnson writes:

> My dear friend, clear your *mind* of cant. You may *talk* as other people do: you may say to a man, "Sir, I am your most humble servant." You are *not* his most humble servant.

You may say, "these are sad times; it is melancholy thing to be reserved to such times." You don't mind the times. You tell a man, "I am sorry you had such bad weather the last day of your journey, and were so much wet." You don't care six-pence whether he was wet or dry. You may *talk* in this matter; it is a mode of talking in Society: but don't *think* foolishly.[43]

In this sense, cant shares much with jargon. Cant, as Johnson refers to it, is a kind of *social jargon*. Both cant and jargon are often empty phraseology or socially pretentious.

Doublespeak

Jargon has been classified by some as a kind of doublespeak, and doublespeak has been classified by others as a kind of jargon.

For example, in *Doublespeak: From 'Revenue Enhancement' to 'Terminal Living'— How Government, Business, Advertisers, and Others Use Language to Deceive You*, William Lutz classifies jargon as one of four kinds of *doublespeak*. According to Lutz's classification scheme, the four kinds of *doublespeak* are *euphemism* ("an inoffensive or positive word or phrase used to avoid a harsh, unpleasant, or distasteful reality"); *jargon* ("the specialized language of a trade, profession, or similar group" and "pretentious, obscure, and esoteric terminology used to give an air of profundity, authority, and prestige to speakers and their subject matter"); *gobbledygook* or *bureaucratese* ("a matter of piling on words, of overwhelming the audience with words, the bigger the words and the longer the sentences the better"); and *inflated language* ("language that is designed to make the ordinary seem extraordinary; to make everyday things seem impressive; to give an air of importance to people, situations, or things that would not normally be considered important; to make the simple seem complex").[44] Except for the sense of jargon as a specialized language of a trade or profession, all of these kinds of doublespeak share an intention on the part of the writer or speaker to deceive the reader or listener in one way or another.

In *The Joys of Jargon*, Tom Fahey discusses doublespeak as one of seven varieties of jargon: *occupational slang, professional/technical terminology, technical metaphors, buzzwords, doublespeak, abbreviations,* and *ex-jargon*.

For Fahey, *occupational slang* is "insider's lingo" of workers in particular jobs.[45] Police in Philadelphia refer to criminals as "*critters*" who may charged with *REAP*, recklessly endangering another person.[46]

Professional/technical terminology refers to the terms of "professional (academic) languages," words found in titles of academic journals such as "Dermatogylphic Patterns and Pattern Intensities of the Genus *Cacagao* (*Cebidae, Platyrrhini*) with Observations on Interspecific Differentiation."[47]

Technical metaphors are metaphors created by "taking a technical concept and applying it to an ordinary situation," words or phrases such as *litmus test, black hole,* and *out of the loop*.[48]

Buzzwords "are the latest 'in' phrases," such as *user-friendly* or *ozone-friendly*.[49]

Doublespeak "is deliberately obscure language intended to mislead."[50] For Fahey, doublespeak becomes jargon "when it is used consciously and professionally, as when a bank disguises a bad loan by calling it a *nonperforming asset*."[51]

Abbreviations are initials (*mph* for miles per hour) or acronyms (*NASA* for National Aeronautics and Space Administration).

Finally, *ex-jargon* occurs "When a term becomes known by virtually everybody, in other words, when it stops being a word 'peculiar to a particular group.'" The word *radar* is a good example, a word that originated as an acronym for **Ra**dio **D**etection **A**nd **R**eceiving."[52]

So is jargon a kind of doublespeak or is doublespeak a kind of jargon? Lutz and Fahey do agree that *doublespeak* deliberately obscures, so perhaps it is more clear to think of jargon as a kind of doublespeak. *Doublespeak* is a useful umbrella term for any kind of deceptive language.

Tech Speak

According to Edward Tenner in his book *Tech Speak or How to Talk High Tech,*

> Tech Speak is a postcolloquial discourse modulation protocol for user status enhancement. It's a referential system for functional-structural, microscopically specific macroscopic-object redesignation. It's a universal semantic transformation procedure. It's a holophrastic technocratic sociolect. It's a meta-semiotic mode for task specific nomenclature.
>
> Tech Speak is an accretive substantive-compound formulation tool. It's a representational Newtonian-solid identification technique. It's an *n*-dimensional matrix of analytical denomination-conventions. It's a recombinant autonomous neologistic paradigm. It's a post-Linnaean taxonomic organic/inorganic classificategory strategy....[53]

With tech speak Tenner goes one step further than jargon and gobbledygook. Tenner suggests, "Tech Speak consists of jargon, but it's more than jargon. It unifies rather than separates people, as jargon does."[54] Tech Speak is equally mystifying to both expert and layperson. Tenner suggests Tech Speak is not simply euphemistic talk nor is it the language of scientists and engineers. And Tech Speak is not just a series of long words. Tech Speak is distinctive because it "brings rigorous structural-functional description" to the things of everyday life.[55] As Tenner suggests, "You can name and describe anything in Tech Speak. You need only talk about it as though you had just invented it. By setting aside all the connotations of a familiar word and introducing others, the Tech Speaker is almost creating a new thing."[56] Finally, Tech Speak "must use complexity up to the limit of our comprehension to reveal the intricate true nature of the things it represents. Of course, it never achieves this."[57]

In Tech Speak a family becomes an *in vivo recombinant genetic system* (IVRGS) or a "dyadic hominid autopropagation unit for intergenerational meiotic chromosome redistribution."[58] A toaster becomes a *dual carbohydrate-oxidation chamber* or, more specifically, "an electroresistive self-limiting Maillard reaction initiator for infrared irradiative thermal preparation of microbially aerated, kiln-stabilized milled cereal-endosperm sections for accelerated absorption of spatula-delivered nonfermented bovine lipid emulsions or hydrogenated carotene-pigmented vegetable glycerides."[59]

Obviously, Tenner's *Tech Speak* is a parody of trends in jargon that make it increasingly complex and confusing to audiences of all knowledge levels.

Technobabble and Computerese

Computerese is the use of computer jargon in your writing. Technobabble is another specialized language because John Barry views it as "an offshoot of *computerese*."[60] Barry suggests technobabble differs from computerese in two key respects: "the heavy influence of marketing and public relations (PR) and the extracomputer contexts in which it is used."[61] Barry argues that computerese or computer jargon devolves into babble under the following circumstances: when it used as filler or decoration, when it is employed intentionally for obfuscatory purposes, when it is employed gratuitously, when it is used obsessively, and when it is used by those unfamiliar with its meanings in an attempt to sound as if they know what they are talking about.[62]

Shop Talk

Shop talk is jargon that has become so specialized that it applies only to one company or subgroup of an occupation. For example, shop talk at Walt Disney World Company includes *turtle* (a slow guest), *419* (a handsome male), *914* (a beautiful female), *ETR* (employee time recording), *ER* (early release, to go home early), and *greeter* (a person who straightens strollers).

Abbreviations, Acronyms, Initialisms

Abbreviations are shortened versions of words and are formed by the first letters of words: *co.* for *company* or *c.o.d.* for *cash on delivery*. Abbreviations save space in technical prose, but abbreviations may also present problems for your readers. There is no guarantee that all of your readers will understand your abbreviations, so you have to analyze your audience when you are deciding whether to use them. If you choose to abbreviate, identifying the abbreviations first will aid your reader: *According to the Drug Abuse Warning Network (DAWN), drug-related emergency admissions are at an all time high. DAWN has completed a new questionnaire to widen its research to include ages and ethnic background.*

Acronyms are abbreviations formed by combining the first letter or letters of several words. An acronym is pronounced as a word and should be written in capital letters without periods: *NASA (National Aeronautics and Space Administration); RAM (Random-access memory)*. Acronyms that have become accepted as nouns are the only exception to the capitalization rule. These acronyms are written with lower case letters without periods: *scuba (self-contained underwater breathing apparatus); sonar (sound navigation ranging)*.

Initialisms are abbreviations that are formed by combining the first letter of each word in a phrase consisting of many words. An initialism is pronounced as separate letters: *STC (Society for Technical Communication); P.M. (post meridiem)*. Initialisms are written using both uppercase and lowercase letters. Use periods for only lowercase initialisms. The two exceptions to this rule are geographic names and academic degrees: *U.S.A. (United States of America); A. A. (Associate of Arts)*. The common rule for plural acronyms and initialisms is to add an *s,* not an apostrophe *s:* DOSs; CRTs.

Consider your reader's knowledge concerning acronyms and initialisms. What might be familiar to you will not necessarily be familiar to your reader. When you first use an acronym or initialism, spell the entire phrase out with the acronym or initialism in parentheses. Spelling the acronym or initialism out when it is first used will guarantee that your reader will know what the abbreviation means.

When writing a long document, you should periodically write out the acronym or initialism in parentheses after the abbreviation. Using this technique will ensure that your reader will not have to refer elsewhere continually to find the meaning of the acronym or initialism. Acronyms and initialisms should also be placed in the glossary of your document.

Filling your prose with abbreviations, acronyms, and initialisms has many names, including *abbrevomania* and *agglomerese:* "The D200 has a dual Centronics parallel/RS 232C serial interface plus IBM PC compatibility and Epson emulation (Smith and Corona ad, 1980s)."[63]

Tips for Using Clear Technical Language

Of course, clear technical language consists of more than just using technical terms your audience understands. An ill-defined purpose, a confusing organization, poorly written sentences, overly long paragraphs, and many other faults contribute to an ineffective document. Yet there are many factors you can consider to make your use of technical terms more appropriate.

First, understand the meanings of jargon—necessary technical terms you must use for many occasions within a discourse community, inappropriate technical terms used for the wrong audience, and the pretentious use of technical terms. Second, understand how jargon is different from related terms and other kinds of specialized languages. Third, in any situation where you have doubts about some of the audience's ability to understand the terms you are using, define or clarify the terms you use. You can define or clarify your terms using a number of strategies:

- Provide a glossary
- Provide definitions for terms within the text (parenthetically or set off as a phrase using commas)
- Provide an analogy or some other kind of comparison
- Provide a graphic to show what the object or concept is
- Provide an anecdote or some other type of narrative to clarify the term
- Provide a description
- Use nontechnical words instead of technical words
- Use specific and concrete words, rather than general and abstract words
- Omit unnecessary modifiers and qualifiers
- Eliminate wordiness
- Use terminology consistently

Chapter Summary

Writers of technical prose are faced with many special challenges, especially when writing for lay people. Technical terms and jargon are particularly difficult to handle well for this kind of audience. You need to know some basic distinctions among words, terms, and terminologies. You also need to know the various meanings of the term *jargon*—as the technical terms of a discourse community, as technical terms used inappropriately, and as pretentious terms. And you need to know how jargon and gobbledygook are related.

Writers of technical prose also benefit from knowing that jargon and gobbledygook are just some of the many specialized languages available to them. Slang, cant, argot, doublespeak, tech speak, technobabble, computerese, and shop talk also present many challenges. Abbreviations, acronyms, and initialisms are also challenging to writers of technical prose. Many strategies can be employed to use a specialized language effectively if your purpose is to be clear, unpretentious, and sincere in your writing.

Case Study 5: Apollo 13 *and Jargon*[64]

Problem

When technical jargon is a necessity due to the highly technical nature of the discourse community, how do insiders communicate clearly and effectively in times of crisis?

Scenario

"At a little after nine central standard time on the night of Monday, April 13, 1970, there was, high in the western sky, a tiny flare of light that in some respects resembled a star exploding far away in our galaxy."[65]

The tiny speck of light noticed by several amateur astronomers signaled the beginning of the greatest rescue in the history of the American manned space program. Thousands of miles away, the crew of *Apollo 13,* James A. Lovell, Jr., John L. Swigert, Jr., and Fred W. Haise, Jr., knew the third planned lunar landing was in serious trouble.

Until encountering onboard problems, *Apollo 13* represented the smoothest flight ever by an Apollo spacecraft. At 46 hours 43 minutes Joe Kerwin, the CapCom (Capsule Communicator) on duty, said, "The spacecraft is in real good shape as far as we are concerned. We're bored to tears down here."[66]

At 55 hours 46 minutes, the crew finished a 49-minute TV broadcast with Lovell stating: "This is the crew of *Apollo 13* wishing everybody there a nice evening, and we're just about ready to close out our inspection of Aquarius (the Lunar Module [LM]) and get back for a pleasant evening in Odyssey (the Crew Module [CM]). Good night."[67]

Nine minutes later, Oxygen Tank No. 2 blew up (later referred to as a "tank failure" by NASA media specialists).[68] This failure also caused Oxygen Tank No. 1 to fail. The *Apollo 13* CM's normal supply of electricity, light, and water was lost. The stranded astronauts were now about 200,000 miles from Earth.

The CapCom from the Black Team now on duty, Jack Lousma, received a message from Swigert, "OK, Houston, we've had a problem here." Swigert said he saw a warning light that accompanied a bang, then another sharp bang and vibration. The time was 2108 hours on 13 April.[69]

What followed until the splashdown of *Apollo 13* on 17 April was a furious effort to save the lives of the three astronauts. But more importantly, the first few minutes of the crisis proved that professionals working under incredible stress must communicate effectively and efficiently. During the first minutes after the explosion, three stranded astronauts knew they were in capable hands—they had to be to survive.

Investigation

What follows is the actual Houston control room transcript of the events from 2.5 minutes before the accident to about five minutes after. Times given are in Ground Elapsed Time (GET): time elapsed since liftoff of *Apollo 13* on 11 April 1970 at 2:13 P.M. (EST).

55:52:31 Master caution and warning triggered by low hydrogen pressure in Tank No. 1.

55:52:58 CapCom (Charlie Duke): "13, we've got one more item for you, when you get a chance. We'd like you to stir up the cryo tanks. In addition, I have shaft and trunnion…

55:53:06 Swigert: "Okay."

55:53:07 CapCom: "…for looking at Comet Bennett, if you need it."

55:53:12 Swigert: "Okay. Stand by."

55:53:18 Oxygen Tank No. 1 fans on.

55:53:19 Oxygen Tank No. 2 pressure decreases 8 psi.

55:53:20 Oxygen Tank No. 2 fans turned on.

55:53:20 Stabilization control system electrical disturbance indicates a power transient.

55:53:21 Oxygen Tank No. 2 pressure decreases 4 psi.

55:53:22.718 Stabilization control system electrical disturbance indicates a power transient.

55:53:22.757 1.2 volt decrease in ac bus 2 voltage.

55:53:22.772 11.1 amp rise in Fuel Cell 3 current for one sample.

55:53:26 Oxygen Tank No. 2 pressure begins rise lasting for 24 seconds.

55:53:38.057 11 volt decrease in ac bus 2 voltage for one sample.

55:53:38.085 Stabilization control system electrical disturbance indicates a power transient.

55:53:41.172 22.9 amp rise in Fuel Cell 3 current for one sample.

55:53:41.192 Stabilization control system electrical disturbance indicates a power transient.

55:54:00 Oxygen Tank No. 2 pressure rise ends at a pressure of 953.8 psia.

55:54:15 Oxygen Tank No. 2 pressure begins to rise.

55:54:30 Oxygen tank No. 2 quantity drops from full scale for 2 seconds and then reads 75.3 percent.

55:54:31 Oxygen Tank No. 2 temperature begins to rise rapidly.

55:54:43 Flow rate of oxygen to all three fuel cells begins to decrease.

55:54:45 Oxygen Tank No. 2 pressure reaches maximum value of 1008.3 psia.

55:54:51 Oxygen Tank No. 2 quantity jumps to off-scale high and then begins to drop until the time of telemetry loss, indicating failed sensor.

55:54:52 Oxygen Tank No. 2 temperature sensor reads –151.3 F.

55:54:52.703 Oxygen Tank No. 2 temperature suddenly goes off-scale low, indicating failed sensor.

55:54:52.763 Last telemetered pressure from Oxygen Tank No. 2 before telemetry loss is 995.7 psia.

55:54:53.182 Sudden accelerometer activity on X, Y, Z axes.

55:54:53.220 Stabilization control system rate changes begin.

55:54:53.323 Oxygen Tank No. 1 pressure drops 4.2 psi.

55:54:53.500 2.8 amp rise in total fuel cell current.

55:54:53.542 X, Y, and Z accelerations in CM indicate 1.17g, 0.65g, and 0.65g.

55:54:53.555 Master caution and warning triggered by DC main bus B undervoltage. Alarm is turned off in 6 seconds. All indications are that the cryogenic Oxygen Tank No. 2 lost pressure in this time period and the panel separated.

55:54:54.741 Nitrogen pressure in Fuel Cell 1 is off-scale low indicating failed sensor.

55:54:55.350 Telemetry recovered.

55:54:56 Service propulsion system engine valve body temperature begins a rise of 1.65 F in 7 seconds. DC main bus A decreases 0.9 volts to 28.5 volts and DC main bus B 0.9 volts to 29.0 volts. Total fuel cell current is 15 amps higher than the final value before telemetry loss. High current continues for 19 seconds. Oxygen Tank No. 2 temperature reads off-scale high after telemetry recovery, probably indicating failed sensors. Oxygen Tank No. 2 pressure reads off-scale low following telemetry recovery, indicating a broken supply line, a tank pressure below 19 psi, or a failed sensor. Oxygen Tank No. 1 pressure reads 781.9 psia and begins to drop.

55:54:57 Oxygen Tank No. 2 quantity reads off-scale high following telemetry recovery indicating failed sensor.

55:55:01 Oxygen flow rates to Fuel Cells 1 and 3 approached zero after decreasing for 7 seconds.

55:55:02 The surface temperature of the service module oxidizer tank in Bay 3 begins a 3.8 F increase in a 15-second period. The service propulsion system helium tank temperature begins a 3.8 F increase in a 32-second period.

55:55:09 DC main bus A voltage recovers to 29.0 volts, DC main bus B recovers to 28.8.

55:55:20 Swigert: "Okay, Houston, we've had a problem here."

55:55:28 Duke: "This is Houston. Say again please."

55:55:35 Lovell: "Houston, we've had a problem. We've had a main B bus undervolt."

55:55:42 Duke: "Roger. Main B undervolt."

55:55:49 Oxygen Tank No. 2 temperature begins steady drop lasting 59 seconds, indicating a failed sensor.

55:56:10 Haise: "Okay. Right now, Houston, the voltage is—is looking good. And we had a pretty large bang associated with the caution and warning there. And as I recall, Main B was the one that had an amp spike on it once before."

55:56:30 Duke: "Roger, Fred."

55:56:38 Oxygen Tank No. 2 quantity becomes erratic for 69 seconds before assuming an off-scale low state, indicating a failed sensor.

55:56:54 Haise: "In the interim here, we're starting to go ahead and button up the tunnel again."

55:57:04 Haise: "That jolt must have rocked the sensor on—see now—oxygen quantity 2. It was oscillating down around 20 to 60 percent. Now it's full-scale high."

55:57:39 Master caution and warning triggered by DC main bus B undervoltage. Alarm is turned off in 6 seconds.

55:57:40 DC main bus B drops below 26.25 volts and continues to fall rapidly.

55:57:44 Lovell: "Okay. And we're looking at our service module RCS Helium 1. We have—B is barber poled and D is barber poled, Helium 2, D is barber pole, and secondary propellants, I have A and C barber pole." AC bus fails within 2 seconds.

55:57:45 Fuel Cell 3 fails.

55:57:59 Fuel cell current begins to decrease.

55:58:02 Master caution and warning caused by AC bus 2 being reset.

55:58:06 Master caution and warning triggered by DC main bus undervoltage.

55:58:07 DC main bus A drops below 26.25 volts and in the next few seconds levels off at 25.5 volts.

55:58:07 Haise: "AC 2 is showing zip."

55:58:25 Haise: "Yes, we got a main bus A undervolt now, too, showing. It's reading about 25 and a half. Main B is reading zip right now."

56:00:06 Master caution and warning triggered by high hydrogen flow rate to Fuel Cell 2.[70]

Some Observations about the Dialogue

During these tense 7.5 minutes, we can make these observations about the dialogue between the stranded astronauts and the Houston flight controllers: a discernible lack of panic and frequent use of technical terms, some so specialized that it's shop talk.

Some shop talk used:

1. "stir up the cryo tanks"—turn on the heating elements within the cryogenic fuel tank to expand and mix the liquid
2. "shaft and trunnion"—the tripod anchoring a camera or telescope
3. "stand by"—please wait
4. "undervolt"—low electrical gauge reading
5. "Roger"—OK
6. "caution and warning"—a flashing panel light on the "Caution and Warning" panel indicating a potential malfunction
7. "amp spike"—a sharp increase in electrical current
8. "button up the tunnel"—close the hatch between the lunar module and the crew module
9. "oscillating down"—gauge needle moving toward zero
10. "full-scale high"—gauge needle off the scale to the high side
11. "RCS Helium high"—Reaction Control System (the small maneuvering jets) helium tank reading high pressure

12. "barber poled"—a thin, vertical gauge containing a black and white striped indicator; upward movement indicates extreme usage; "barber poled" would mean the indicator fills the entire gauge
13. "showing zip"—no reading

Some Conclusions about Case Study 5

Henry Cooper believes that "what he [Swigert] felt may in fact have been not unlike the disconcerting shudder that first puzzled some of the passengers aboard the *Titanic* as the ship scraped against an iceberg."[71] Unlike the passengers of the *Titanic,* all the passengers of the *Apollo 13* spaceship were successfully rescued. In any other profession, the official outcome of *Apollo 13*'s mission—"Classed as 'successful failure' because of experience in rescuing crew"—would be viewed as a ridiculous oxymoron.[72] The 1500 passengers and crew who lost their lives when the *Titanic* went down in 1912 clearly hoped for such an oxymoron.

Questions

1. What, if anything, does this case study illustrate about the efficiency of technical language?
2. Provide a list of what you consider the most unusual technical terms from the dialogue. Why are these technical terms unusual?
3. Are the thirteen examples given above shop talk or jargon? Defend your answer.
4. What parts of the dialogue are in plain English? Why?
5. Could more of the dialogue have been in plain English? If so, which parts?
6. What kinds of backgrounds can you assume for the audience that was receiving the transmissions from *Apollo 13*?

Questions/Topics for Discussion

1. What was the most recent situation where you found yourself listening to someone use technical terminology you could not understand? What could have been done differently to help you understand?
2. What are the different meanings of jargon offered in this chapter? Is it important to make a distinction in these meanings? Why or why not?
3. In what key ways does scientific English differ from everyday English?
4. Do you think it's necessary, as John Barry apparently does, to distinguish *technobabble* from *computerese*? Why or why not?
5. John Tenner is obviously offering a parody in *Tech Speak*. What do you think is the subject of his parody and does parodying this subject have any value?
6. Do you think the use of abbreviations, acronyms, and initialisms has gone too far in some companies? Why or why not?
7. What do you think are the best strategies for handling technical terms for an audience that is unfamiliar with these terms?
8. Have you ever read something recently written in what Lanham calls The Official Style? If so, what was the document? Was the style "self-conscious put-on" or due to "real ineptitude, genuine system sickness"?
9. In what ways do the media—television, film, radio, and print—contribute to the spread of pretentious jargon?
10. What are some of the technical terms concerning the Internet you find strange and unfamiliar?

Exercises

1. Find a technical document aimed at the home user—for example, a manual for a camera, computer, audio system, home theater system, television, appliance, or tool—and identify the technical terms in the document. Are the terms handled effectively? If so, how? Are there any confusing technical terms? If so, why are they confusing? Are there any pretentious words?
2. Find a technical document aimed at owners of small businesses—for example, manuals to fax machines or photocopies, computer hardware manuals, software manuals, procedural guides, or employee handbooks. Discuss how appropriate the technical terms are for the intended audience.
3. Visit a nearby hardware, computer, audio systems, television, appliance, or automobile repair shop. Interview a sales clerk and ask what kinds of terminology are regularly used with customers.
4. Interview an expert or technician. You may want to interview a professor in any discipline, for example. Or you may want to interview a doctor, lawyer, accountant, dentist, or other professional. Or visit an electrician, automobile mechanic, or appliance repair person. Ask what this person does to clarify technical concepts for a lay audience. Also ask if this person can identify the occasions when he or she is not concerned about defining or clarifying terms.
5. What kinds of shop talk have you used on your job or some other social group (family, church, club, organization, and so on)? What has been the value of this shop talk?
6. What characteristics does jargon share with slang? How is jargon different from slang?
7. Find a technical report in the government documents section of your local community or university library. Identify some of the technical terms used. What makes the terms difficult to understand?
8. Find a recent book about the Internet and identify a dozen technical terms used in the book. Are the terms adequately defined in the book? Why or why not?
9. Access the World Wide Web using your favorite browser. Visit five sites and characterize how effectively you think these sites handle technical terms.
10. Find the *McGraw-Hill Dictionary of Scientific and Technical Terms* in the library. Characterize the usefulness of this dictionary to writers of technical prose.
11. Read the following description of the Marburg and Ebola viruses published in the *Review of Medical Microbiology*. Make a list of any terms which are jargon to you.

> Marburg virus disease is an acute hemorrhagic febrile infection first recognized in 1967 during an epidemic among laboratory workers exposed to the infected tissues of imported African green monkeys (*Cercopithecus aethiops*) in Germany and Yugoslavia. Person-to-person transmission occurred from the hospitalized cases to medical personnel. No further epidemics of sporadic cases have appeared among persons exposed to nonhuman primates in America or Europe. Nevertheless, serologic surveys have indicated that the virus is present in East Africa (Uganda and Kenya) and causes infections in monkeys and humans. Two cases occurred in 1975 in tourists traveling through Africa.
>
> Marburg virus has been isolated in guinea pigs and various cell culture systems. The virus particle contains RNA and has a cylindric or filamentous shape by electron microscopy. Although this appearance and the cytoplasmic inclusions in infected cells superficially resemble those of rabies, the basic morphologic structure of Marburg virus is distinctive. It shares no antigenic properties with rhabdoviruses or with any other known viruses.
>
> Monkeys of various species, including *C aethiops,* have been experimentally inoculated with Marburg virus. These animals developed a uniformly fatal infection that pathologically resembled the disease in humans.

Marburg virus disease is an acute infectious disease, with fever, malaise, headache, myalgia, vomiting, and diarrhea. Clinical jaundice is not seen, but chemical evidence of liver damage appears during the second week of illness. Proteinuria and renal failure occur in the severely ill. The mortality rate is about 30% in hospitalized patients. Complications include bacterial pneumonia, orchitis, and acute psychosis. Convalescence is prolonged, with lethargy, fatigue, and loss of hair continuing for 3–4 weeks. Complement-fixing antibodies are present in the serum by the second or third week of illness.

Treatment is symptomatic, with maintenance of electrolyte balance and renal function. Severe hemorrhage may occur. Transfusion of convalescent plasma may be of help. Extreme care should be taken to prevent exposure of medical personnel to blood, saliva, and urine from the patient. Infectious virus can persist for up to 3 months in semen.

In 1976 two severe epidemics of hemorrhagic fever occurred in Sudan and Zaire. The virus responsible, provisionally named Ebola virus after a river in Zaire, resembles Marburg virus morphologically but is antigenically distinct. The outbreaks involved over 500 cases and at least 350 deaths. The illness was marked by a sudden onset of severe headache, fever, muscle pains, and prostration, quickly followed by profuse diarrhea and vomiting. In each outbreak, whether of Marburg or Ebola fever, hospital staff became infected, often with tragic results. In one hospital in Sudan, 76 staff members became ill and 41 died. Direct human-to-human transmission was responsible for all the secondary infections, which usually occurred after close and prolonged contact with sick patients and particularly through contact with patients' blood.

Once Marburg and Ebola viruses have been transmitted to humans through contact with unknown reservoir, they can adapt themselves to direct human-to-human transmission, and these severe infections may thus be introduced from their country of origin to regions where the natural host does not exist.[73]

12. For the following examples, read the text and notice the technical terminology. In each case, how would you characterize the word choices: jargon, slang, shoptalk, argot, something else? Is the terminology used appropriately? Explain your answer.

Example 1

Figure 1 is set in 10-pica justified line lengths with every second line indented 2 ems. Because the lines are short and the type is justified, the type size is small so that there will be enough characters on each line to avoid excessively open word spacing.

Figure 2 shows two columns of type shuffled together like playing cards. The two columns are of different widths and use different type sizes; one is text and the other a caption. To spec this, keep the leading the same for both blocks of copy. The type's baselines occur every 12 points where the two areas of type overlap. So spec the type 8/24 and 10/24 with the instruction to skip every other line and insert a line of the opposing text. Then spec the amount of indentation for the lines of text in the right-hand copy block.[74]

Example 2

Helios's ColorSync 2 QuarkXTension lets XPress designers tap into ColorSync's color-matching capabilities. For use in conjunction with the Helios EtherShare OPI software, this XTension provides complete and consistent color control throughout prepress. The XTension also underscores Helios's strategy to make EtherShare OPI capable of working with third-party software add-ons.

Archetype is taking a similar approach to color management. ColorBlind (developed with Color Solutions) adds color-separation capabilities to InterSep, while providing consistent, predictable color from input to output. Archetype is also staking a claim in another area of development: OPI and image databases. OPI servers have limited archiving functions and don't have sophisticated query tools for searching through hundreds of stored images. A database like Archetype's new MediaBank addresses those shortcomings: it finds specific high-resolution images quickly and provides a vehicle for job tracking.[75]

Example 3

If you've studied the example programs in the time-honored Windows programming references, you've probably noticed that the traditional way to handle the various window messages passed to the application's callback routine is to decode the messages with elaborate (and sometimes nested) switch {} statements. Because of the incredible variety of possible messages, particularly in the case of an application with a complex menu bar, these switch statements often result in a single callback function that drags on for literally hundreds of lines. This antimodular, antistructured approach to handling messages flies in the face of good programming style in C or any other language; it interferes with code readability, code maintenance, and code reuse. Even the Windows CASE tools, which have no particular obligation to adopt this technique over any other, generate source code based on the same old ugly, convoluted switch statements.[76]

Example 4

Shy Vi, the princess of text editors
    The vi text editor is head and shoulders above ed in every way. It is a screen editor rather than a line editor; it shows you as much of the file as it can fit on-screen. You don't have to beg it to display bits and pieces of your file—definitely a step forward.
    The bad news is that command-wise and mode-wise, vi works just like ed because deep down it's just a souped-up version of ed. Two modes with no clue about which one you are using, cryptic one-letter commands—the works. You're still better off learning emacs, but vi really is better than ed.[77]

Example 5

As we sit at the computer and shift our attention from the screen to the printed copy or to look around the office, we must refocus our eyes. Each time we change our focusing distance the ciliary muscles must modify the shape of the lens to insure that a clear image appears on the retina. Shifting from a dimly lit display screen to the brightly [lit] background of the office requires changing the pupil size. For this to happen, within the iris, radial muscles dilate or circular muscles constrict to alter the pupil size.[78]

Notes

1. Tom Fahey, *The Joys of Jargon* (New York: Barron's, 1990) 10.
2. John A. Barry, *Technobabble* (Cambridge, MA: MIT P, 1991) 57–58.
3. Richard Lanham, *Revising Prose,* 3rd ed. (New York: Macmillan, 1992) 72.

4. Tom McArthur, ed. *The Oxford Companion to the English Language* (New York: Oxford UP, 1992) 1119.
5. McArthur, p. 1033.
6. McArthur, p. 1032–1033.
7. McArthur, pp. 1032–1033.
8. From *WinNews, Volume 2, No.21* (December 5, 1995), an online newsletter.
9. Barry, p. 38.
10. David Crystal, *The Cambridge Encyclopedia of the English Language* (New York: Cambridge UP, 1995) 372.
11. J. D. Watson, and F. H. C. Crick, "A Structure for Deoxyribose Nucleic Acid," *Nature* April 25, 1953: 737–738, reprinted in James D. Watson's *The Double Helix: A Personal Account of the Discovery of the Structure of DNA,* ed. Gunther S. Stent (New York: Norton, 1980) 240.
12. John Harris, "The Naming of Parts: An Examination of the Origins of Technical and Scientific Vocabulary," *Journal of Technical Writing and Communication* 14.3 (1984): 184–188.
13. Robert Hays, "What is Technical Writing?" *Word Study* (April 1961): 1–4.
14. Walter Nash, *Jargon: Its Uses and Abuses* (Cambridge, MA: Blackwell, 1993) 5–6.
15. McArthur, p. 543.
16. Irving Howe, ed., *Orwell's Nineteen Eighty-Four: Text, Sources, Criticism,* 2nd ed. (New York: Harcourt Brace Jovanovich, 1982) 37–38.
17. John Ciardi, *A Browser's Dictionary and Native's Guide to the Unknown American Language* (New York: Harper and Row, 1980) 154.
18. Ciardi, p. 154.
19. Lanham, pp. 68–69.
20. Lanham, p. vii.
21. Lanham, p. 60.
22. Lanham, p. 60.
23. Lanham, p. 60.
24. Lanham, p. 64.
25. Lanham, p. 62.
26. Lanham, p. 63.
27. Lanham, p. 63.
28. Lanham, p. 63.
29. Lanham, p. 65.
30. Fahey, p. 14.
31. Fahey, p. 14.
32. Fahey, p. 56.
33. Fahey, p. 56.
34. Fahey, p. 60.
35. Fahey, p. 64.
36. Fahey, p. 67, p. 69.
37. Ron Schroeder, Posted on newsgroup bit.listserv.techwr-l on July 20, 1995.
38. Lanham, p. 98.
39. Lanham, p. 103.
40. Lanham, p. 104.
41. McArthur, p. 188.
42. Anthony Burgess, *A Clockwork Orange* (New York: Ballantine, 1986) 3.
43. Quoted in Nash, 87.

44. William Lutz, *Doublespeak: From 'Revenue Enhancement' to 'Terminal Living'—How Government, Business, Advertisers, and Others Use Language to Deceive You* (New York: Harper & Row, 1989) 2–8.

45. Fahey, p. 21.
46. Fahey, p. 22.
47. Fahey, p. 23.
48. Fahey, p. 24.
49. Fahey, p. 24.
50. Fahey, p. 25.
51. Fahey, p. 25.
52. Fahey, p. 25.
53. Edward Tenner, *Tech Speak or How to Talk High Tech* (New York: Crown, 1986) 5–6.
54. Tenner, p. 5.
55. Tenner, p. 9.
56. Tenner, p. 9.
57. Tenner, p. 10.
58. Tenner, p. 16.
59. Tenner, p. 20.
60. Barry, p. 3.
61. Barry, p. 3.
62. Barry, pp. 4–5.
63. McArthur, p. 26.
64. All references found within this case study are taken from these sources: Henry S. F. Cooper, Jr., *Thirteen: The Flight That Failed.* (New York: Dial, 1973); National Aeronautics and Space Administration (NASA) World-Wide Web servers; and *STS and Associated Payloads: Glossary, Acronyms, and Abbreviations. NASA Reference Publication, 1981.* Where applicable, direct quotes are annotated parenthetically, for example, (Report 12).

65. Cooper, p. 3.
66. http://www.ksc.nasa.gov/history/apollo/apollo-13/apollo-13.html
67. apollo-13.html
68. Cooper, p. 21.
69. apollo-13.html
70. http://nssdc.gsfc.nasa.gov/planetary/lunar/ap13chrono.html
71. Cooper, p. 22.
72. apollo-13.html
73. "Marburg (Green Monkey) Virus Disease (African Hemorrhagic Fever)" in *Review of Medical Microbiology,* 13th ed. (Los Altos, CA: Lange, 1978): 460.

74. Alex White, *How to Spec Type* (New York: Watson-Guptill, 1987).

75. Barry Green, "At Your Service: OPI Servers Streamline Printing Process." *Publish* March 1996: 86.

76. Ray Duncan. "The Windows File-Oriented Common Dialog Functions." *PC Magazine* 26 May 1992: 386.

77. John Levine and Margaret Levine Young. *UNIX for Dummies,* 2nd Ed. (San Mateo, CA: IDG, 1995) 124.

78. Edward C. Godnig and John S. Hacunda. *Computers and Visual Stress: Staying Healthy* (Grand Rapids, MI: Abacus, 1991).

Chapter 7

Creating Sentences with Style

A perfectly healthy sentence, it is true, is extremely rare.[1]
—HENRY DAVID THOREAU

For many writers the pleasure of crafting a good paragraph, even a single sentence, lies mainly in its simple achievement. What it finally comes down to, I think, is the responsibility we feel for our language and the private satisfaction we take in meeting that responsibility.[2]
—JOSEPH WILLIAMS

Chapter Overview

Sentences in Technical Prose
A Brief Review of Sentence Basics
Sentence Combining Techniques
Emphasis
Rhythm
Sentence Variety
Sentence Elegance in Technical Prose
Sentence Faults
Chapter Summary
Questions/Topics for Discussion
Exercises
Notes

Sentences in Technical Prose

The art of writing clear technical prose lies in knowing more than the most appropriate diction to use, including appropriate technical terminology. The art of writing clear technical prose also depends on understanding how to construct the most effective sentences. So the focus here is on how an effective technical sentence is achieved. After a brief review of sentence basics (structure, forms, functions, and styles), more complex sentence elements are discussed: combining techniques, emphasis, rhythm, variety, and elegance. The chapter concludes with a review of common sentence faults. Of course, cohesion and coherence could also be discussed in this chapter because both are major concerns at the sentence level. However, because cohesion and coherence are also main concerns of effective paragraphing, they are discussed in Chapter 8.

A Brief Review of Sentence Basics

Much of the discussion in this section is brief because you probably already know a great deal about sentences. Also, a good grammar handbook will cover most of the points discussed here in far more detail. For example, phrases and clauses are described only briefly, and much can be said about both. Still, some basic discussion is necessary to provide a point of departure for the later sections that, for the most part, discuss more sophisticated sentence features.

Structure

A *sentence* is typically defined as a word or a group of words that expresses and conveys a complete thought. But the problem of defining a sentence is more complex than this definition suggests. Some sentences seem to express a complete thought, but do not: *Great! Okay. Yeah, right.* Some sentences express more than one thought: *One day soon I plan to buy a cordless phone, a home theater sound system, a cd player, and a laptop computer.* Some writers have suggested a "sentence is something which begins with a capital letter and ends with a full stop."[3] This definition falls short, however, for two reasons. First, it does not allow for question marks and exclamation marks, or for constructions that do not use punctuation but that are sentences (as in many ads and newspaper headlines). Second, there is considerable disagreement about the best ways to punctuate text.[4]

While recognizing the limitations of sentence definitions, we can still begin with some kind of definition. Many writing guides, for example, define a sentence according to its parts. Typically, we are told a sentence must have a *subject* and a *predicate,* either expressed or implied. A sentence may also have an *object complement* or other *complements, modifiers, connectives,* and *independent elements.*

A *subject* is the topic of a sentence; a *predicate* is what is said about the subject. For example, in the sentence *The computer has multimedia capabilities, the computer* is the subject and *has multimedia capabilities* is the predicate. (The receiver of the action denoted by the simple predicate is the *direct object: The programmer wrote the **code**.*)

An *object complement* is a noun or adjective referring to the direct object: *He called the man a liar* or *He called the man deceptive.*

A *modifier* is a word or group of words that qualifies and characterizes the meaning of another word: *The new computer worked **smoothly**.*

Prepositions (sometimes called *connectives* or *connectors*) serve to join the sentence parts: *The new computer came **with a modem**.*

Various *independent elements* include: *interjections* (Great! That's good news!), *direct addresses* (Fred, I want you to attend the meeting.), *exclamations* (Oh no! The computer crashed again!), *parenthetical limiting expressions* (He attended the meeting, I have no doubt, to please his boss.), *responsives* (No, he was not at the meeting.), and *nominative absolutes* (The meeting over, we went to lunch.).

Phrases

The most common major elements of sentences are phrases and clauses. A phrase is a group of related words used as part of a sentence but not having a subject or a predicate: *The computer **on the table** is a Compaq.*

Clauses

A clause is a group of words which contains a subject and a predicate: ***Phil repairs computers*** *after he attends his classes.*

As parts of a sentence clauses are classified as *independent* (also called *principal* clause or *main* clause) and *dependent* or *subordinate.* An independent clause makes complete sense when standing alone as in the above example. A dependent clause is used as a part of speech in a sentence and usually does not make sense when standing alone: *Phil repairs computers **after he attends classes**.*

Forms

Sentences may be classified into one of four forms or constructions: *simple, compound, complex,* and *compound-complex.* The form of the sentence depends on the type and number of clauses.

Simple

A simple sentence has only one independent clause and no dependent clause: *The Internet is a vast network of computers.* Here's another example.

> Multimedia is any combination of text, graphic art, sound, animation, and video delivered to you by computer or other electronic means.[5]

Compound

A compound sentence is equivalent to two or more simple sentences connected by coordinating conjunctions or by punctuation: *The Internet is a vast network of computers, and every day thousands of people are subscribing to one online service provider or another.* Here's another example of a compound sentence.

> Sound and movies are usually created with editing tools dedicated to these media, and then the elements are imported into the authoring system for playback.[6]

Complex
A complex sentence contains one independent clause and one or more dependent clauses: *After Web browsers were created in the early 1990s, the number of people accessing the World Wide Web increased at a phenomenal pace.* Here's another example of a complex sentence.

> To check the spelling of these words, and to prevent Word from questioning them if they are spelled correctly, add them to a custom dictionary.[7]

Compound-Complex
If either independent clause of a compound sentence has a subordinate clause, the sentence is called compound-complex or complex-compound: *After Web browsers were created in the early 1990s, the number of people accessing the World Wide Web increased at a phenomenal pace, and the Internet almost had more traffic than it could handle.* Here's another example of a compound-complex sentence.

> Furthermore, you are not limited to using union talent, and if your call is posted on bulletin boards in public places (in the theater department of a local university, for example), you may find yourself with many applicants, both union and nonunion, who are eager for the work.[8]

Functions

Sentences are divided into four classes according to function: *declarative, imperative, interrogative,* and *exclamatory.*

Declarative Sentences
The most common sentence in technical prose, the declarative sentence makes a statement or asserts a fact: *A good virus detection program may save your computer from potentially damaging computer viruses.*

Imperative Sentences
The imperative sentence expresses a command, request, or entreaty. The subject of an imperative sentence is *you,* but this subject is usually omitted: *Create a new directory on the hard drive and save the virus detection program to this directory.*

Interrogative Sentences
The interrogative sentence asks a question: *Is your virus detection program current or several months old?*

Exclamatory Sentences
The exclamatory sentence expresses surprise or a strong emotion. Exclamatory sentences can be useful in technical prose to motivate novices who, for example, are becoming familiar with a software program for the first time. Many programs have online messages which read *Congratulations! The installation was successful!*

Styles

Sentences are also classified as to the arrangement of their material as *periodic, balanced,* or *loose*. Periodic and balanced sentences are far more carefully arranged than loose sentences and are therefore particular favorites of prose stylists, but these sentences also occur in technical prose.

Periodic Sentences

Periodic sentences are complex sentences in which the main clause is delayed until the end of the sentence. These sentences are tightly structured so that they wait until the last word or phrase to make their point. Such sentences are called periodic sentences because they do not make much sense until the reader reaches the period at the end. Richard Lanham observes that "in a periodic sentence, things don't fall into place until the last minute, and when they do, they do with a snap, an emphatic climax."[9] Look at the following example:

> When you weave together the sensual elements of multimedia—dazzling pictures and animation, engaging sounds, compelling video clips, and raw textual information—you can electrify the thought and action centers of people's minds.[10]

Periodic sentences are more difficult for people to read because readers have to hold all of the earlier information in memory. When they reach the end of the sentence, they can fit the subordinate elements to the main point. Periodic sentences may be used in a technical memo or letter, for example, to help make a point more emphatically or persuasively. In general, however, you should choose a less sophisticated sentence style. Tightly structured periodic sentences help authors achieve elegance in their prose, but, as we discuss in a section later in this chapter, elegance is not the primary goal of writers of technical prose.

Balanced Sentences

Because balanced sentences also help writers achieve elegance in their prose, they are also less common in technical prose than in other kinds of prose. Balanced sentences use parallel structure, antithesis, or symmetry to achieve their effect. Basically, a pattern that is used at the beginning of the sentence is repeated elsewhere in the sentence. Like the periodic sentence, the balanced sentence also has a tight structure, but the balanced sentence is not as structured as the periodic sentence. However, a balanced sentence may also be a periodic sentence.

Here is an example of a balanced sentence that is also periodic.

> **The writer who** learns the knack of balance or of deliberate imbalance; **the writer who** understands how to quicken his tempo with short words, quick darting words, words that smack and jab; **the writer who** learns to slow his composition with soft and languorous convolutions; **the writer who** practices the trick of sentence-endings, striving deliberately for syllables that are accented a particular way, for the long vowel sound or the **short—such a writer is on his way toward mastery of a marvelous tool.**[11]

Note the repetition of the phrase "the writer who" leading to the statement of the main idea at the end: "such a writer is on his way toward mastery of a marvelous tool."

Some rhetoricians prefer to call both the periodic and balanced sentence styles just one style: the periodic style. Richard Lanham, for example, observes, "The periodic stylist works with balance, antithesis, parallelism, and careful patterns of repetition."[12]

The following discussion points out the essential differences among parallel structure, antithesis, and symmetry:

Parallel Structure. Writers achieve parallel structure in their sentences by creating a series of nouns, verbs, phrases, clauses, or sentences. The example below shows a series of clauses:

> This is the greater danger for our species, to try to pretend **that we** are another kind of animal, **that we** do not need to satisfy our curiosity, **that we** can get along somehow without inquiry and exploration and experimentation, and **that the human mind** can rise above its ignorance by simply asserting that there are things it has no need to know.[13]

In technical prose balance, when it is aimed for by writers, is often achieved through the parallel structure of lists in the form of nouns, verbs, phrases, clauses, or sentences. The following is an example of nouns in a list:

With multimedia authoring software, you can make

- Video productions
- Animations
- Demo disks and interactive guided tours
- Presentations
- Interactive kiosk applications
- Interactive training
- Simulations, prototypes, and technical visualizations[14]

Antithesis and Symmetry. Less common in technical prose for achieving balance are antithesis and symmetry. In *antithesis* one half of a sentence is pitted against the other. Sentences achieving this kind of balance are usually wise or witty. *He hoped for the best product release, but expected the worst.* To use *symmetry* to achieve balance in a sentence, the order in the second half of an expression reverses that of the first. This technique is also commonly referred to as *chiasmus: When the going gets tough, the tough get going.* These examples of antithesis and symmetry show authors achieving a certain grace or elegance in their prose, qualities that are not the primary goals of sentences in technical prose.

Loose Sentences

In contrast to periodic sentences, loose sentences are complex sentences in which the main clause comes first, followed usually by several subordinate clauses. Such sentences are generally easier for readers to understand because the main clause is at the beginning, and because of this easier comprehension, loose sentences are very common in technical prose.

Periodic and balanced sentences may be more tightly structured, but loose sentences have a definite structure as well. They are also called *cumulative* sentences because they cumulate subordinate clauses after the main clause. Note the structure of the following loose sentence: "You can connect to the Internet in a number ways, for example, by subscribing to a local service provider, by subscribing to a large commercial service such as America Online or Prodigy, or by subscribing for free through a university or company."

Of course, some writers prefer to play with the style of loose sentences, making them appear less structured by using add-on phrases and clauses almost as though additional thoughts just occurred to the writers. This is a common technique in a literary style. Note the freewheeling structure of the examples below:

> The vacuum cleaner turned on in the apartment's back bedroom emits a high-pitched lament indistinguishable from the steam alarm on the teakettle in the kitchen, and the only way of judging whether to run to the stove is to consult one's watch: there is a time of day for the vacuum cleaner, another time for the teakettle.[15]

> The futuristic sheen of virtual reality—a simulation technology that employs TV goggles and quadraphonic sound to immerse users in 3-D, computer-graphic worlds—briefly captured the media's magpie eye in 1991 with the promise of a tomorrow where virtual thrill seekers, like Arnold Schwarzenegger in *Total Recall,* roam the red seas of Mars without leaving their armchairs.[16]

Technical prose style, like other prose styles, presents a mixture of periodic and loose sentences.

Sentence Combining Techniques

Sentence combining is a technique for helping you write more effective sentences. The approach calls for you to reduce complex ideas to their simplest parts. Then you write these parts into a series of short, simple sentences. Then you take these simpler parts and blend them into a good, readable sentence.

Research shows that sentence combining works at least as well as other methods of focusing on writing at the sentence level. As a writer using this approach, you can worry less about concentrating on underlying grammatical concepts, and, instead, focus on "the innate ability of language users to tinker with expression until they get it right."[17] Practicing combining sentences in this way makes you more aware of the many choices you have for writing sentences differently. What follows is a brief discussion of sentence elements that will aid you in sentence combining.

Coordination

Coordination is achieved by combining simple sentences using conjunctions. These conjunctions give each sentence or independent clause equal weight. Typical coordinating conjunctions are: *and, but, or, nor, for, yet, so.*

> The Briefcase application lets you speak with the same file, or document, on more than one computer—**but** without having to worry about whether you're working with the most current version. This may sound like no big deal, **but** here's an example that shows otherwise.[18]

Subordination

Subordination is achieved by using conjunctions that turn one sentence into a *dependent clause*. Typical subordinating conjunctions are: *if, when, whenever, since, as, because, although, whereas, where, before, after.*

> **When** you choose the Shut Down command from the Start Menu, you see a dialog box that asks, "Hey, what do you mean, 'shutdown'?[19]

Relative Clauses

A relative clause is a type of dependent clause. It refers to a particular noun in an independent clause through a relative pronoun (*who, whom, whose, which*). This kind of sentence combining is also called combining with the Wh-connection. These clauses also include *that*, so *that* is included as another possible Wh-connection.[20]

> The people **who weave multimedia into meaningful tapestries** are *multimedia developers.*[21]

Participial Phrases

A participle is a verb ending in *-ed* or *-ing* (for example, *walked* or *walking, talked* or *talking*). You may use a participial phrase when you want to compress two sentences that have the same subject into one sentence. You create one sentence by dropping the subject from one of the sentences and using only the present or the past participle of the verb. Consider the following two sentences:

> The keyboard wristpad is designed to help reduce the risk of carpal tunnel syndrome. Wristpads are easy to find in computer stores and are inexpensive.

These two sentences may be rewritten into one sentence:

> Designed to help reduce the risk of carpal tunnel syndrome, keyboard wristpads are easy to find in computer stores and are inexpensive.

Appositives

Appositives are phrases that identify or define something, as, for example, in the sentence: "Windows® 95, **Microsoft's new operating system,** was released in August of 1995." Another more complex use of appositives occurs in this example:

> Supported by Kodak and Apple, Plugged In, **a nonprofit organization dedicated to bringing new technologies to children of low-income communities,** gave cameras to six students in East Palo Alto, California, who made the award-winning "Escapes from the Zoo," **a Portfolio format Photo CD that includes poignant first person views of their everyday lives.**[22]

Absolutes

An absolute is a simple sentence that is subordinated to an independent clause by removing or altering its verb form to establish a closer connection between clauses. Note the simple sentence below: "He finished the chapter and put it in the mail." This sentence may be revised in the following manner: "The chapter finished, he put it in the mail."

Noun Substitutes

One kind of noun substitute is a simple sentence that has been converted into a noun clause by beginning the sentence with either the word *that* or an altered form of the verb (participial phrase):

You will attend the meeting. It's essential.
That you will attend the meeting is essential.

They pleaded for an extension on the deadline. It didn't help their cause.
Pleading for an extension on the deadline didn't help their cause.

Addition/Deletion

This approach requires you to delete some material from some short sentences and to add to the main sentence. Consider the following independent clauses:

All glassware should be cleaned and rinsed with acid.
The glassware is nondisposable.
The cleaning and rinsing should be thorough.
The acid should be nitric.
The acid should be a 50 percent solution.[23]

Written like this the sentences create an abrupt, choppy style. Rewritten into one sentence, these clauses may look like this:

All nondisposable glassware should be thoroughly cleaned and rinsed with a 50 percent nitric acid solution.

Notice that the following words were deleted:

All glassware should be cleaned and rinsed with acid
The glassware is nondisposable.
The cleaning and rinsing should be thorough.
The acid should be nitric.
The acid should be a 50 percent solution.

Creating Sentences with Style   153

The only part of a word added is the suffix **-ly:**

> All nondisposable glassware should be thorough**ly** cleaned and rinsed with a 50 percent nitric acid solution.

Rearrangement

Rearrangement requires you to place the emphasis in a sentence where it will get the best response from the reader. The end of the sentence is usually considered to be the most emphatic part, the beginning less emphatic, and the middle least. Note the three separate ideas in the three sentences below:

> Windows® 95 is a user-friendly operating system. That's what a vendor told us. We attended the computer show.

Now notice how these sentences can be rearranged into one sentence placing the emphasis on the most important point:

> When we attended the computer show, a vendor told us that Windows® 95 is a user-friendly operating system.

Notice how the emphasis shifts when different ideas are placed at the end of the sentence:

> A vendor told us that Windows® 95 is a user-friendly operating system when we attended the computer show.

> When we attended the computer show, a vendor told us that a user-friendly operating system is Windows® 95.

Emphasis

Sentence emphasis is the art of knowing how to end a sentence. Strategies for ending sentences effectively include trimming the end, shifting less important ideas to the front of the sentence, and shifting more important ideas toward the end of the sentence.[24] The idea of emphasis has already been discussed under the topic of rearrangement, but this sentence quality is so important it deserves more detailed attention.

Some Strategies for Ending Sentences

Trimming the end:

> Many Internet service providers such as America Online and Compuserv offer a trial membership for fifteen hours or more of unlimited Internet access in the hope that customers will want to continue the service after the trial period of fifteen hours or more is complete.

Rewrite as:

> Many Internet service providers such as American Online and Compuserv offer a trial membership for fifteen hours or more of unlimited Internet access hoping customers will want to continue the service after the trial period.

Shifting less important ideas to the front of the sentence:

> Commercial Internet service providers offer many of the same kinds of services for the most part.
>
> For the most part, commercial Internet service providers offer many of the same kinds of services.

Shifting more important ideas toward the end of the sentence:

> You will see that each of the commercial Internet service providers offers some unique services if you take the time to look at each provider carefully.
>
> If you take the time to look at the commercial Internet service providers carefully, you will see that each offers some unique services.

Emphasis and Technical Terms

In technical prose you should locate technical terms in the ending of the sentence when you first introduce these terms. In the paragraphs below, notice how the authors place technical terms that are new to their readers at the end of sentences:

> In order to complete this chapter and our overview of the Internet, we need to spend a few moments talking about **TCP/IP.** As you know, the Internet is built on a collection of networks that cover the world. These networks contain many different types of computers, and somehow, something must hold the whole thing together. That something is **TCP/IP.**
>
> The details of TCP/IP are highly technical and are well beyond the interest of almost everybody, but there are a few basic ideas you will want to understand.
>
> To ensure that different types of computers can work together, programmers write their programs using standard ***protocols.*** A protocol is a set of rules that describes, in technical terms, how something should be done. For example, there is a protocol that describes exactly what format should be used for a mail message. All Internet mail programs follow this protocol when they prepare a message for delivery.
>
> TCP/IP is the command name for a collection of over 100 protocols that are used to connect computers and networks. We have already mentioned two of the TCP/IP protocols, **Telnet** and **FTP** (file transfer protocol). [25]

Rhythm

All sentences, whether written or spoken, have a certain rhythm. Rhythm is a recurring pattern of strong and weak elements or stresses. The strong elements are the peaks of the sentence; the weak elements are the valleys. Good sentences have a balance, a flow, and a rising and falling rhythm.

Note the effective rhythm the author achieves in the following sentence:

> Biology is something else again, another matter, quite another matter indeed, in fact very likely another form, or at least another aspect of matter, probably not glimpsed, or anyway not yet glimpsable, by the mathematics of quantum physics.[26]

The sentence begins with an independent clause—"Biology is something else again," and is followed by a series of phrases. There is a rising and falling in the independent clause, matched by a parallel tempo in each of the phrases, a rhythm of weak and strong stresses.

Sentence Variety

Sentence variety is essential for achieving an effective style. Using too many simple sentences together may create a primer style; using too many compound-complex styles may create a turgid style. Sentence variety is not only a matter of sentence form, but of sentence function, style, combining techniques, and length. Varying sentence length is especially important. Place too many short sentences together, and you will create a choppy style; place too many long sentences together, and you will create a wordy style.

Others have commented on the importance of length:

> Many writers feel that long sentences are inevitable if complex interactions have to be expressed. This is a mistake. Any subject can be broken up into longer or shorter items of information at will, and the determining factor is how much the reader can comfortably absorb, not how much information is "logically" joined together.... Sentence length should be determined by what the reader can effectively decode.[27]

It's impossible to set a limit on the length of a sentence. The average length of a sentence is eighteen words, but many longer sentences are often necessary. The complexity of the information is also a factor in determining sentence length. A long sentence may have too much complex information and would be better rewritten in several shorter sentences.

There are some helpful factors to consider in determining the length of a sentence. You should consider, for example, the familiarity of readers with the material, the importance of the material, whether the sentence is offering new information or repetitive information, and how you will achieve a logical flow of ideas. Christopher Turk and John Kirkman offer this additional helpful guideline: "Anything unfamiliar, complex, and new will require stating in shorter sentences, while a review of familiar information can be coded in longer sentences."[28]

Sentence Elegance in Technical Prose

> *Sentences which suggest far more than they say, which have an atmosphere about them, which do not merely report an old, but make a new impression; sentences which suggest as many things and are durable as a Roman aqueduct; to frame these, that is the* art *of writing.*[29]
>
> —HENRY DAVID THOREAU

In *Style: Ten Lessons in Clarity and Grace,* Joseph Williams devotes his last chapter to elegance in prose and how to achieve it. For Williams, elegance is more than clarity in prose: "A touch of class, a flash of elegance can mark the difference between unremarkable clarity and a thought so elegantly shaped that it not only fixes itself in the mind of our readers forever, but gives them a moment of pleasure when they recall it."[30] Williams offers a variety of helpful strategies for achieving what he calls "artful prose," strategies including balance and symmetry, climactic emphasis at the end of the sentence, varied length and rhythm, and metaphor.[31]

Should writers of technical prose aim for this same artful prose? As we have seen from various discussions in earlier chapters, technical prose is often emotive, expressive, and written with much skill. We have seen that the history of scientific and technical communication is filled with many classics. And in this chapter we have seen that technical prose relies on balanced and periodic sentences as well as loose sentences and depends on emphasis, varied length, and rhythm.

Yet the primary aim of technical prose is functionality or usefulness. Technical prose does not usually aim to inspire, illuminate, or entertain as does much artful prose. Of course, nonfiction may inspire, illuminate, or entertain as much as fiction, but this is not an issue of nonfiction and fiction; rather, it's an issue concerning the purposes of a special kind of nonfiction—technical prose. Turk and Kirkman perhaps state the point too strongly when they suggest, "Elegance may or may not be a by-product; but it can never be an intention."[32] Writers of technical prose have given us and can continue to give us elegant and memorable passages, but they do not generally aim to entertain or inspire us with the same richness of figurative language found in much artful prose.

Sentence Faults

Crafting effective sentences requires more than knowing the basics of structure, functions, and styles. And writing good sentences demands more than knowing techniques of sentence combining and techniques for achieving emphasis, rhythm, and variety. You must also be alert to basic sentence faults. Again, a good grammar handbook will guide you through the do's and don'ts in far more detail. Common sentence faults are reviewed only briefly here.

Comma Splices

Comma splices occur when you join together two independent clauses using a comma rather than a coordinating conjunction, a period, or a semicolon. Essentially, the writer is splicing two sentences together with a comma: "Insert the first program disk into the disk drive, run the setup.exe file."

This error may be avoided in a number of ways:

"Insert the first program disk into the disk drive. Run the setup.exe file."

"Insert the first program disk into the disk drive and run the setup.exe file."

"Insert the first program disk into the disk drive; run the setup.exe file."

Fragments

Contrary to what some teachers of writing would have you believe, fragments or incomplete sentences are widely used by professional writers in all kinds of writing. Over the centuries some of the best essayists have routinely used fragments for particular stylistic effects. Intentional fragments are also commonplace in technical prose. Yet inexperienced writers do use fragments when they should not be used. For example, you might accidentally write: "While waiting for the computer to boot up." Instead, you should write: "While waiting for the computer to boot up, make sure you have the program disks available."

Fused Sentences

Fused sentences occur when two independent clauses are fused together with no punctuation or coordinating conjunctions. Essentially, the writer is trying to make two sentences one sentence: "You'll need a good virus detection program in case your computer becomes infected with a virus a virus detection program will detect and delete any viruses."

Run-on Sentences

Some view run-on sentences and fused sentences as the same kind of sentence error. Both kinds of sentences are unnecessarily long. But run-on sentences are sentence errors that are more than merely fusing together two independent clauses with no punctuation. Run-on sentences are often complex or compound-complex sentences that are unnecessarily long sentences and often awkward constructions: "After installing the program, you will be prompted to restart your computer, and after restarting your computer, you double-click the icon for the program to open the program perhaps discovering that the program was not successfully installed after all."

Dangling Modifiers

Modifiers should not be left dangling with nothing to modify. Dangling modifiers occur when the sentence is worded in such a way that the modifying phrase does not modify the subject: "Opening the software package, the program disks and assorted manuals should be checked to see if anything is missing." In this sentence the modifier, "opening the software package" modifies "the program disks and assorted manuals." The sentence needs to be rephrased: "After opening the software package, check the program disks and assorted manuals to see if anything is missing." The phrase, "After opening the software package," now modifies the implied subject, "you."

Faulty Parallelism

One particularly common sentence error consists of making the verbs or verb phrases not parallel to each other. For example, a typical nonparallel sentence may be the following: "Taking all of the parts out of the boxes and check them with the parts list to make sure you received all the parts required to set up your new computer properly." The basic rule ignored here concerns parallelism. The verb "taking" is not parallel with the verb "make." The sentence may be revised in many different ways. The easiest way is to change the verb "taking" to "take": "Take all of the parts out of the boxes and check them with the parts list to make sure you received all the parts required to set up your new computer properly." Another possible revision is the following sentence: "To set up your new computer properly, take all of the parts out of the boxes and check them with the parts list to make sure you have received all of the parts required."

Awkward Sentences

Most college students have seen the abbreviation **awk** written in the left or right margin of a paper at one time or another. Simply defined, an awkward sentence is a sentence that often makes no sense because it ignores some basic rule of syntax. An awkward sentence may be due to a combination of sentence errors discussed above, combining a run-on sentence, for example, with nonparallel sentence elements, and perhaps even omitting essential information.

For example, you might encounter a sentence such as the following: "To attach the external portable hard drive to your pc, you will first need to install a SCSI adapter card, and then connecting the external drive using a cable and power adapter, then installing the software so your computer recognizes the new hardware." This sentence should be revised into several shorter sentences to make sense: "To attach the external portable hard drive to your pc, you will first need to install a SCSI adapter card. Then you will need to connect the external drive to the adapter card using a SCSI cable. Plugging in a power adapter will also be necessary to supply power to the external drive. Finally, you will need to install software so your computer will recognize the new hard drive."

Chapter Summary

Writing effective sentences is more difficult than most people think. Whether you are writing a personal letter, an essay, a memorandum, an e-mail message, or a technical manual, you can learn ways to improve your sentences. You cannot begin to construct better sentences unless you know and understand essential sentence terminology. First, you must have a good understanding of sentence basics: structure, forms, functions, and styles. You must study how different writers use periodic, balanced, and loose sentence styles for varying effects. Second, you must know about various sentence combining techniques involving coordination, subordination, relative clauses, participial phrases, appositives, absolutes, noun substitutes, addition/deletion, and rearrangement. Third, once you understand the basics of sentences and combining techniques, you also must know strategies for achieving sentence emphasis, rhythm, and variety. Finally, you must know the important differences between the aims of technical prose and artful prose. Your technical sentences will not succeed if they are unnecessarily stylish; but, at the same time, you should not aim to avoid nicely crafted, even memorable, technical prose.

Questions/Topics for Discussion

1. How are sentence structure, form, function, and style different from each other?
2. What do subordination and coordination contribute to a sentence?
3. What is the difference between appositives and absolutes?
4. How is loose sentence style different from periodic and balanced sentence styles?
5. When is a periodic sentence style appropriate in technical writing?
6. What is sentence combining?
7. What is sentence emphasis?
8. What is sentence rhythm?
9. How often do you use one, two, or three word sentences in your writing? Why?
10. What is artful prose? Should the aim of technical writing be artful prose? Why or why not?
11. What is a good example of a metaphor in technical prose? What is a good example of a metaphor in artful prose? How do the two metaphors differ, if at all?
12. Would you list any other sentence faults in the discussion of sentence faults? If so, which one(s) and why?

Exercises

1. In the following passage, identify the sentences by their form: simple, compound, complex, and compound-complex.

    > Everyone needs an owner's manual for the car and the house. Now we have one for the body, but if it shares the fate of most owners' manuals, a place at the bottom of a drawer, you will not profit very much. If the owners of such manuals read about the care, repair and maintenance of their possessions, there would be fewer ripoffs, less disaster, not to mention totaled mechanisms. Where the human body is concerned, this is trebly true. Most diseases are inflicted by less than healthy life-styles, but most pain could be eased or erased if there were a thorough understanding of how pain progresses as well as how to prevent it in the first place.[33]

2. In the following passage, identify the sentences by their function: declarative, imperative, interrogatory, or exclamatory.

    > If you have the Capitalize First Letter of Sentences box checked, WinWord looks to see if you've typed a period or exclamation point, followed by a space. If so, the WinGods light

up: this must be the beginning of a sentence! The next letter is automatically capitalized. Something this squirrelly should not be an option.

As the preceding cap "O" in "Officials" so amply demonstrates, a period followed by a space is not synonymous with "end of sentence" in many situations. This bogus setting will only cause you grief. Keep the box unchecked.[34]

3. Find a technical document—a memo, letter, or a part of a report—and identify and list the sentences according to their style: loose, periodic, or balanced.
4. In the following two paragraphs, identify what kinds of transitions are used to connect the sentences and the paragraphs. Are the transitions used effectively?

> GIF is nominally a lossless compression scheme; for grayscale images, it truly is lossless. Color, however, is a different story. GIF works only on indexed color images, and a huge amount of information is lost when you convert a 24-bit color image to 8-bit indexed color—you go from a possible 16.7 million colors to a mere 256. Images destined for the Web never contain anything close to 16.7 million colors, for the simple reason that they never contain anywhere near that number of pixels. But even a small, 320 × 240-pixel image can contain 300 times more colors than indexed color can represent, which can result in an 8-bit or 5-bit GIF that looks grainy and unsharp.
>
> Nevertheless, GIF still has some advantages that make it an important format. First and foremost, it's a de facto standard, supported by every graphical Web browser known to mankind. If you use GIF, you can confidently expect that everyone, everywhere, will be able to download your image.[35]

5. Find a technical prose example that effectively illustrates sentence rhythm.
6. Find a nontechnical prose example from a well known essayist that effectively illustrates sentence rhythm.
7. Find a technical prose example that effectively illustrates the repetition of key words and ideas.
8. Find a nontechnical prose example that effectively illustrates the repetition of key words and ideas.
9. Consider the following passage in which the author describes her grandmother. Would you consider it artful prose? Discuss briefly what the author is attempting to do with the language.

> It was partly the black hair, so improbably black and glossy. It was partly the mascara and the eye shadow surrounding her black narrow watchful eyes, though these aids to beauty were not applied carelessly but with an infinite discretion. It was the rouge, perhaps, most of all, the rouge and the powder and the vanishing cream underneath. When she perspired, on a warm day, the little beads of sweat on her eagle nose under her nose veil and on her long upper lip would produce a caked look that seemed sad, as though her skin were crying. Yet not even her cosmetics and the world of consummate artifice they suggested could account for the peculiarly florid impression she made as she moved across the store, peering through her lorgnon at the novelties and notions, and vanished into the elevator, up to the lending library or the custom-made or the hat department—her favorite purlieus—where elderly salespeople, her salespeople, would hurry up to greet her, throwing their arms around her, just as though they had not seen her the day before.[36]

10. In the following passage, discuss what qualities make the prose well crafted. Would you consider this an example of artful prose? Why or why not?

Bartering is the natural method of goods distribution; the introduction of money is, in fact, an enormous abstraction of the process. Material having no intrinsic value is used instead of goods to make trading easier. Just as a bitmap (a stored sequence of ones and zeroes) makes no sense in itself, money derives its value solely from the accepted conventions of our economy. Another example is in the art of painting. An abstract painting makes no sense whatsoever when the person looking at it is not in a position to interpret—in other words, to "concretize"—what he or she has seen. Hence, one able to apply a real concretion technique might find a work by Picasso more expressive, more meaningful than a realistic, concrete painting.

The second abstraction in font storage, the outline format, is the equivalent of a cashless bank transaction. A checkbook is an abstract, compact expression of assets, as the outline character is the abstract, compact expression of type. With a check, we can obtain a specific amount of money when needed. The special point is, that we do not receive from the bank the very same notes and coins paid in, rather an equivalent sum—we might even choose to take the money in a foreign currency.[37]

Notes

1. Eva M. Burkett and Joyce S. Steward, *Thoreau on Writing* (Conway, AR: U of Central Arkansas P, 1989) 185.
2. Joseph Williams, *Style,* 4th Ed. (New York: HarperCollins, 1994) 225.
3. David Crystal, *The Cambridge Encyclopedia of the English Language* (New York: Cambridge UP, 1995) 215.
4. Crystal, p. 215.
5. Tay Vaughan, *Multimedia: Making It Work,* 2nd ed. (New York: Osborne McGraw-Hill, 1994) 4.
6. Vaughan, p. 6.
7. *User's Guide: Microsoft Word* (Microsoft, 1994), 84.
8. Vaughan, p. 433.
9. Richard Lanham, *Analyzing Prose* (New York: Scribner's, 1983) 55.
10. Vaughan, p. 4.
11. James Kilpatrick, *The Writer's Art* (New York: Andrews, McMeel & Parker, 1984) 54.
12. Lanham, p. 54.
13. Lewis Thomas from "The Hazards of Science," *The Medusa and the Snail: More Notes of a Biology Watcher* (New York: Bantam, 1980) 60.
14. Vaughan, p. 148.
15. Lewis Thomas from "On Smell," *Late Night Thoughts on Listening to Mahler's Ninth Symphony* (New York: Bantam, 1984) 40.
16. Mark Dery, *Escape Velocity: Cyberculture at the End of the Century* (New York: Grove, 1996) 7.
17. James DeGeorge, Gary A. Olson, and Richard Ray, *Style and Readability in Technical Writing: A Sentence-Combining Approach* (New York: Random, 1984), v.
18. Stephen L. Nelson, *Field Guide to Microsoft Windows® 95* (Redmond, Washington: Microsoft, 1995) 28.
19. Nelson, p. 128.
20. DeGeorge, Olson, and Ray, p. 13.

21. Vaughan, p. 6.
22. Vaughan, pp. 14–15.
23. This extended example is borrowed from DeGeorge, Olson, and Ray, pp. 6–7.
24. These strategies are suggested by Joseph Williams, pp. 147–148.
25. Harley Hahn and Rick Stout, *The Internet Complete Reference* (New York: Osborne McGraw-Hill, 1994) 30.
26. Lewis Thomas, *Late Night Thoughts on Listening to Mahler's Ninth Symphony,* p. 70.
27. Christopher Turk and John Kirkman, *Effective Writing: Improving Scientific, Technical and Business Communication,* 2nd ed. (London: SPON, 1989) 94.
28. Turk and Kirkman, p. 97.
29. Burkett and Steward, p. 124.
30. Williams, p. 209.
31. Williams, p. 218.
32. Turk and Kirkman, p. 90.
33. Bonnie Prudden, *Myotherapy: Bonnie Prudden's Complete Guide to Pain-Free Living* (New York: Ballantine, 1984) 44.
34. Woody Leonhard, *The Underground Guide to Word for Windows:*™ *Slightly Askew Advice from a WinWord Wizard* (Reading, MA: Addison-Wesley, 1994) 45.
35. Bruce Fraser, "Graphic Images." *Publish* Mar. 1996: 52+.
36. Mary McCarthy, *Memories of a Catholic Girlhood* (San Diego: Harcourt, 1985) 218–219.
37. Peter Karow, *Font Technology: Methods and Tools* (Berlin: Springer-Verlag, 1994) 81.

Chapter 8

Structuring Paragraphs and Other Segments

What matters more than length is, as in the sentence, good internal form. A paragraph is ideally one piece. If anything is a brick in composition—and there is some doubt about that—the paragraph is it. [1]
—JACQUES BARZUN

A paragraph is a way of grouping closely related sentences that deal with a thought or division of thought. The sole purpose of a paragraph is to make the reader's job easier by "packaging" the information and ideas conveyed. [2]
—DAVID EWING

When...an entire page (typed or printed) contains no paragraph breaks, some readers are repelled. A friend of mine calls such a paragraph a "paragiraffe." [3]
—DAVID EWING

Paragraphs in academic works, works of reference, religious scriptures, specialist journals, consumer magazines, quality newspapers, and tabloid newspapers all follow different rules of thumb in their construction. [4]
—TOM MCARTHUR

Chapter Overview

Writing Technical Paragraphs
Writing on One Topic
Developing the Paragraph Topic
Achieving an Effective Flow
Providing Adequate Details
Larger Segments in Technical Prose

Common Paragraph Faults
Chapter Summary
Questions/Topics for Discussion
Exercises
Notes

Writing Technical Paragraphs

Whether you are writing fiction or nonfiction, essays or journal articles, personal letters or e-mail, business letters or technical instructions, the paragraph is the major building block for what you want to express to your audience. Some inexperienced writers give little thought to paragraphs while they write. They don't worry much about focusing on one major idea, about developing the topic, about achieving a smooth flow, or providing adequate details. And, of course, their paragraphs reflect their lack of concern. Many other writers, however, know it's as important to give as much thought to effective paragraphing as to diction and sentences.

This chapter focuses on the strategies necessary for achieving effective paragraphs in your technical prose. Writing paragraphs of technical prose often presents special challenges. For example, while many technical paragraphs begin with a topic sentence (this general-to-particular pattern is the most common in technical prose), many technical paragraphs have no topic sentence. Instead, they often use a section heading or summary in place of a topic sentence. And finally, technical prose consists of more than the words, sentences, and paragraphs of your text. Technical prose occurs in headings, headers, footers, and numerous other reader aids. All of these aids are important factors for indicating to your readers what topics you are discussing.

Writing on One Topic

That your paragraphs should each focus on one major idea may seem like a simple and straightforward suggestion. Problems arise, however, when you realize that an idea may be expressed in one sentence or in hundreds of pages.

One Topic and Appropriate Paragraph Length

So how do you know when you have expressed a complete thought and are ready to move on to another paragraph? There's no easy answer to this question. Jacques Barzun offers some useful advice: "But if you ask yourself after writing for some 10, 12, 15, 18, 20 lines: 'Have I settled some point and am I now ready to take up the next?' you will have a rough guide to paragraphing. Do not be afraid to make, from time to time, one paragraph six lines long and the next twenty-two."[5]

Bill Stott tells us paragraphing is "a matter of tone. You want to have a proper paragraph length for your subject, your audience, and your degree of seriousness (or frivolity)."[6] Stott takes a more casual approach to paragraphing than many others who offer advice on the subject:

> How long is a paragraph?
> As short as that.
> Shorter.
> Or as long as it needs to be to cover a subject....
> When the topic changes, a new paragraph starts.[7]

Stott goes even further by suggesting that paragraph length also depends on the kind of writing you're doing. Stott believes writing that "aims to be inviting"—newspapers, popular magazines—uses predominantly shorter paragraphs. In this kind of writing

> New paragraphs are begun before a topic is exhausted.
> Anytime.
> For no reason at all.
> Because each new paragraph lightens the tone, encourages readers, offers a foothold down the page.[8]

Yet even Stott is aware that "when paragraphs are short, writing does seem easier. Less happily, it also seems disjointed and superficial—as though the writer can't concentrate on a subject."[9] As a writer of technical prose, sometimes you will write disjointed or superficial short paragraphs, but if you are careful, you will choose to use numerous short paragraphs effectively.

Although we have no consensus on what constitutes an appropriate length, we can point to several factors that help determine length in technical prose. The trend in most kinds of technical prose is to keep paragraphs shorter rather than longer. In many kinds of technical instructions, for example, paragraphs are generally no longer than three or four sentences. Part of the reason for this brevity is to aid the reader in dealing with the complexity of the information. Few things are more annoying to readers than a two- or three-page paragraph discussing a complex technical subject. The writer is quite simply asking the reader to do too much work. In these instances the writer should do some of the work for the reader by breaking the text up into much smaller paragraphs.

Page Layout and Design and Paragraph Length

> *Paragraph construction is...as much a matter of layout and visual balance as of content and logical relationship between preceding or subsequent paragraphs.*[10]
> —TOM MCARTHUR

Another reason for the trend in shorter paragraphs in technical prose lies in the principles of effective page layout and design. Pages in a technical manual, for instance, are more appealing if the paragraphs are short and introduced by headings. One common fault of novice writers of technical prose is the tendency to provide one long paragraph after another with no headings and subheadings to help readers determine what they are reading. Readers expect and need a clear heading hierarchy (one that clearly differentiates first, second, third, and fourth level headings), and paragraphs that appropriately discuss only the information indicated by these headings.

Because headings (or the lack of headings) often indicate a technical writer's awareness of paragraph topics, a little more discussion on the issue is necessary here. When novice writers of technical prose use headings, they often make one or more of the following mistakes. Sometimes they create only one second-level heading for a section instead of at least two. The rule about dividing first-level headings into second-level (and second-level headings into third-level, and so on) is simple. You always have to divide into at least two subheadings.

For example, in an outline for a report on fingerprint patterns, an appropriate outline for a section of the report might be as follows:

Arches
    Plain Arch
    Tented Arch
Loops
    Radial Loop
    Ulnar Loop
Whorls
    Plain Whorl
    Central Pocket Whorl
    Double Loop Whorl
    Accidental Whorl

A less experienced writer may create an outline such as the following:

Arches
    Plain Arch
Loops
    Radial Loop
Whorls
    Plain Whorl
    Central Pocket Whorl
    Double Loop Whorl
    Accidental Whorl

The mistake here is to divide the topic arches into one subtopic, plain arch, and to divide the topic loops into one subtopic, radial loop.

Novice writers also often forget about the importance of parallel structure in their headings. Your headings should be all nouns or noun phrases (notice all of the parallel nouns in the example above), or all verbs or verb phrases, or even, if appropriate, complete sentences. You should never provide a combination of all of the above, for example, providing noun phrases in some headings and verb phrases in others. Parallel structure must be achieved not only at the sentence level but at the heading level and in any lists throughout a technical document.

Finally, many novice writers do not provide enough headings in a technical document. Although there are many exceptions, in many technical documents, especially in instructions, you should at least consider using a heading above almost every paragraph.

Developing the Paragraph Topic

In addition to being limited to one complete thought, good paragraphs must have a discernible pattern or order. This pattern or order is how you have chosen to develop the topic, whether, for example, you have chosen to begin with your topic sentence, or ask a question, or contrast or define.

Achieving this order is not as easy as it seems because paragraphs do not always fall into easily recognizable patterns. Many paragraphs begin with a topic sentence, but many paragraphs do not. A topic sentence may occur at any point in the paragraph. And often no one sentence expresses the topic, but the paragraph still concerns itself with only one topic.

The patterns or orders discussed below are actually strategies for developing the topic of a paragraph. How do you tell what is the best paragraph strategy to use for a particular subject and a particular audience? Of course, there is no easy answer to this question either. Different paragraph patterns have different purposes and often affect audiences in different ways.

A few guidelines are provided to show why some methods of development may be preferred over others in some cases, but, in many cases, you may be free to choose from a wide variety of methods of paragraph development to make the same point clear. After all, all of these methods of development have the common purpose of helping to make the unfamiliar familiar to readers.

Examples

Providing examples to your readers is often one of the most effective ways to illustrate a point. Notice in the following paragraphs how the author uses several examples to illustrate the concept of hypertext:

> The World Wide Web has more names than any other Internet resource. We like to call it the Web, but you will also see it referred to as *WWW* or *W3*. To understand the Web, we need to start with the idea of hypertext.
>
> *Hypertext* is data that contains *links* to other data. A simple example of hypertext is an encyclopedia. Say that you are reading the entry on "Trees." At the end of the article, you see a reference that says, "For related information, see Plants." This last line is a link, from the "Trees" article to the "Plants" article.
>
> Of course, this is a simple example. The Web is based on hypertext that is a lot more complex. In particular, there may be links anywhere within a document, not just at the end.

Here's an imaginary example. Say that you are using the Web to read a hypertext article about trees. Every time the name of a new tree is mentioned there is a link. Each link is marked in some way so that it stands out. For example, a word that has a link may be highlighted or underlined, or it may be identified by a number.

If you follow that link, you will jump to an article about that particular type of tree. Within the main article, there are also links to other related topics such as "rain forests" or "wood." These links lead to complete articles. You will also find links to technical terms such as "deciduous" and "coniferous." When you follow one of these links, you will find a definition.[11]

If you want your readers to make immediate connections, if you want your readers to see links between the unfamiliar and the familiar, and if you want to demonstrate a point by showing, then use examples wherever appropriate.

General-to-Specific

A general-to-specific paragraph begins with a topic sentence and provides specific supporting details to illustrate the topic sentence. This paragraph pattern is the most common in technical prose because it helps readers recognize immediately what the main idea is:

> **A keyboard is the most common method of interaction with a computer.** Keyboards provide various tactile responses (from firm to mushy), and they have various layouts depending on your computer system and keyboard model. Most provide the common QWERTY typewriter layout (in the U.S.), large keys with roman letter labels and raised dots on the F, J, and 5 keys so that number-processing software can use these and the surrounding keys to emulate a calculator pad (on the Macintosh, the raised dot keys are D, K, and 5). For users who spend substantial time doing numeric entries and accounting, a numeric keypad is an essential part of the keyboard. Function keys let users perform special operations or *macros* with a single keystroke. Keyboards are typically rated for at least 50 million cycles (the number of times a key can be pressed before it might suffer breakdown).[12]

Readers know from the outset what your point is and now are looking at your supporting details for information to support the topic. Much technical prose is skimmed by readers, and providing topic sentences at the beginning of paragraphs helps readers to skim your documents more quickly.

Specific-to-General

A specific-to-general paragraph begins with supporting details and ends with the topic sentence. This pattern can be an effective way to prepare an audience for an idea. This type of paragraph is often used in circumstances where you want to be persuasive.

John Barth makes good use of this pattern to convince us of the value of public school teachers:

> For a superachiever in the U.S.A., public-school teaching is a curious choice of profession. Salaries are low. The criteria for employment in most districts are not notably high; neither is the schoolteacher's prestige in the community, especially in urban neighbor-

hoods and among members of the other professions. The workload, on the other hand, is heavy, in particular for conscientious English teachers who demand a fair amount of writing as well as reading from the hundred or more students they meet five days a week. In most other professions, superior ability and dedication are rewarded with the five P's: promotion, power, prestige, perks, and pay. Assistant professors become associate professors, full professors, endowed-chair professors, emeritus professors. Junior law partners become senior law partners; middle managers become executives in chief; doctors get rich and are held in exalted regard by our society. Even able and ambitious priests may become monsignors, bishops, cardinals. But the best schoolteacher in the land, if she has no administrative ambitions (that is, no ambition to get out of the classroom), enters the profession with the rank of teacher and retires from it decades later with the rank of teacher, not remarkably better paid and perked than when she met her maiden class. Fine orchestral players and repertory actors may be union-scaled and virtually anonymous, but at least they get, as a group, public applause. Painters, sculptors, poets may labor in poverty and obscurity, but, as Milton acknowledged, "Fame is the spur." The condition of the true artisan, perhaps is most nearly akin to the gifted schoolteacher's: an all but anonymous calling that allows for mastery, even for a sort of genius, but rarely for fame, applause, or wealth; whose chief reward must be the mere superlative doing of the thing. **The maker of stained glass or fine jewelry, however, works only with platinum, gemstones, gold, not with young minds and spirits.**[13]

When readers reach this last sentence, they have been well prepared for the observation. The conclusion or topic sentence seems completely appropriate. Writers of technical prose may use this particular method of development for some paragraphs when the audience needs more subtle persuasion.

Question-to-Answer

A question-to-answer paragraph pattern simply asks a question, usually at the beginning of the paragraph. The answer may be a simply stated "yes" or "no," may be unstated but implied, or may be the remainder of the paragraph. Placing a question at the beginning of a paragraph is often an effective way to provide a transition from the preceding paragraph and is also often an effective way to catch your reader's attention. Notice how the author in the example below asks a common question before answering it with several brief paragraphs:

> **How does the Net know where your data is going?** If you want to send a letter, you can't just drop the typed letter into the mailbox and expect delivery. You need to put the paper into an envelope, write an address on the envelope, and stamp it. Some addressing goes at the beginning of your message; this information gives the network enough information to deliver the *packet* of data.
>
> Internet addresses consist of four numbers, each less than 256. When written out, the numbers are separated by periods, like this:
>
> 192.112.36.5
> 128.174.5.6
>
> (Don't worry, you don't need to remember numbers like these to use the network.) The address is actually made up of multiple parts. Since the Internet is a network of

networks, the beginning of the address tells the Internet routers what network you are part of. The right end of the address tells that network which computer or *host* should receive the packet. Every computer on the Internet has a unique address under this scheme....[14]

Definition

Definition appears in all kinds of writing for all kinds of purposes. Definition is especially commonplace in technical prose and is often necessary to clarify new terms, concepts, or ideas. A definition may be in one sentence or may occupy many pages. In the example below, the author uses definition to clarify several basic concepts so his readers will have a much better understanding of bulletin board systems.

> **A computer bulletin board (also known as a *bulletin board system,* or *BBS* for short) is a computer that uses a special program that allows other computers to call it over standard telephone lines.** A BBS acts as a storage facility, where people calling from their computers can post messages and send and receive programs and files.
> *Modems* (short for *mo*dulator-*dem*odulator) make communication between BBSes and computers possible. When you call a BBS, the modem attached to your computer translates the electrical impulses generated by your computer into tones. The tones are then sent over telephone lines to a modem attached to the BBS which translates the tones back into electrical impulses that the BBS computer can understand.[15]

Description

Description may be the main purpose of an entire book, an essay, a laboratory report, or even a paragraph or part of a paragraph. In the example below the author gives his readers a good overview of what disks look like and what they are made of before he provides more detailed discussion later of the role of disks in creating multimedia:

> Hard disks are the most common mass-storage device used on computers. A *hard disk* is actually a stack of hard metal platters coated with magnetically sensitive material, with a series of recording heads or sensors that hover a hairsbreadth above the fast-spinning surface, magnetizing or de-magnetizing spots along formatted tracks using technology similar to that used by floppy disks and audio and video tape recording. Hard disks range from 20 megabytes (20,000,000 bytes) to more than three gigabytes (3,000,000,000 bytes) of storage capacity. For making multimedia, you need a large-capacity hard disk drive.[16]

One kind of common descriptive paragraph is organized spatially. In a technical description, for example, you might need to describe the dimensions as well as the texture, weight, and other features of an object.

Narration

Narrative paragraphs have all kinds of uses in technical prose. For example, you may want to report what happened, as in an accident report or travel report. A kind of narrative paragraph is a chronological paragraph. This type of paragraph may discuss the time necessary for performing various steps of a procedure or the time during which certain events occurred, for example, in a laboratory report.

A common type of narrative paragraph is the anecdote. Below is an example of a brief but effective anecdote Lewis Thomas uses to discuss ways nature has to help us deal with death:

> The worst accident I've ever seen was on Okinawa, in the early days of the invasion, when a jeep ran into a troop carrier and was crushed nearly flat. Inside were two young MP's, trapped in bent steel, both mortally hurt, with only their heads and shoulders visible. We had a conversation while people with the right tools were prying them free. Sorry about the accident, they said. No, they said, they felt fine. Is everyone else okay, one of them asked. Well, the other one said, no hurry now. And then they died.[17]

Comparison

A comparison paragraph can be effective in helping readers see similarities more readily:

> The life of the earth resembles that of an embryo, and the life of our species within the life of the earth resembles that of a central nervous system. The earth itself is an organism, still developing and differentiating.[18]

Here is another example of a comparison paragraph that also effectively uses a question at the beginning of the paragraph:

> Ever see the television show "Star Trek"? If you did, you may remember the transporter room. It let the Starship *Enterprise* move Captain Kirk, Mr. Spock, and nearly anyone or anything else just about anywhere. The Clipboard is the Windows 95 equivalent of the *Enterprise's* transporter room. With the Clipboard, Windows 95 easily moves just about anything anywhere. When working with a Windows-based **application,** you can use the Clipboard to move chunks of text, tables, and even graphic images to and from different **files.** You can also use the Clipboard to move text, tables, and graphic images between Windows-based applications—such as from Microsoft Word to Microsoft Excel.[19]

Contrast

Of course, you may use both comparison and contrast in the same paragraph, or you may make contrast the chief purpose of a paragraph, as in this example:

> In production and manufacturing industries, estimating costs and effort is a relatively simple matter. To make chocolate-chip cookies, for example, you need ingredients such as flour and sugar and equipment such as mixers, ovens, and packaging machines. Once the process is running smoothly, you can turn out hundreds of cookies, each tasting the same and each made of the same stuff. You then control your costs by

fine-tuning known expenses, negotiating deals on flour and sugar in quantity, installing more efficient ovens, and hiring personnel at a more competitive wage. **In contrast, making multimedia is not a repetitive manufacturing process.** Rather, it is by its very nature a continuous research and development effort characterized by creative trial and error. Each new project is somewhat different from the last, and each may require application of many different tools and solutions.[20]

Analogy

An analogy is a comparison of something unfamiliar to something familiar or a comparison of something to something else that is similar. For example, you may choose to describe a cylindrical object in terms of how it compares to a scuba diving tank. Analogies and analogy paragraphs are used often in technical prose. In fact, analogies may extend over many paragraphs. Recall the example from the "Conclusions" of the report *Spill: The Wreck of the Exxon Valdez* discussed in Chapter 4. There, the report's authors created an analogy between the nation's passenger air transport system and the maritime transport system.

An analogy paragraph is used effectively by Lewis Thomas in the following example:

> We can imagine three worlds of biology, corresponding roughly to the three worlds of physics: the very small world now being explored by the molecular geneticists and virologists, not yet as strange a place as quantum mechanics but well on its way to strangeness; an everyday, middle-sized world where things are as they are; and a world of the very large, which is the whole affair, the lovely conjoined biosphere, the vast embryo, the closed ecosystem in which we live as working parts, the place for which Lovelock and Margulis invented the term "Gaia" because of its extraordinary capacity to regulate itself. This world seems to me an even stranger one than the world of very small things in biology: it looks like the biggest organism I've ever heard of, and at the same time the most delicate and fragile, exactly the delicate and fragile creature it appeared to be in those first photographs taken from the surface of the moon. It is at this level of things that I find meaning in Wallace Stevens, although I haven't any idea that Stevens intended this in his "Man with the Blue Guitar": "they said, 'You have a blue guitar,/you do not play things as they are.' The man replied, 'Things as they are/are changed upon the blue guitar.'" It is a long poem, alive with ambiguities, but it can read, I think, as a tale of the earth itself.[21]

Classification

A classification paragraph is also often called an analysis paragraph. You divide the subject into parts and show how these parts are related, as in this example:

> **There are three general categories of buttons: text, graphic, and icon. Text buttons** and their fonts and styles are described in Chapter 9. **Graphic buttons** can contain graphic images or even parts of images—for example, a map of the world where each country is color coded, and a mouse click on a country yields further information.

Icons are graphic objects designed specifically to be meaningful buttons and are usually small (although size is, in theory, not a determining factor). Icons are fundamental graphic objects symbolic of an activity or entity. Figure 15-7 shows a selection of clip-art icon buttons available to users of ToolBook and Figure 15-8 shows buttons supplied with Passport Producer. Most authoring systems provide a tool for creating text buttons of various styles (radio buttons, check boxes, push buttons, animated buttons, and spin buttons), as well as graphic and icon buttons.[22]

Enumeration

An enumerative paragraph is one that usually states the supporting points as *first, second, third*...or in some other variation:

> TrueType is Microsoft Corporation's scalable font technology. If you're working with a Windows-based application, using TrueType fonts in your documents delivers two benefits. **First,** Windows and Microsoft Windows-based applications come with some cool TrueType fonts. (OK. Maybe that shouldn't count as a benefit, but you don't get any PostScript fonts with Windows or with Microsoft applications. PostScript is the competitive scalable font product.) **Second,** because of the way a scalable font is created, it's easy for Windows and Windows-based applications to change the point size in a way that results in legible fonts. Windows identifies TrueType fonts in the various Font list boxes with the TT prefix.[23]

This next example is a more subtle use of the same principle:

> For the students, life is not so easy. **They must learn three or four natural languages of God at least; they must learn a bureaucratized, difficult, highly Latinate language,** rich in impersonal and hence guiltless passives, called "soc-sci"; **they must learn a mathematicized language,** studded with charts and graphs, called "nat-sci"; and **they must learn a computer language or two**—which we might consider, I suppose, the natural languages of the gods-in-the-machine. And if through some curricular shipwreck they should stray into English Department territory and land on the island called Critical Theory, **they will have to master a language of priestly complexity and anarchistic terror,** one composed of magical spells, secret handshakes, ritualized genuflections before freshly deified Continental heroes, and more words for "text" than the Eskimos have for "snow." Naturally enough, student-travelers wandering in this world often don't write very well, and sometimes indeed go crazy, breaking into a hyperkinetic babble of "I mean," "like you know," and "I goes and he goes," which they schizophrenically and quite pathetically claim to be a "language of our own." It is not surprising that students, in such a world, come to hunger for a major subject and select one prematurely, simply to find a place of rest and belonging, a constant conceptual universe and a professional language which they can begin to master. Who wants to be—*horresco referens*—an "undeclared freshperson"?[24]

Cause and Effect

The following paragraphs discuss the process of making compact discs. Each new step is a cause for the creation of discs:

> **Compact discs are made in what is generally referred to as a *family* process.** The glass master is made using the well-developed photolithographic techniques created by the microchip industry: First an optically ground glass disk is coated with a layer of photoresist material 1/10 micron thick. A laser exposes (*writes*) a pattern of pits onto the surface of the chemical layer of material. The disc is developed (the exposed areas are washed away) and is silvered, resulting in actual pit structure of the finished master disc. The master is then electroplated with layers of nickel one molecule thick, one layer at a time, until the desired thickness is reached. The nickel layer is separated from the glass disc and forms a metal negative, or *father*.
>
> In cases where low runs of just a few disks are required, the father is used to make the actual discs. Most projects, though, require several *mothers*, or positives, to be made by plating the surface of the father.
>
> In a third plating stage, *sons* or stampers are made from the mother, and these are the parts that are used in the injection molding machines. Plastic pellets are heated and injected into the mold or stamper, forming the disc with the pits in it. The plastic disc is coated with a thin aluminum layer for reflectance and protective lacquer for protection, given a silkscreened label for marketing, and packaged for delivery. Most of these activities occur in a particle-free cleanroom, because one spec of dust larger than a pit can ruin many hours of work. The mastering process alone takes around 12 hours.[25]

Of course, writers of technical prose have other methods of developing topics in a paragraph, but this discussion has focused on many of the most commonly used ones.

Achieving an Effective Flow

Not only must paragraphs focus on one complete thought and develop the topic often using some common pattern, but they must also flow smoothly from the first sentence to the last. This smooth flow is achieved chiefly by the principles of cohesion and coherence.

Cohesion

In *Style: Ten Lessons in Clarity and Grace,* Joseph Williams distinguishes cohesion from coherence. For Williams, *cohesion* is "only the first step toward creating in your readers a sense of a whole."[26] Cohesion is the way you create a sense of flow in your sentences. You create this flow by including information that is old and information that is new in each sentence. More specifically, to enhance the flow or cohesion of a sentence you should do the following: (1) "Begin sentences with ideas that your readers will readily recognize, ideas that you have just mentioned, referred to, or implied, or with concepts that you can assume they know"; (2) "End sentences with information that your readers cannot anticipate or with information that is more difficult to understand: lists, technical words, complex conditions."[27]

Martha Kolln defines *cohesion* as "the ties that connect each sentence to what has gone before—the glue that gives a paragraph unity."[28] She informs us, like Williams, that the "**known,** or given, **information** generally fills the subject slot; the new information—the real purpose of the sentence—generally comes in the predicate."[29]

An example of cohesion can be seen in the following paragraph:

> Multimedia elements are typically sewn together into a project using ***authoring tools.*** **These software tools** are designed to manage individual **multimedia elements** and provide **user interaction. In addition to providing a method for users to interact** with the project, **most authoring tools** also offer **facilities for creating and editing text and images,** and they have **extensions** to drive videodisc players, videotape players, and other relevant hardware peripherals. **Sounds and movies** are usually created with editing tools dedicated to these media, and then the elements are imported into the authoring system for **playback. The sum of what gets played back** and how it is presented to the viewer is the ***human interface.*** **This interface** is just as much the rules for what happens to the user's input as it is the actual graphics on the screen. The hardware and software that govern the limits of what can happen are the multimedia *platform* or *environment.*[30]

The author introduces the reader to the new information of "**authoring tools**" in the first sentence. In the next sentence the now old information, "**These software tools,**" begins the sentence as the author introduces the reader to the new information of "**multimedia elements**" and "**user interaction.**" The next sentence begins with the now old information of "**In addition to providing a method for users to interact**" and now discusses "**multimedia elements**" more specifically as "**facilities for creating and editing text and images,**" and "**extensions.**" The next sentence begins with the now familiar concept of "**sounds and movies**" and introduces the reader to the topic of "**playback.**" Then the next sentence begins with "**The sum of what gets played.**" Finally, the author introduces the reader to the "**human interface**" and begins the next paragraph with "**This interface.**"

Coherence

Whereas cohesion consists of linking each sentence to the one before by the principle of old–new chaining, *coherence* requires linking sentences so that they "merge into a unified passage."[31] To achieve coherence in a paragraph, you need to see the string of topics in your sentences the way your readers will. The topic of a sentence is the subject of a sentence. To write more coherent prose, limit the number of topics you discuss in a paragraph. In effect, try to provide a consistent topic string.[32] You make topics evident in a number of ways: using transitional words and phrases, repeating key words, and repeating key ideas.

Using Transitional Words

Transitional words are essential for achieving a more coherent style. Without appropriate transitional words and phrases, your prose may seem choppy or disjointed. Good writers make frequent use of all kinds of transitional words and phrases: to show addition (*also, again, and, next, moreover, first*); to show contrast (*however, otherwise, yet, but*); to show comparison (*similarly, likewise*); and to show a result (*consequently, therefore, then*).

Below is an example of the appropriate use of a few key transitional words:

> An interactive multimedia project typically consists of a body of information through which a user navigates by pressing a key, clicking a mouse, or pressing a touchscreen. The simplest menus consist of text lists of topics. You choose a topic, click it, and go there. As multimedia and graphical user interfaces become more pervasive in the computer community, certain intuitive actions are being widely learned. **For example,** if there are three words on a computer screen, the typical response from the user, without prompting, is to click one of these words to evoke activity. This click-to-act function is becoming widely understood by computer users. Sometimes the menu items are surrounded by boxes or made to look like push buttons. **Or,** to conserve space, text such as Throw Tomatoes, Play Video, and Press to Quit is often shortened to Tomatoes, Video, and Quit. **Nonetheless,** the intention remains clear to the user.[33]

Repeating Key Words

One helpful strategy is repeating key words or phrases. Doing so helps readers to follow your discussion more easily or may give your writing a stronger stylistic impact as in the example below:

> Multimedia is an eerie wail as two cat's eyes appear on a dark screen. **It's** the red rose that dissolves into a little girl's face when you press "Valentine's Day." **It's** a small window of video, showing an old man recalling his dusty journey to meet a rajah, laid onto a map of India. **It's** a catalog of fancy cars with a guide to help you buy one. **It's** a real-time video conference with three colleagues in Paris, London, and Hong Kong on your office computer. At home, **it's** an algebra or geography lesson for a fifth grader. At the arcade, **it's** goggled kids flying fighter planes in sweaty virtual reality.[34]

Repeating Key Ideas

Readers of technical prose are often lost by the complexity of the subject matter, the length of the paragraphs, the technical terms, and many other elements. By repeating key ideas in your paragraphs, you are giving your readers signposts that help them follow you more easily. Repeating key ideas will give individual paragraphs a much stronger unity and a stronger sentence rhythm:

> The conventional way of looking at the human brain is to see it as an intricately wired **calculating machine,** receiving inputs of information from those regions of the outside world for which it has been structurally equipped with receptors, then either storing the information for later retrieval or sending it along from one center to another for immediate processing and action, depending on the circumstances. In recent years, the brain has become immensely more complicated. It is not just a **hard-wired device,** but turns out now to turn its inner affairs by an exchange of chemical messages, hundreds of

them, among its billions of neurons. It has become the custom in recent years to refer respectfully to **this awesome device** as the most complicated and elaborate of all structures in the universe, perhaps even including the universe itself; indeed, there are some who suggest that the universe itself would not exist without it; that the only reality is what the human brain perceives as reality, like that falling tree in that earless forest. Black holes only exist because we observe them into existence.[35]

Providing Adequate Details

In addition to having unity, order, cohesion, and coherence, good paragraphs must also be complete. At the beginning of this chapter, we discussed the problem of paragraph length. Paragraphs may be as brief as one word or may occupy several pages. It's important to remember that there is no set length for paragraphs. Your subject, purpose, audience, and the context in which you are writing are all factors in determining how long is long enough. The question addressed here concerns knowing when you have written enough.

One helpful strategy is to remember that providing many details is one of the most common ways to develop the topic of a paragraph. Notice how the writer in the paragraph below provides good details about the machines, software, capacity, and limitations of writing to your own CD-ROMs:

> With a special compact disc recorder, you can make your own CDs using special CD-recordable (CD-R) blank optical discs to write a disc in most of the formats of CD-ROM and CD-Audio (see Chapter 18). The machines are made by Sony, Philips, Ricoh, Kodak, JVC, Yamaha, and Pinnacle. Software such as TOPIX from Optical Media, Inc., lets you organize files on your hard disk(s) into a "virtual" structure, then writes them to the CD in that order. CD-R disks are made differently than normal CDs but can play in any CD-Audio or CD-ROM player. They are available in either a "63-minute" or "74-minute" capacity—for the former, that means about 560MB, and for the latter, about 650MB. These *write-once* CDs make excellent high-capacity files archives, and are used extensively by multimedia developers for premastering and testing CD-ROM projects and titles. Once the data is written onto these CDs, that part of the disc cannot be overwritten or changed.[36]

The writer is specific about various manufacturers of compact disk recorders (Sony, Philips, Ricoh, Kodak, JVC, Yamaha, and Pinnacle), specific about software (TOPIX) for organizing files on your hard disk, specific about playing time (63-minute or 74-minute), and specific about many features relevant to disk recorders. You are provided a great deal of useful and necessary information in this one paragraph.

In contrast to this nicely detailed paragraph, many paragraphs fail to provide essential or adequate information to illustrate the topic sentence. A common fault in much technical prose is undeveloped paragraphs.Consider the example below from an undergraduate forensic science major's report aimed at explaining fingerprint identification to lay people:

> Fingerprints are impressions made by the end joints of the fingers, and, therefore, are reversed reproductions of the skin surface details. These surface details are produced by the palmar surfaces of the hand and the plantar surfaces of the feet. These surfaces have skin formations called papillary or friction ridges.

Is this paragraph complete? The answer is "it depends." The topic of fingerprints is defined, but not in much detail. The lay reader is left wondering what *palmar* and *plantar* surfaces and *papillary* or *friction* ridges are. Additional discussion is necessary for these terms. In technical prose you may decide to provide additional information on and definitions of these terms by providing additional short paragraphs instead of adding more sentences to the same paragraph.

The following paragraph from the same student report more adequately discusses one topic:

> The knowledge of fingerprint ridge patterns dates back to the late seventeenth century. Scientists knew that ridge detail existed, but it did not occur to them that those ridge patterns were unique and individual to each person. In 1880 an article published by Dr. Faulds documented the possibilities of identifying criminals by the fingerprint patterns left at crime scenes. He demonstrated his theory by identifying an individual through a greasy print left behind on some laboratory supplies. This case was one of the earliest fingerprint identifications. It was not until 20 years later that a man was convicted based on fingerprints in the case People versus Jennings. **Fingerprints have come a long way since 1880, and they are now the most positive means of identifying individuals.**

This paragraph is a specific-to-general paragraph with the topic sentence coming at the end of the paragraph. The student could have provided more details or examples, but the information provided adequately illustrates her point that fingerprints have come a long way since 1880.

The trend in technical prose is toward shorter paragraphs. For example, look at almost any user's manual accompanying most software programs, and you will see paragraphs that are at most three or four sentences long. Additionally, these short paragraphs are increasingly accompanied by illustrations, bulleted or numbered lists, headings, and numerous other reader aids. Note the following discussion of a path in *Introducing Windows 95,* the brief manual accompanying the CD-ROM version of this new operating system:

> **What is a path?**
>
> A path is a more direct way to describe where a file, such as a document or program, is located on your computer or the network. It lists the drive, such as the hard disk, floppy disk, CD-ROM drive, or shared network folder, that contains the document. It also lists all the folders that you need to open to find the document.
>
> To specify the full path for a document, type the drive letter, followed by a colon (:) and backslash (\). Then list the folders in the order you open them. If there are more than one, separate the names by backslashes. Then type the filename.
>
> Windows 95 supports long filenames, which can contain up to 250 characters. If you use long filenames, enclose the path in quotation marks.

Here are some examples of paths:

- To specify the location of the Readme file, which is located on drive C in the Windows folder, you would type:
  c:\windows\readme.txt

- To specify the location of a document named Party List.doc, located in the Holiday folder, which is in the Social Events folder on drive C, you would type:
  "c:\social events\holiday\party list.doc"

- To specify the location of a bitmap (drawing) named Canyon, which is located on the network in a shared folder named \\Pictures\Scenic, you could type:
  \\pictures\scenic\canyon.bmp

- Or, if the folder is mapped to drive D, you could type:
  D:\canyon.bmp[37]

As you can see, some of the paragraphs in this example are very brief, but each is on one topic, and each paragraph adequately discusses this topic for the audience of first-time users of Windows 95.

Larger Segments in Technical Prose

There are larger segments than paragraphs in technical prose. The major genres of technical prose—proposals, manuals, and reports—often consist of standard sections specific to that genre. For example, a proposal often includes sections for discussing your schedule, qualifications, and estimated costs. Sometimes these topics can be covered in a few sentences as part of a letter or memo. Sometimes, for much longer proposals, each of these sections may occupy many pages in book format.

Whether your manuals, reports, and proposals are provided in letter, memorandum, or book format, you must give your readers a good sense of the same qualities of a good sentence or a good paragraph: unity, order, cohesion and coherence, and completeness. Of course, you can achieve these qualities, in part, with reader aids such as executive summaries, summaries, or abstracts; tables of contents, introductory sections, sectional tables of contents, a consistent and clear heading hierarchy throughout the document; a consistent and appealing page layout and design with helpful headers and footers; and appropriate back matter such as glossaries, supplementary data, notes, references, and indexes. In technical prose, because so much page layout and design as well as technical illustration depends on text, it's not easy to separate style from these concerns.

Common Paragraph Faults

Consistency is one of the most important qualities of effective technical prose. Writers of technical prose often violate the consistency of their prose with one or more of the following errors.[38]

Inconsistency in Verb Tense

When the computer technician *removed* the cover from the computer, he *installs* the modem.

Inconsistency in Mood

Use the diagnostics disks to find any problems. The problems *are found* most quickly using these disks.

Inconsistency in Voice

Note the shift from passive to active voice in this example:

The program *was installed* successfully. Next *install* the supplementary programs.

Inconsistency in Person and Number

You can also move the cursor in other ways. *People* often hold the Ctrl key down and press the right arrow key, allowing you to move the cursor one word to the right. If *a person* wants to move quickly to the top of the screen, *he or she* should press the Home key and then the up arrow.

Inconsistency in Tone and Point of View

This example shifts from a third-person point of view and formal tone to a second-person point of view and informal tone:

During the detailed system investigation, the *analyst* should have gained a thorough understanding of the proposed new system. So *you* might want to *get going* on a prototype of the new system as you see the system design phase *starting up*.

Chapter Summary

Writing effective paragraphs requires more strategies and techniques than most people think. You must fully understand the qualities of topic unity, topic development or order, cohesion and coherence, and completeness. You must know when your audience and the technical prose genre in which you are writing will support longer or shorter paragraphs. You must know that many elements of prose in your page layout and design and technical illustrations work on the same principles of consistency as do paragraphs. You must understand that effective paragraphing in technical prose is not completely arbitrary, but it is not bound by too many rules either. You must also keep in mind that paragraphs are just one of many kinds of segments that occur in technical prose. The genres of technical prose often determine what these segments or sections will be, and the format you have chosen—a letter, memorandum, or book—often helps determine the length of these sections. In sum, learn how to suit your paragraphs and their length to your subject, purpose, audience, genre, format, and communication context.

Questions/Topics for Discussion

1. What is a paragraph and how does it differ from a sentence?
2. How do you keep a paragraph focused on one topic?
3. How is it helpful to know different strategies for developing the topic of your paragraphs?
4. What strategies for developing paragraph topics do you use most often in your writing? Why?
5. What is the difference between cohesion and coherence?
6. What are some common transitional words or phrases not mentioned in the section on coherence?
7. In your own writing, what kinds of decisions do you make to determine that your paragraphs are long enough?
8. In what ways do effective sentences and effective paragraphs share the same qualities?
9. In what ways can a paragraph be too long? Too short?
10. How do some elements of page layout and design and technical illustration constitute textual concerns?
11. What are some other common paragraph faults not mentioned in this chapter?

Exercises

1. In the following paragraph, discuss what the writer should have done to achieve better unity.

   It may be a surprise to some collectors to learn that the silver dollar had not circulated to any great extent in the United States after 1903. The coin had been turned out steadily since 1840, but for various reasons such as exportation, melting, and holding in bank vaults, the dollar was virtually an unknown coin. The Law of 1873 in effect demonetized silver and committed our country to a gold standard. The silver mining interests came to realize what had occurred a little later, and the ensuing quarter century of political and monetary history was filled with their voluble protests. There was a constant bitter struggle for the return to bimetallism.[39]

2. Read the following paragraph. Discuss the level of detail and why you think it is appropriate or inappropriate.

   Medicine has always been under pressure to provide public explanations for the diseases with which it deals, and the formulation of comprehensive, unifying theories has been the most ancient and willing preoccupation of the profession. In the earliest days, hostile spirits needing exorcism were the principal pathogens, and the shaman's duty was simply the development of improved techniques for incantation. Later on, especially in the Western world, the idea that the distribution of body fluids among various organs determined the course of all illnesses took hold, and we were in for centuries of bleeding, cupping, sweating, and purging, in efforts to intervene. Early in this century the theory of auto-intoxication evolved, and a large part of therapy was directed at emptying the large intestine and keeping it empty. Then the global concept of focal infection became popular, accompanied by the linked notion of allergy to the presumed microbial pathogens, and no one knows the resulting toll of extracted teeth, tonsils, gallbladders, and appendixes: the idea of psychosomatic influences on disease emerged in the 1930s and, for a while, seemed to sweep the field.[40]

3. For these examples, discuss why you think the use of example is effective or not effective.

Example 1

Something else that needs to be watched is the habit of overgeneralizing from the speaker's remarks. If a speaker is critical of, let us say, the way in which design is taught at a particular school, some persons in the audience seem automatically to assume that the speaker is saying that design shouldn't be taught at all. When I speak on the neglected art of listening, as I have done on many occasions, I am often confronted with the question, "If everybody listened, who would do the talking?" This type of misunderstanding may be called the "pickling in brine fallacy," after the senior Oliver Wendell Holmes's famous remark, "Just because I say I like sea bathing, that doesn't mean I want to be pickled in brine." When Alfred Korzybski found himself being misunderstood in this way, he used to assert with special forcefulness, "I say what I say; I do not say what I do not say." Questions of uniqueness, properly chosen, prevent not only the questioner but everyone else present from projecting into a speaker's remarks meanings that were not intended.[41]

Example 2

Besides prefixes and suffixes, some languages have infixes. An infix is a morpheme that is inserted within another morpheme instead of being affixed to an end of it. If English had morphemes TTH meaning "tooth" and GSE meaning "goose" (which it doesn't!), then we could say the -OO- was an infix meaning "singular" and -EE- an infix meaning "plural." English speakers find any interpretation calling for singular and plural infixes in words like tooth/teeth and goose/geese to be counterintuitive, but other languages do exploit this morphological possibility. In Tagalog (the most widely spoken language of the Philippines), infixing does exist. The word gulay meaning "greenish vegetables" can take the infix -IN-, creating the word ginulay, meaning "greenish blue." Compared to prefixes and suffixes, infixes are relatively rare in the languages of the world.[42]

4. In the following paragraph, what is the topic? Does the paragraph have a topic sentence? If so, which sentence functions as a topic sentence? If not, can the paragraph be considered effective without one? Why?

To install LANtastic for Windows, first be sure you have installed both LANtastic as explained in Chapter 5 and Windows as explained in this chapter. Then put the LANtastic for Windows distribution floppy disk in your A drive, type win a:winstall and press <enter>. This command starts Windows and runs the WINSTALL program. Select the Install option to copy the necessary files into your LANTASTIC directory, or another directory if you choose, and into the WINDOWS directory. Be patient. For the first one or two minutes the floppy disk spins but nothing else appears to happen. Eventually the "0–100%" meter shows some action and installation completes soon after.[43]

Structuring Paragraphs and Other Segments **183**

5. Examine the following examples. For each paragraph, determine the method the writer uses to develop the topic: specific-to-general, general-to-specific, comparison, contrast, enumeration, definition, description, narration, and classification. Discuss why you think these methods of development are effective for the paragraph topic.

Example 1

Kodak Photo CD—Why stick with 8-by-10 glossies? Some photo developers can store your pictures on a compact disc. A Kodak Photo CD-compatible drive can read these discs and display their pictures on your computer's monitor. The more expensive drives are multisession—they let you add additional photos to the same disc. The single-session drives cannot read the multisession discs, so the photo developer gives you a new discful of pictures every time you take a batch in for developing.[44]

Example 2

NOTE In Windows 3.1 you could get an idea of how much system memory an item on the Clipboard was occupying by switching to Program Manager or File Manager and choosing Help → About. The resulting dialog box told you how much memory and User Resources were available. Windows 95 isn't as informative. Many About dialog boxes that under Windows 3.x reported free memory now simply report the total amount of RAM your system has. Windows 95 utility programs that monitor and report memory usage will probably begin to appear on the market. In the meantime, you can run the System Monitor program from the Accessories → System Tools folder if you really need to keep track of memory (and many other system resource) usages.[45]

Example 3

The lawn had reached its optimal height (I was having problems finding the house from the street), so I decided it was time to mow. I went into my garage to retrieve the lawn mower. As I opened the door to the garage, I was struck by that wonderful garage smell. A smell of insecticide, fungicide, herbicide, weed killer, gasoline, and other remembrances of my youth. I closed my eyes and inhaled deeply. I must have blacked out momentarily, because I began to think of all the detritus in the garage—boxes, furniture, yard care products—as files. The garage was a file system. I began to explore.[46]

Example 4

I've watched students generate an initial draft of a letter, print it, revise the text, reprint it, and then revise a third or fourth time, each time producing an improved version. That they were willing to write and revise their text once surprised me (I'm innately jaded); watching them revise numerous times made me rejoice. The computer allowed them to do so. I've never had students revise so often or so efficiently when their correspondence was typed. The physical act of retyping the text deters them. Computers, in contrast, make revision possible, if not actually enjoyable. There's something similar to video game playing inherent in word processing—writing complete with blinking cursors, whirring noises, beeping, and the colorful text seen on a colored monitor. *Writing on a computer is fun.*[47]

Example 5

The other three new ones, "Inner Bevel," "Carve," and "Cutout," are all extremely useful for creating 3-D effects on your flat surfaces. "Cutout" creates a look that can feel like one sheet of paper floating over another. "Carve" is exactly as described, although it would be fun to have an option to choose the texture of the carving as stone or wood. And finally, "Inner Bevel" makes elegant, smooth curves that nicely raise the surfaces of your backgrounds into buttons or frames.[48]

Example 6

Not all social animals are social with the same degree of commitment. In some species, the members are so tied to each other and interdependent as to seem the loosely conjoined cells of a tissue. The social insects are like this; they move, and live all their lives, in a mass; a beehive is a spherical animal. In other species, less compulsively social, the members make their homes together, pool resources, travel in packs or schools, and share the food, but any single one can survive solitary, detached from the rest. Others are social only in the sense of being more or less congenial, meeting from time to time in committees, using social gatherings as ad hoc occasions for feeding and breeding. Some animals simply nod at each other in passing, never reaching even a first-name relationship.[49]

6. Find an example of a paragraph that is not adequately developed. Discuss why you think the paragraph needs more development to be complete.
7. Find an example of a paragraph you think is too long. Photocopy the paragraph and use lines and arrows to indicate where you think new topics begin. Discuss why you think new paragraphs should begin in the places you have marked.
8. Photocopy a page or two from a technical document that you think has unified, ordered, cohesive and coherent, and complete paragraphs. Discuss why you think the paragraphs have these qualities.
9. In the following paragraph, indicate the elements that contribute to unity, order, cohesion, coherence, or completeness. Discuss why you think the paragraph has some of these qualities.

You do character sketches to practice pulling useful traits from the real world onto the page, where you combine them with character qualities from your reading experience. You might begin with a concrete observation about externals—our Stranger, wearing a designer tennis dress and a Rolex watch, strides to her Jaguar—and expand the sketch with an assumption based on behavior: as your Stranger races to the aid of a victim being harassed in the parking lot, you conclude that she is courageous. Working the sketch, you jot down external details and internal qualities of your Stranger and begin your creative guesswork. For example, your Stranger, a modern-day Cinderella who witnesses the victimization of another, puts aside what she's attained to rush to the aid of the victim.[50]

Notes

1. Jacques Barzun, *Simple and Direct: A Rhetoric for Writers,* rev. ed. (New York: Harper & Row, 1985) 163.
2. David W. Ewing, *Writing for Results in Business, Government, the Sciences and the Professions,* 2nd ed. (New York: Wiley, 1979) 307.

3. Ewing, p. 307.
4. Tom McArthur, ed., *The Oxford Companion to the English Language* (New York: Oxford UP, 1992) 749.
5. Barzun, p. 163.
6. Bill Stott, *Write to the Point and Feel Better about Your Writing* (Garden City, NY: Anchor, 1984) 130.
7. Stott, p. 130.
8. Stott, p. 130.
9. Stott, p. 130.
10. McArthur, p. 749.
11. Harley Hahn and Rick Stout, *The Internet Complete Reference* (New York: Osborne McGraw-Hill, 1994) 496.
12. Tay Vaughan, *Multimedia: Making It Work,* 2nd ed. (New York: Osborne McGraw-Hill, 1994) 80.
13. John Barth, "Teacher," *The Best American Essays 1987,* ed. Gay Talese (New York: Ticknor & Fields, 1987) 6.
14. Ed Krol, *The Whole Internet User's Guide and Catalog,* 2nd ed. (Sebastopol, CA: O'Reilly, 1994) 25–26.
15. John Hedtke, *Using Computer Bulletin Boards* 2nd ed. (New York: MIS, 1992) 1–2.
16. Vaughan, p. 77.
17. Lewis Thomas, "On Natural Death," *The Medusa and the Snail: More Notes of a Biology Watcher* (New York: Bantam, 1980) 85.
18. Lewis Thomas, "Science and the Health of the Earth," *The Fragile Species* (New York: Collier, 1992) 117.
19. Stephen L. Nelson, *Field Guide to Microsoft Windows 95* (Redmond, WA: Microsoft, 1995) 37.
20. Vaughan, pp. 375–76.
21. Lewis Thomas, "Things Unflattened by Science," *Late Night Thoughts on Listening to Mahler's Ninth Symphony* (New York: Bantam, 1984) 74–75.
22. Vaughan, pp. 396–97.
23. Nelson, p. 120.
24. Richard A. Lanham, *The Electronic Word: Democracy, Technology, and the Arts* (Chicago: U of Chicago P, 1993) 142.
25. Vaughan, p. 462.
26. Joseph Williams, *Style: Ten Lessons in Clarity & Grace,* 4th ed. (New York: HarperCollins, 1994) 120.
27. Williams, p. 129.
28. Martha Kolln, *Rhetorical Grammar: Grammatical Choices, Rhetorical Effects* (New York: Macmillan, 1991) 16.
29. Kolln, p. 16.
30. Vaughan, pp. 6–7.
31. Williams, p. 120.
32. Williams, p. 124.
33. Vaughan, p. 202.
34. Vaughan, p. 4.
35. Lewis Thomas, *The Fragile Species,* p. 28.
36. Vaughan, p. 79.
37. *Introducing Microsoft Windows 95* (Redmond, WA: Microsoft, 1995) 16.

38. These common paragraph faults and some of these examples are from Philip Rubens, ed., *Science and Technical Writing: A Manual of Style* (New York: Henry Holt, 1992) 33–35.

39. R. S. Yeoman, *A Guide Book of United States Coins,* 49th ed. (Racine, WI: Whitman, 1995) 13.

40. Lewis Thomas, "On Magic in Medicine," *The Medusa and the Snail* (New York: Bantam, 1980) 16.

41. S. I. Hayakawa, "How to Attend a Conference," *The Use and Misuse of Language,* ed. S. I. Hayakawa (Greenwich, CT: Fawcett, 1962) 70–76.

42. Edward Finegan, *Language: Its Structure and Use,* 2nd ed. (Fort Worth, TX: Harcourt, 1994) 88.

43. Tom Rugg, *LANtastic Made Easy* (Berkeley, CA: Osborne McGraw-Hill, 1992) 297.

44. Dan Gookin and Andy Rathbone, *PCs for Dummies,* 2nd ed. (San Mateo, CA: IDG, 1994) 180.

45. Robert Cowart, *Mastering Windows® 95: The Windows 95 Bible* (San Francisco: Sybex, 1995) 272.

46. John Montgomery, *The Underground Guide to UNIX* (Reading, MA: Addison-Wesley, 1995) 113.

47. Steven M. Gerson, "Commentary: Teaching Technical Writing in a Collaborative Computer Classroom." *Journal of Technical Writing and Communication* 23.1 (1993): 26.

48. Linn Susanne DeNesti, "A Practical Box of Fun." *Adobe Magazine* 7.4 (Mar./Apr. 1996): 31.

49. Lewis Thomas, "Social Talk," *The Lives of a Cell* (New York: Bantam, 1974) 102.

50. Robert J. Ray, *The Weekend Novelist* (New York: Dell, 1994) 31.

Chapter 9

Establishing an Appropriate Tone

Style is largely a matter of tone. The writer uses a style; the reader infers a communication's tone. Tone comes from what a reader reads into the words and sentences a writer uses.[1]
—JOHN S. FIELDEN

The best tone is the tone called plain, unaffected, unadorned. It does not talk down or jazz up; it assumes the equality of all readers likely to approach the given subject; it informs or argues without apologizing for its task; it does not try to dazzle or cajole the indifferent; it takes no posture of coziness or sophistication. It is the most difficult of all tones, and also the most adaptable. When you can write plain you can trust yourself in special effects.[2]
—JACQUES BARZUN

If the tone of your letter, memorandum, or report is inappropriate, your missive may misfire disastrously.[3]
—DAVID EWING

Thoreau chose carefully those words that added a general earthiness of tone and color to his writing. As the wind rose and the branches of the trees waved stiffly, he heard a brattling *sound. When the wind blew strong and the atmosphere cleared brightly, it was a* washing *day, and the wind was a* washing *wind. Rain that fell thinly as a light mist was a* mizzling *rain, or* drisk. *September, the first month after the summer heat, a month of early frost, had a* burr *to it. Wild apples were* knurly; *cold days were* snapping *cold; the note of the toad was a* sprayey *one; the slate-colored snowbird had a* jingling *note; sparrows had* lisping *notes; the grasshopper's flight was a* crackling *one; and so on and on. Always the prose craftsman was finding the word that suited the particular matter about which he wrote.[4]*
—REGINALD COOK

Chapter Overview

Tone in Technical Prose
The Key Emphases of Tone
Strategies for Developing an Appropriate Tone
Humor in Technical Writing
Chapter Summary
Case Study 6: Computer Viruses and Tone
Questions/Topics for Discussion
Exercises
Notes

Tone in Technical Prose

A Definition of Tone

Perhaps some people think the tone of technical prose is always formal, impersonal, and serious. After all, much technical prose *is* formal, impersonal, and serious. But the truth is that technical prose, like nontechnical prose, conveys a wide range of tones, tones that must be appropriate to the subject, purpose, audience, and context. As we will see in this chapter, much technical prose is also informal, personable, and humorous.

Tone is usually defined as your attitude toward your subject and your audience. However, *tone* is more accurately defined as your attitude toward your subject, your audience, and yourself. You must carefully consider all three to convey the desired attitude in your documents.

The Importance of Tone

Your writing may have a clear purpose, be well organized, be free of errors in grammar, punctuation, spelling, and so on, but unless you also convey an appropriate tone your document may fail to communicate your message to your intended audience. Tone is especially important in persuasive technical documents. You may have put together a logical and fact-filled document, but unless you also establish the proper tone for your audience, your argument will fail more often than not. It's probably fair to say that the most common reason many technical documents fail is an inappropriate tone. Perhaps no other element of style is so difficult to master or so important.

Persona, Person, Point of View, Voice

Persona, person, point of view, and voice are also important style elements, but some of these elements are used as synonyms for tone. As you will see, some writers and teachers of writing use *point of view* or *voice* to refer to tone. *Tone* is the preferred term in this chapter and throughout this book. This section briefly discusses definitions for these terms and suggests that there is no need for you to worry about a lack of consensus concerning *point of view, voice,* or *tone.*

Persona

Persona means "mask" and is the Latin term for a character (in drama). In classical Greek drama actors wore masks to assume their roles. The term *persona* is now used in literary criticism to refer to the role constructed by the narrator of a text, for example, the persona of a poem. Of course, writers of nonfiction may use a persona for stylistic effect as much as writers of fiction, drama, or poetry. See the discussion of **ethos** later in this chapter for some important distinctions between ethos and persona, and for some discussion on how both significantly shape the tone of a piece of writing. A persona, then, often helps to shape the tone of your writing, but many other elements do so as well.

Person

Person concerns the personal pronouns you use in your writing to refer to yourself, your readers, and the people you write about.

Person	*Singular*	*Plural*
First	I, me, my	we, us, our
Second	you, your	you, your
Third	he, him, his	they, them, their
	she, her, hers	
	it, its	

Choosing to address the reader as "you" contributes significantly to making instructions, for example, more informal and reader-friendly. Like persona, person helps shape the tone of a document, but it too is only one of many contributing elements.

Point of View

Now we move to elements more commonly associated with tone. *Point of view* has several major meanings. It may refer to the narrative point of view from which a story is told, for example, first-person point of view or third-person omniscient. Point of view may also be defined as it relates to the larger discourse situation of narrator-text-reader. As Katie Wales suggests, "on the ideological level the implied author may communicate explicitly or implicitly to the readers a sense of his or her attitude to the story narrated (Dickens, for instance on the social abuses of his time); and this affects tone: the use of satire or irony, for example. Or the narrator may reveal a particular attitude to the readers: of deference, friendliness, etc. In this sense point of view clearly applies to any kind of discourse, non-literary as well as fictional."[5] Wales's definition here makes point of view very similar to tone.

Defined simply, point of view concerns your relationship to the information you are writing about in terms of your use of person. Usually, you express point of view in first person, second person, or third person. When you think of point of view as your attitude toward your readers, you are essentially making tone and point of view the same stylistic element.

Voice

Voice has several important meanings. Voice in grammar concerns the relationship of subject and object in a sentence or a clause, whether, for example, you are using active voice (*Jack bought the computer*) or passive voice (*The computer was bought by Jack*). A narrative voice is a term in literary criticism for the person who tells the story, third person or first person, for example. In this sense, narrative voice is the same as point of view.

Complicating matters further, some authors equate voice with style or tone. For example, William Zinsser comments:

> Relax and say what you want to say. And since style is who you are, you only need to be true to yourself to find it gradually emerging from under the accumulated clutter and debris, growing more distinctive every day. Perhaps the style won't solidify for several years as *your* style, *your* voice—and it shouldn't. Just as it takes time to find yourself as a person, it takes time to find yourself as a stylist, and even then, inevitably, your style will change as you grow older.[6]

Dona Hickey also defines voice as broadly as Zinsser does.

> By voice, I mean the writer's relationship to subject, audience, and occasion as it is revealed through the particular blend of speech patterns you hear as you read.
>
> Voice is the sum effect of all the stylistic choices a writer makes to communicate not only information about a subject but also information about himself or herself to a particular audience. In this sense, all writing can be said to be expressive....[7]

Interestingly, Hickey also suggests "all writing can be said to be expressive—except for highly technical discourse, in which the convention is the absence of voice. To audiences of technical discourse, the felt presence of the writer is undesirable. That is not to say that the technical writer does not have a 'style.' Rather, the style—a writer's habitual language choices—is not voiced. We don't hear the strong intonation patterns of personal speech. We don't hear the person behind the information."[8] Much technical discourse is impersonal and lacks a sense of the writer (or writers) behind the prose. However, much technical prose is quite informal and personable. The writer may project a strong presence in the prose.

As you can see, some writers and teachers of writing prefer to use one term instead of another. If you prefer to see the elements in this chapter as matters of voice rather than tone, then that decision is fine. Knowing how to strengthen or shape your tone for an intended effect is more important than matters of semantics.

The Key Emphases of Tone

Tone is, of course, achieved in many ways, but three essential elements to establish tone in your documents are the level of formality, the attitude or emotion you want to convey, and the ethos you want to project.

Levels of Formality

Tone is, first of all, a matter of distance or formality, the distance you want to establish between yourself, your subject, and your audience. Do you, for example, want to be informal or formal with your audience? Your decision to be one or the other or something in between determines, for example, what kind of pronouns you use to refer to yourself and your readers or if you will use formal or informal diction (see Chapter 5).

Formal and Informal Technical Prose

Tones are usually broadly discussed as either formal or informal. Review, for instance, the examples of formal, informal, colloquial, and slang diction provided at the beginning of Chapter 5. Below is an example by Lewis Thomas providing a formal treatment of the AIDS virus:

> A third line of research involves the human immune system itself, the primary victim of the AIDS virus. Actually, most if not all patients with AIDS die from other kinds of infection, not because of any direct lethal action of the virus itself. The process is a subtle one, more like endgame in chess. What the virus does, selectively and with exquisite precision, is to take out the population of lymphocytes that carry responsibilities for defending the body against all sorts of microbes in the world outside, most of which are harmless to normal humans. In a sense, the patients are not dying because of HIV, they are being killed by great numbers of other bacteria and viruses that can now swarm into a defenseless host. Research is needed to gain a deeper understanding of the biology of the immune cells, in the hope of preserving them or replacing them by transplantation of normal immune cells. This may be necessary even if we are successful in finding drugs to destroy the virus itself; by the time this has been accomplished in some patients, it may be that the immune system has already been wiped out, and the only open course will be to replace these cells.[9]

Of course, Thomas could have written about this subject more informally. Many writers have, but it suits Thomas's larger purpose in this essay to treat the subject more formally.

In the following example Thomas chose to treat a far less threatening virus more informally:

> Warts are wonderful structures. They can appear overnight on any part of the skin, like mushrooms on a damp lawn, full grown and splendid in the complexity of their architecture. Viewed in stained sections under a microscope, they are the most specialized of cellular arrangements, constructed as though for a purpose. They sit there like turreted mounds of dense impenetrable horn, impregnable, designed for defense against the world outside.
>
> In a certain sense, warts are both useful and essential, but not for us. As it turns out, the exuberant cells of a wart are the elaborate reproductive apparatus of a virus.
>
> You might have thought from the looks of it that the cells infected by the wart virus were using this response as a ponderous way of defending themselves against the virus, maybe even a way of becoming more distasteful, but it is not so. The wart is what the

virus truly wants; it can flourish only in cells undergoing precisely this kind of overgrowth. It is not a defense at all; it is an overwhelming welcome, an enthusiastic accommodation meeting the needs of more and more virus.

The strangest thing about warts is that they tend to go away. Fully grown, nothing in the body has so much the look of toughness and permanence as a wart, and yet, inexplicably and often very abruptly, they come to the end of their lives and vanish without a trace.[10]

Impersonal and Personal Technical Prose

Human beings, we need to remind ourselves here, are social beings. Our reality is a social reality. Our identity draws its felt life from our relation to other people. We become uneasy if, for extended periods of time, we neither hear nor see other people. We feel uneasy with The Official Style for the same reason. It has no human voice, no face, no personality behind it. It creates no society, encourages no social conversation....the closer you get to the impersonal essence of The Official Style, the more distant any felt human reality becomes.[11]

—RICHARD LANHAM

Another aspect of distance in tone is your decision to be impersonal or personal in your technical prose. Recall the examples of pretentious jargon and gobbledygook in Chapter 6 for when writers choose to be impersonal. Notice in the examples of formal and informal prose above that both are still personable prose. Thomas chooses first-person plural, *we,* to discuss the AIDS virus, and he uses second-person plural, *you,* as well as first-person plural, *us,* to discuss the subject of warts.

However, whether your prose is impersonal or personal depends on far more than your choice of person. William Zinsser refers to this quality as both "warmth" and "humanity" in your prose. Commenting on warmth, Zinsser writes: "Another way of making science accessible is to write like a person and not like a scientist. It's the same old question of being yourself. Just because you're dealing with a scholarly discipline that's usually reported in a style of dry pedantry is no reason why you shouldn't write in good fresh English."[12] On the element of "humanity," Zinsser comments: "Just because people work for an institution they don't have to write like one. Institutions can be warmed up. Administrators can be turned into human beings."[13] For Zinsser, adding warmth and humanity to your writing is to put people in your writing using first- and second-person pronouns, using examples and details that people can relate to or that involve people, and "it's a question of remembering that readers identify with people, not with abstractions like 'profitability,' or with Latinate nouns like 'utilization' and 'implementation,' or with inert constructions in which nobody can be visualized doing something ('prefeasibility studies are in the paperwork stage')."[14]

Of course, we must also recognize that much technical prose must be impersonal, for example, many types of abstracts, parts of some license agreements for software programs, many types of contracts, and certain kinds of scientific, engineering, or medical reports.

Look at the following warranty and license agreement for a typical shareware program. This document discusses three topics: the shareware concept, the license agreement, and the disclaimer. Notice how the disclaimer uses much more legalese (as well as all caps) and is much more impersonal than discussions of the shareware concept and the license agreement:

Warranty and License Agreement

In the following agreement "software" refers to ALL FILES included and listed in the PACKING.LST file with this product.

The Shareware Concept

The software is distributed through a means known as Shareware. Shareware is not a type of software, but a means of distribution. SHAREWARE IS NOT FREE SOFTWARE. You are granted a right to use the software for an evaluation period of no more than 30 days. If you continue to use the software after the evaluation period you are required to either register the software or remove the software from your computer. While the software is not registered you may freely distribute it to anyone you want to, provided you distribute the entire software package (all files). The software is considered "registered" once the author has received the appropriate payment for the software and the "registration ID" has been entered in to the program along with your name. Once registered the software does not display the message explaining the Shareware concept and therefore can not be distributed through the Shareware channels. Once the software is registered you are responsible for adhering to the license agreement below.

License Agreement

You the original purchaser, are granted a non-exclusive license to use this Software under the terms stated in this Agreement. You may use this software on any computer you own or use. You may physically transfer the software from one computer to another provided that there is NO POSSIBILITY that the software is being used on one computer while it is being used on another. YOU MAY NOT use the software with multiple terminals or on multiple computers simultaneously in a network nor on more than one computer at a time. YOU MAY NOT distribute to others copies of the software, any file copied from the distribution diskette, or the documentation. For information about using this software on a network, please contact the author regarding a site license.

    The software and documentation are copyrighted. United States Copyright law provides civil and criminal penalties for the unauthorized use, reproduction, distribution, and/or sale of copyrighted material.

Disclaimer

THE AUTHOR DISCLAIMS ALL WARRANTIES RELATING TO THIS SOFTWARE WHETHER EXPRESS OR IMPLIED, INCLUDING BUT NOT LIMITED TO ANY IMPLIED WARRANTIES OF MERCHANTABILITY AND FITNESS FOR A PARTICULAR PURPOSE, AND ALL SUCH WARRANTIES ARE EXPRESSLY AND SPECIFICALLY DISCLAIMED. NEITHER THE AUTHOR NOR ANYONE

ELSE WHO HAS BEEN INVOLVED IN THE CREATION, PRODUCTION, OR DELIVERY OF THIS SOFTWARE SHALL BE LIABLE FOR ANY INDIRECT, CONSEQUENTIAL, OR INCIDENTAL DAMAGES ARISING OUT OF THE USE OR INABILITY TO USE SUCH SOFTWARE EVEN IF THE AUTHOR HAS BEEN ADVISED OF THE POSSIBILITY OF SUCH DAMAGES OR CLAIMS. IN NO EVENT SHALL THE AUTHOR'S LIABILITY FOR ANY DAMAGES EVER EXCEED THE PRICE PAID FOR THE LICENSE TO USE THE SOFTWARE. REGARDLESS OF THE FORM OF CLAIM, THE PERSON USING THE SOFTWARE BEARS ALL RISKS AS TO THE QUALITY AND PERFORMANCE OF THE SOFTWARE.

SOME STATES DO NOT ALLOW THE EXCLUSION OF THE LIMIT OF LIABILITY FOR CONSEQUENTIAL OR INCIDENTAL DAMAGES, SO THE ABOVE LIMITATION MAY NOT APPLY TO YOU.

THIS AGREEMENT SHALL BE GOVERNED BY THE LAWS OF THE STATE OF NEW YORK AND SHALL INURE TO THE BENEFIT OF THE AUTHOR AND ANY SUCCESSORS, ADMINISTRATORS, HEIRS AND ASSIGNS. ANY ACTION OR PROCEEDING BROUGHT BY EITHER PARTY AGAINST THE OTHER ARISING OUT OF OR RELATED TO THIS AGREEMENT SHALL BE BROUGHT ONLY IN A STATE OR FEDERAL COURT OF COMPETENT JURISDICTION LOCATED IN SUFFOLK COUNTY, NEW YORK. THE PARTIES HEREBY CONSENT TO IN PERSONAM JURISDICTION OF SAID COURTS.

Although many technical documents have an impersonal tone, many impersonal technical documents would achieve their purpose far more effectively if they were written in a less impersonal tone. Writers of technical prose can use first- and second-person pronouns in many types of technical documents, and they can use analogies, examples, details, and even personal anecdotal information people can relate to. And often writers of technical prose can avoid the institutional passive and write in a straightforward active voice style. One of the major reasons for the virtual renaissance of science writing in the 1970s, '80s, and '90s, is the ability of so many gifted scientists and science journalists to humanize their subjects for their readers. Lewis Thomas, Oliver Sacks, Stephen Jay Gould, Paul Davies, Stephen Hawking, Peter Medawar, Loren Eisely, and numerous others have found just the right balance of scientific and technical information and a personable tone.

Attitude and Emotion

Tone is, in part, a matter of attitude and emotion, the attitudes and emotions you want to convey about your subject and to your audience.

What is an attitude? The word *attitude* has many meanings. One meaning is "a position assumed for a specific purpose," such as a threatening attitude. Other meanings are "a mental position with regard to a fact or state," or a "feeling or emotion toward a fact or state." Yet other relevant meanings for our purposes here are "a negative or hostile state of mind," or "a cocky or arrogant manner."[15]

What is an emotion? An *emotion* is defined as "the affective aspect of consciousness" or "a state of feeling," or "a psychic and physical reaction (as anger or fear) subjectively experienced as strong feeling and physiologically involving changes that prepare the body for immediate vigorous action." A feeling is a "subjective response to a person, thing, or situation. FEELING denotes any partly mental, partly physical response marked by pleasure, pain, attraction, or repulsion; it may suggest the mere existence of a response but imply nothing about the nature or intensity of it."[16]

Examples of attitudes are trust, distrust, sincerity, insincerity, seriousness, lightheartedness, hope, despair, belief, incredulity, acceptance, denial, desire, repulsion, and attraction or rejection. Examples of emotion are anger, love, hate, desire, envy, lust, joy, excitement, disgust, contempt, revulsion, obsession, and indifference. As you can see, attitudes and emotions overlap. What's important is to realize that all of the attitudes and emotions listed above (and many more) can be an appropriate tone in a piece of writing.

Before you write, you need to decide how you want to treat both your subject and your audience. Do you, for example, want to treat the subject seriously and convey this seriousness to your audience? Or do you want to treat your subject with anger and convey this anger to your audience? To convey a serious tone, you have to show readers that you respect them and take them seriously, that you take the subject matter seriously, and that you also take yourself seriously. To convey this serious tone, you have to choose your words carefully, phrase your sentences and paragraphs with skill, and organize the document so that it will achieve the desired effect. To convey anger about a subject, all of your style choices must help you suggest this anger appropriately.

Note the tone in the following excerpt from a letter written by a company executive:

> Enclosed is your invoice 2924, which you note replaces invoice 2900. Since we have already paid invoice 2900, we have no intention of paying 2924. In addition, since invoice 2924 is for $550.00 and represents a $60.00 a month increase from the $490.00 we were paying, we dispute your claim for a higher rent, absent prior notice, which we have not received. Your comment that the increase was to move us to "market rate" is not an acceptable explanation. "Market rate" is what you negotiate for, not what is unilaterally set by one party to a transaction; and there has been no negotiation between [company A] and [company B].
>
> For the future, we accept your proposal and the renewal of $550.00 per month with one month free.
>
> Please send us a lease and an invoice for October and we will continue our stay with you.

Notice the negative phrases—"we have no intention," "we dispute your claim," and "not an acceptable explanation." This letter, of course, has other problems, for example, a cluttered opening paragraph and poor organization, but its chief failure is in tone. Not only is the tone too angry, but it is also inconsistent. Note that at the end of the letter the author agrees to accept the proposal for an increased monthly rent, and the author ends with several conciliatory sentences.

The following tones are some of the most common ones in professional writing. Notice what kinds of words or strategies are used to help create these tones:

Friendliness, Warmth, Graciousness

To convey a friendly tone, use words such as *glad, pleased, delighted, happy, benefit, pleasure, fine, privilege, grateful, welcome, successful, progressive, generous,* and *rewarding.*

The following is a friendly opening paragraph in a user's guide:

> Welcome to **LView Pro 1.6,** an interactive image viewer that enables you to view, edit, and retouch computer images in many popular formats. With **LView Pro 1.6** you can improve images and save them in a format that can be incorporated into many word processing programs, including WordPerfect and Microsoft Word. If you are interested in adding professional quality graphics to your text documents, **LView Pro 1.6** is the key. If you want a top-quality program for viewing, editing, and retouching photographs, paintings, cartoons, or any other images you may find, then you have made the wise choice in selecting **LView Pro 1.6.**[17]

Empathy

To convey a sympathetic tone, use what is commonly called the "you attitude." As David Ewing comments,

> In its most obvious form the "you attitude" simply means choosing the words *you* and *your* wherever possible instead of *I, me,* and *mine* or impersonal words. But the "you attitude" should go deeper. It should mean stating the advantages, disadvantages, and implications of an idea or fact in terms of your reader's interest rather than your own. It also should mean being honest with the reader and truly considering his or her viewpoint as faithfully as you can.[18]

Conveying empathy does not mean conceding everything to your audience. Empathy is balancing your interests with those of your audience's.

Look at the following example of effective empathy in a letter.

> I am writing to see if we can agree on a date for you to finish the remaining work in our technical writing laboratory. I know you have been very busy these past few months, but there are still a few small tasks that need to be taken care of before our lab can open for the students.
>
> I'll remind you what the remaining problems are. First, the printers still do not work well with the network. It takes much too long to print. We know this is a network problem because when we remove any of the printers from the network, they print very quickly. Second, we are still missing a cable to connect our projector to our faculty pc. Without this cable we are unable to demonstrate various programs and student work to the classes that meet in our lab. Third, several of the pc's still do not work properly on the network, and we don't know why. Fourth, you still haven't given our lab director the training session she needs to become familiar with the network.
>
> I anticipate you will counter with the views that networks quite often slow down the printing process, that the cable connection is the responsibility of another department, that it's not unusual for a few pc's to have a few problems with a new network, and that you haven't any time yet to set up a training session with our lab director. All

of these responses are valid, and I sympathize with how busy you must be. It's been a very busy summer for me, too.

However, these problems have to be addressed immediately so we can open the lab on time. Please make every effort to meet with me and our lab director in the lab this Thursday at 3:00 P.M. to address these remaining problems. I hope this meeting won't interfere with your plans.

Anger and Accusation

Although letters and memos written in a tone of anger will seldom gain you positive results, sometimes it is necessary to be angry in your communications. Of course, with care and attention to detail, you may control your anger to achieve results. You may blame others for errors and do so in strong terms without totally alienating the reader. Or you may be more extreme and be too blunt and accusatory. An effective way of achieving this latter tone is to use many *I*-words, treat the reader with condescension, and use many expressions such as "your fault," "your mistake," "your oversight," and "you are to blame." Below are two examples, the first example written in less inflammatory language than the second:

> I am writing once again concerning remaining tasks in our technical writing laboratory. I have written to you and called you numerous times over the past few months, but you have still not finished the work that remains to be done. I know you have been very busy, but I think the delays concerning our lab have been unreasonable. This letter is my final request for you to finish your work before complaining to your supervisor about your failure to respond to my repeated requests.
>
> I'll remind you what the remaining problems are. First, the printers still do not work well with the network. It takes much too long to print. We know this is a network problem because when we remove any of the printers from the network, they print very quickly. Second, we are still missing a cable to connect our projector to our faculty pc. Without this cable we are unable to demonstrate various programs and student work to the classes that meet in our lab. Third, several of the pc's still do not work properly on the network, and we don't know why. Fourth, you still haven't given our lab director the training session she needs to become familiar with the network.
>
> The other technical writing faculty and I have decided to give you two more days to solve the problems described here. If these problems are not addressed within this time, I have no choice but to appeal to your supervisor.

This letter obviously has an angry tone, but its anger is not misdirected. It is still possible that the intended reader will finish the remaining tasks. The following letter, however, is a different matter:

> After five months of trying to convince you to finish the work you are responsible for in our technical writing laboratory, I have given up trying. I no longer have any patience to deal with you. Nor do I have any confidence that you are able to do the job responsibly. I am calling your supervisor tomorrow morning and complaining in the strongest terms possible about your incompetent behavior concerning our lab.

You know as well as I do that you have no valid excuse for not finishing over a month ago all of the remaining work in this lab. I refuse to remind you in this letter what still needs to be done. I don't think doing so would matter because I don't think you care about how well you do your job. I have politely reminded you over a dozen times what needed to be finished, and yet you have not made the time to finish your work for our lab. I understand that you are busy tending to matters in other labs as well, but that's no excuse for not finishing the few tasks in our lab.

I am relieved to know that either your supervisor will take care of the lab issues or he will have someone else do so. Quite honestly, I would much rather that you never walk in to our lab again.

It is possible that this letter may lead to the intended audience finishing what needs to be done in the lab, but it is not likely. The letter makes it perfectly clear that no future dealings with the audience are wanted. And the letter very clearly states that the supervisor will be contacted to look into this matter.

The point here is for you to be careful about writing e-mail, letters, or other documents in which you express anger. You may regret what you have put into writing, and expressing your anger in writing may come back to haunt you later. Often a face-to-face meeting or telephone call are better ways to vent anger. Finally, if you do decide to express your anger in writing, keep in mind you can do so and still achieve results without totally alienating your audience.

Familiarity and Cuteness

There are many times when it's appropriate to stress a close, personal relationship with an audience. This tone is often established in e-mail, for example. You need to write very informally, even conversationally. You especially need to sound like yourself if you want to establish this tone effectively.

Note how the author in the example below achieves an informal rapport with his readers:

So there you have it, pilgrim—my own personal rendition of a WinWord workin' stiff's Toolbars and desktop. I'm not about to claim it's the best for you—far from it! But I *will* claim that it's a hell of a lot better than the version that ships with WinWord. If you use it as a starting point, you won't be far off.[19]

The *RoboHELP User's Guide* is another example of a technical manual that effectively uses familiarity and cuteness. The author uses sentences such as the following:

RoboHELP provides amazing hotspot graphics capabilities which you are really going to enjoy.

Choose the Choose button (heh, heh) to browse for a destination topic or to define a Help macro.

I had always been annoyed at the number of inconvenient steps required to capture and tweak screen images for use in printed documentation and Help systems.

Screen Capture Utility is amazingly good at capturing pull-down menu images. You have to experience it to believe it.[20]

Ethos

> *Thoreau contended that one reason why writers used torpid or wooden or lifeless words like* humanity, *words which he said had a paralysis in their tails, was because the writer did not speak out of a full experience. There was no life in the writing because there had been an incomplete realization of life in the writer.*[21]
>
> —REGINALD COOK

Finally, tone is a matter of ethos or character and how you establish this ethos with your audience. You may recall that in Chapter 4 we discussed Aristotle's three modes of persuasion: *ethos, logos,* and *pathos. Ethos,* precisely defined, is "the distinguishing character, sentiment, moral nature, or guiding beliefs of a person, group, or institution."[22] Ethos comes from the Greek word *ethos,* meaning custom or character. Ethos is the character of the writer, the writer's integrity, competency, objectivity. Of course, like poets and fiction writers, nonfiction writers may create a persona in their prose that is the opposite of their true selves. Humorists, for example, do this all the time. Ethos is different from a persona. A persona, in keeping with its Latin origins as a "mask," is a mask for the self whereas ethos is the authentic self.

For example, do you want to be perceived as credible or knowledgeable? Trustworthy? In control? Deferential? Friendly? You may make numerous stylistic decisions to help you create an ethos in your communication, but, to some degree, the ethos of your style is beyond your control. You are who you are, and, often, your writing reflects your identity. Exceptions, of course, are when you choose to be satiric or ironic or when you choose to be deceptive. See Chapter 11 for more discussion of deliberately deceptive prose.

In Chapter 1 style was partially defined as "who you are and how you reflect who you are, intentionally or unintentionally, in what you write." What follows is a brief discussion of some tones your ethos helps you establish in your prose.

Establishing Authority or Credibility

You can establish your authority or credibility in many ways. The most important way, of course, is to know your subject extremely well and to show this expert knowledge throughout the document in your attention to detail, your technical accuracy. Richard Eastman reminds us that

> The writer's knowledge hardly shows to best advantage through a flaunting of degrees earned, offices held, or big names known. Rather, one proves one's knowledge by use of a vocabulary and syntax obviously capable of illuminating a topic to the full scope of the reader's interest. Knowledge can also appear in one's continuous but quiet respect for substantiation, a modest familiarity with the sweep of one's subject.[23]

However, you may also establish your credibility by simply calling attention to your credentials. Notice, for example, how William Zinsser establishes his credibility early in his book *On Writing Well: An Informal Guide to Writing Nonfiction:*

> This book grew out of a course in nonfiction writing that I taught at Yale in the 1970s. I taught out of my own experience: 13 years as a writer and editor on the *New York*

Herald Tribune and many more as a free-lance writer for magazines. I was lucky in my apprenticeship: the *Herald Tribune* was an education in high standards. The older editors who made us rewrite what we had written and rewritten were custodians of a craft; writing well was a point of honor. That notion of nonfiction writing as honorable work was what I wanted to pass along in my course. I wanted to try to help college students to write about the world they were living in.[24]

Howard Rheingold establishes his credibility in a similar manner by describing his extensive experience.

Since the summer of 1985, for an average of two hours a day, seven days a week, I've been plugging my personal computer into my telephone and making contact with WELL (Whole Earth 'Lectronic Link)—a computer conferencing system that enables people around the world to carry on public conversations and exchange private electronic mail (e-mail). The idea of a community accessible only via my computer screen sounded cold to me at first, but I learned quickly that people can feel passionately about e-mail and computer conferences. I've become one of them. I care about these people I met through my computer, and I care deeply about the future of the medium that enables us to assemble.[25]

Establishing Trust

If you convince your audience that you are very knowledgeable about a subject, you successfully develop one important kind of trust. But, often, an audience wants to know more than the fact that you are an authority on the subject. The audience wants to know you are honest and sincere about the information you provide. The audience wants to know you have no hidden agendas, no ulterior motives, nothing to be gained from the information you provide.

Establishing trust in a memo, letter, or report is no easy task. You cannot simply say "Trust me" and leave it at that. Trust is established by all of your stylistic choices, and trust is established by your authenticity. Simply stated, you must be trustworthy to be trusted.

Notice how Thoreau slowly begins to win our trust beginning on the first page of *Walden*.

I should not talk so much about myself if there were any body else whom I knew as well. Unfortunately, I am confined to this theme by the narrowness of my experience. Moreover, I, on my side, require of every writer, first or last, a simple and sincere account of his own life, and not merely what he has heard of other men's lives; some such account as he would send to his kindred from a distant land; for if he has lived sincerely, it must have been in a distant land to me. Perhaps these pages are more particularly addressed to poor students. As for the rest of my readers, they will accept such portions as apply to them. I trust that none will stretch the seams in putting on the coat, for it may do good service to him whom it fits.[26]

Thoreau maintains his high standard of critical self-examination throughout *Walden*, building on the sincerity he begins to establish on the first page.

Numerous other writers successfully establish a similar trust with their honesty about themselves and their consistency. Montaigne is consistent in his three volumes of essays about both his and our inconsistencies:

> Our ordinary practice is to follow the inclinations of our appetite, to the left, to the right, uphill and down, as the wind of circumstances carries us. We think of what we want only at the moment we want it, and we change like that animal which takes the color of the place you set it on. What we have just now planned, we presently change, and presently again we retrace our steps: nothing but oscillation and inconsistency.[27]

How does an audience know when a writer or speaker is being sincere? There are, of course, no easy answers to this question. It is important to remember, however, that your written prose and your manner of speaking are far more revealing then you think they are. Readers can pick up on the slightest shift in tone, a poorly chosen word, a weak organization, an unconvincing argument, a lack of conviction, a lack of details, poor examples, and so on to detect insincerity. Listeners, of course, can pick up on all of these signs as well as inappropriate gestures, frowns, eye movement, voice intonations, and so on to detect insincerity of one kind or another.

Establishing Control

Sometimes it is necessary to project the image of power or control, that you have the upper hand, in a communication. This situation, for example, is often the case in much legal correspondence, particularly letters from lawyers representing one client written to lawyers representing another client. In the two letters with the angry tone earlier in this chapter, the author is establishing control by demanding that certain tasks be completed and by being serious about what will happen if these tasks are not. The writer could have reminded the reader that because the reader is a member of the staff of the university his primary responsibilities are to tend to the needs of students on campus. He could have been reminded that he is not serving the students well by not completing the setup of the lab. In fact, these points and others could have been made in a letter with a professional and courteous tone instead of an angry tone.

Establishing Deference

Sometimes you convey that you are aware the audience has the upper hand, that you are perhaps a subordinate, an employee, or a client who is dependent on the person to whom you are writing. Often it is necessary to convey a tone of deference if you are, for example, a subordinate writing to your boss or another senior person. You should use words and phrases indicating your awareness that the reader, not you, has the responsibility to decide or act upon what you are discussing. Make it clear that you are only offering advice and that you hope your advice will be helpful.

Once again the letters concerning the technical writing laboratory can be used to illustrate the issue of deference. The letter writer could have opened by stating how much he and the students depend on the services provided by the readers' department. The writer could have stressed that he depends now more than ever on the ability of the reader's staff to finish a job that has been done well up to this point. And in addition to covering other points in the middle of the letter, the writer could close by pointing out how much the future success of the lab depends on the talents of the reader to assure its continued smooth operation.

Establishing an Awareness of a Delicate Situation

Often you are challenged to write when the situation is very delicate. The reader does not want to hear the news you have to convey, or you know the reader will oppose what you have to convey, but you also know you have to convince the reader. These situations demand some of the most careful phrasing in professional writing. Imagine, for example, you have to write to a loyal customer rejecting her request for a complete refund on a product, but, at the same time, you want to continue receiving business from this customer. You have to provide convincing reasons in the proper order, and you have to show you understand your reader's concerns. Or imagine you have to reject a computer company's competitive and solid proposal and accept a bid from a rival of the company. You want to maintain a good relationship with the company that has lost the bid, but you have to convince this company that another bid was in your better interests. Conflicts of this kind are frequent in the corporate world.

All of the letters discussed concerning the computer lab show an awareness of a delicate situation, but some are more sensitive than others to this situation. In this situation you have someone who has certain job responsibilities but who is not meeting these responsibilities. How do you motivate him to do what he is paid to do if he does not want to do his job well? Perhaps the most successful letter in this particular situation will be one that is very sensitive to the difficulties faced by both writer and reader.

Strategies for Developing an Appropriate Tone

Now that you have a better understanding of some of the basic elements of tone, you need to learn some strategies for conveying a tone that is appropriate to your purpose. Essentially, there are six strategies for helping you achieve an appropriate tone.[28]

Determine Your Relationship with Your Audience

First, decide what kind of relationship you want to have with your audience. Do you want your audience to regard you as a friend? Do you want your audience to regard you as highly informed and knowledgeable? As trustworthy? You can see how all of these questions are important to the person writing the letters concerning the lab. Here the writer has to decide whether the reader will be treated as a friend or an enemy, whether the writer will appear very knowledgeable about remaining lab tasks, and whether the writer is trustworthy concerning this information.

Consider How Your Communication Will Be Read (Received)

Second, visualize your audience and put yourself in their shoes. Does, for example, your document convey good news or bad news? If you have good news to convey, you should state it early in your memo or letter. If you have bad news, you should take a more indirect route and save the bad news until the middle and close with good or neutral news. The writer of letter three, the letter with the strongest tone of anger, probably would not mail the letter if he more carefully considered how the reader will read and react to it. And there are always many other variables to consider. Suppose that the reader's supervisor thinks the reader can do no wrong. Any threat on the part of the writer to see the supervisor may appear as an idle threat.

Choose Appropriate Diction for the Best Tone

Third, after you decide what tone you want to convey and after you have reviewed the possible effect of the situation on your communication, you need to choose the appropriate words for your message. If you want to establish a friendly tone, for example, avoid accusatory words and phrases. You can see in the various examples provided in this chapter that tone is very much established by the words you choose. Also remember that readers can recognize and appreciate your efforts to be friendly, courteous, concerned, or insistent without being too angry.

Express Ideas That Convey the Intended Relationship

Fourth, state the ideas and thoughts that show clearly what kind of relationship you want with your audience. For example, if you have reservations, state that you have reservations. If you have confidence in the situation, state that you have confidence. The writer of the two angry lab letters has clearly lost all confidence in the reader, and in the third letter the writer states this point emphatically. Remember before you begin to write that you can add all kinds of phrases and sentences to influence the way your readers will perceive you.

Use an Appropriate Amount of Detail

Fifth, sometimes it is necessary to write richly detailed paragraphs to make your tone convincing. As David Ewing suggests, "in numerous writings we can succeed in developing the right tone only if we elaborate, define, qualify, document, and illustrate our points and ideas."[29] There are many occasions for being brief, but there are also times when careful attention to detail is necessary. One effective way to write an angry letter is to provide lots of details to make your points convincingly and, of course, to avoid any personal attacks on the reader's abilities or integrity. Often the reader will see that your anger is well justified.

Focus on How Your Document Looks

Sixth, don't forget that the page layout and design of your letter, memo, or report also help to establish tone. Hard to read fonts, too many fonts, fonts that are too small, the wrong color of paper, poor margins and spacing, the wrong binder, and many other physical properties of your document all make an impression on the reader. You can make all kinds of careful decisions about your tone, diction, opening, middle, and ending, but unless you also give careful attention to the appearance of your document, you are not doing all you can to communicate well.

Humor in Technical Writing

> *Humor is the secret weapon of the nonfiction writer. It's secret because so few writers realize that humor is often their best tool—and sometimes their only tool—for making an important point.*[30]
>
> —WILLIAM ZINSSER

The Advantages of Humor

Does humor have a role to play in technical prose? Of course, it does. Humor in technical prose, if effectively established, is often a great tone and motivator for novices who are intimidated by the subject matter. Witness the success of the *Dummies* series of books with titles such as *DOS for Dummies, Windows for Dummies,* and *Word for Dummies.* These books rely on humor and a very simple approach to the subject to help walk novices through the steps of using many different kinds of computer programs. Humor, then, is just one of many potentially effective stylistic elements often necessary for appealing to and motivating a lay audience. Technical prose primarily informs or persuades, but it may also entertain at the same time.

Humor, of course, does not have to dominate a technical document. Many technical documents use some subtle or slight humor to help make points more effectively.

> **Curser** A curser is someone who curses, who frequently shares his (or her) vocabulary of obscenities. This isn't a book about manners or linguistics. So I'm not going to share my thoughts with you—nor what my parents told me the first time I brought home one of these special words from grade school. I do want you to know there's a difference between a curser and a cursor.
>
> **Cursor** People sometimes called the **insertion point** a "cursor." You can call it that if you want. One of the fun things about being an adult is that you usually get to make your own decisions. In this book, I'm going to call the insertion point an "insertion point."
>
> Whatever you do, however, don't confuse the terms insertion point and **selection point.**[31]

Here is another example from a technical book on LANtastic.

> Get ready. You're about to undergo a rite of passage by learning the Sneakernet joke. You can't claim to have any significant LAN exposure unless you know the Sneakernet joke.
>
> As you saw earlier in this chapter, one of the most popular LAN card types follows the standard called Ethernet. If you don't have a LAN (Ethernet or another kind), and you want to share data with fellow PC users, how do you do it? You simple copy files to floppy disks, put on your sneakers, and run around to the other users and pass out floppy disks. This not-so-high-tech system has been dubbed *Sneakernet.* Some people who think they're clever like to joke that a company that has no LAN performs data distribution using a Sneakernet network. In fact, some of these people like to use this joke over and over until no one can stand to hear it again. Many of them sell LANs for a living.
>
> There. Now you've heard it, too. If you're lucky, you won't hear it more than fifteen thousand times a year.[32]

Of course, a lot of prose, technical and nontechnical, is humorous unintentionally. Notice the following examples from various church bulletins:

Actual Announcements Taken from Church Bulletins

Don't let worry kill you—let the church help.

Thursday night—Potluck supper. Prayer and medication to follow.

Remember in prayer the many who are sick of our church and community.

For those of you who have children and don't know it, we have a nursery downstairs.

The rosebud on the altar this morning is to announce the birth of David Alan Belzer, the sin of Rev. and Mrs. Julius Belzer.

This afternoon there will be a meeting in the South and North ends of the church. Children will be baptized at both ends.

Tuesday at 4:00 P.M. there will be an ice cream social. All ladies giving milk will please come early.

Wednesday, the ladies Liturgy Society will meet. Mrs. Jones will sing, "Put me in My Little Bed" accompanied by the pastor.

Thursday at 5:00 P.M. there will be a meeting of the Little Mothers Club. All wishing to become little mothers, please see the minister in his study.

This being Easter Sunday, we will ask Mrs. Lewis to come forward and lay an egg on the altar.

The service will close with "Little Drops of Water"; one of the ladies will start quietly and the rest of the congregation will join in.

Next Sunday a special collection will be taken to defray the cost of the new carpet. All those wishing to do something on the new carpet will come forward and do so.

The ladies of the church have cast off clothing of every kind and they may be seen in the church basement Friday.

A bean supper will be held on Tuesday evening in the church hall. Music will follow.

At the evening service tonight, the sermon topic will be "What is Hell?"; come early and listen to our choir practice.

Of course, sometimes intentional humor is not effective. Look at the following example from a local newspaper. The example intends to be humorous about recruiting volunteers to help the elderly prepare their taxes.

Uncle Sam Needs YOU!

"IRS seeks local volunteers to help others with taxes"

As our volunteer forces continue to ensure peace in Bosnia, this in no way diminishes the needs for patriotic efforts within our borders. **Volunteers are needed to help low-income, elderly and disabled taxpayers prepare their federal income tax returns.** Your friends and fellow Americans at the IRS have been decimated by government shutdowns, downsizings, family leave programs, and extended coffee breaks.

We call upon the generosity of right-thinking Americans everywhere to help those with the greatest need at this special time of the year. Some of these unfortunate Americans were even veterans. What better gift could you give to those who have risked their lives to defend your life, liberty and pursuit of happiness?

Volunteers will be trained by the Internal Revenue Service to prepare basic federal tax returns as part of a national program called the Volunteer Income Tax Assistance program. Please specify if you would prefer working with the low-income or the disabled, where our needs are the greatest. Some of these unfortunate Americans do not realize that just because they receive assistance payments from their government, they are still obligated to pay their fair share of taxes. We at the IRS are not heartless. It is at the very last resort that we confiscate their property, freeze their assets, and place them in federal correction institutions. Most of these services require more funds that the meager collections that result. Nonetheless, it is the Law, and we are charged by congress to service all Americans with equality and fairness.

The IRS offers free tax assistance to older taxpayers through a volunteer program called Tax Counseling for the Elderly. These elderly are a particular problem. Many have Alzheimer's disease, and don't know the day of the week, let alone something as simple as a deferred investment tax credit. They also tend to have extended illnesses, and die frequently without fulfilling their obligations to their fellow Americans. Securing their estates as quickly and efficiently as possible is one of the keys to balancing the federal budget.

Free training will be provided for volunteers willing to learn to prepare short forms, 1040A and 1040EZ, as well as the basic 1040 forms with Schedules A and B. We would be particularly interested in attorney volunteers who would like to assist the elderly with their estate planning, and negotiated fees for your "additional" services would be between you and the senior citizen. (Just don't let us catch you billing for the completion of that 1040EZ, okay?)

Those interested in working as VITA or TCE volunteers can call the IRS Taxpayer Education Office at 1-800-829-1040. We guarantee that this IRS phone line will be answered by an actual human being, who will help to arrange for your training and work schedule. Call today! Uncle Sam needs YOU!

Some Do's and Don'ts for Using Humor

Do's

1. Consider your audience. Will your audience appreciate having humor in your document or see the humor as an obstacle against understanding the information? Remember, your readers are primarily reading your document to be informed, not entertained.

2. Consider the genre of your document. Many kinds of documents readily lend themselves to humor—e-mail, letters, and memos on many kinds of pedestrian subjects, user guides and tutorials for novices, and often some progress or status reports on matters not of vital importance. Some genres do not lend themselves well to humor—abstracts, summaries, reference manuals, feasibility reports, and many other kinds of reports. Of course, if you look hard enough you can find humorous examples of all of these kinds of documents, too.

3. Consider the purpose of the document. If your purpose is to explain something of vital interest to your audience, this seriousness in your purpose should override any inclination to be humorous. If your purpose is to show your awareness of a delicate situation, you do not want to undermine your tone with frivolity. But if your purpose is to show an audience how it needs to relax before it can learn, then you might well consider using humor to achieve your goals.

4. Consider the subject matter. Some subjects just do not lend themselves to humor, topics such as AIDS, a death in the family, the possibility of serious injury, and other life threatening matters.

5. Be aware that humor is very subjective. What is humorous to some may not be to others. There are so many variables at play when it comes to deciding on what is or isn't funny. Keep in mind that you can have something in a document that is truly funny, but a reader may be too tired or under too much pressure to appreciate the effort at humor.

Don'ts

1. Don't use humor just for the sake of using humor. Your humor should have a point in your documents. Your humor should enhance or underscore your more serious points.

2. Don't provide too much humor. Humor is fine for motivating some readers and for getting their attention. But ultimately a technical document should inform or persuade, and to do so you must be accurate, precise, and concise. If humor is undermining the larger seriousness of your purpose, then you need to edit out some of your humor.

3. Avoid offensive humor or humor that is too subtle. Sexist jokes, jokes about the physically or mentally disadvantaged, and many other kinds of humor will undermine all of your purposes in a technical document. And don't count on your readers picking up on understated humor. Sarcasm, understatement, and parody may go unnoticed by your readers.

4. Avoid being inconsistent with your humor. You can have a tone of seriousness that is supplemented by humor, but you can't usually have a tone of humor supplemented by seriousness. Don't forget to control your humor and to use humor consistently in some way. In a user guide, for instance, don't just use humor in the introduction but nowhere else in the manual.

5. Don't forget your audience, your purpose, the subject, content, genre, and the communication situation. As we discussed in the "do's" about humor, all of these factors determine what you can and can't do with humor.

6. Keeping in mind all of the points above, don't be afraid to try. Many technical documents on pedestrian subjects are unnecessarily boring or dry. Many documents aimed at instructing beginners are unnecessarily intimidating. Don't be afraid to liven things up a little with appropriate humor.

Chapter Summary

Tone is as important in technical prose as it is in nontechnical prose. Of all of the many elements of style, tone is perhaps the most difficult to master. Mastering tone is the ability to read, and write, between the lines, the ability to understand your audience thoroughly and to know well what will and will not work with your audience, and the ability to choose the words, phrases, clauses, sentences, paragraphs, and organization that will achieve the desired effect.

Complicating discussions of tone is the fact that some terms are used synonymously with tone, particularly *voice* and *point of view*. You need to understand the ways these terms overlap in meaning.

Tone has at least three major emphases: the level of formality you choose, the attitude or emotion you choose to convey, and the ethos you project. You need to know when you should be formal or informal and impersonal or personable. You need to know how to convey a wide range of attitudes and emotions in your prose. And you need to know you can control and shape the ethos that is projected in your writing.

There are some basic strategies for establishing the tone you need. You must understand these strategies and know how to use them for advantage in your prose.

Finally, humor has a role to play in technical prose as it does in nontechnical prose. Technical prose writers use humor sometimes to motivate and reward a rapidly growing audience of home computer users, for example. Learn the do's and don'ts of using humor, and don't be afraid to try using humor.

Case Study 6: Computer Viruses and Tone

Computer viruses are a serious subject to anyone who uses computers frequently. Each year viruses cost large and small companies millions of dollars. The following document is typical of many discussions you will find of viruses. This example comes from a text document that accompanies the anti-virus program F-Prot. Read the document and answer the questions at the end of the case study.

Computer Viruses: An Introduction[33]

A very simple definition of computer viruses is:

A program that modifies other programs by placing a copy of itself inside them.

This definition is somewhat simplified, and does not cover all virus types, but is sufficient to show the major difference between viruses and so-called "Trojan" programs, which is that the virus replicates, but the Trojan does not. (The definition does not cover the so-called "companion"-type viruses, however).

Trojan is a program that pretends to do something useful (or at least interesting), but when it is run, it may have some harmful effect, like scrambling your FAT (File Allocation Table) or formatting the hard disk.

Viruses and Trojans may contain a "time-bomb," intended to destroy programs or data on a specific date or when some condition has been fulfilled.

A time bomb is often designed to be harmful, maybe doing something like formatting the hard disk. Sometimes it is relatively harmless, perhaps slowing the computer down every Friday or making a ball bounce around the screen. However, there is really no such thing as a harmless virus. Even if a virus has been intended to cause no damage, it may do so in certain cases, often due to the incompetence of the virus writer or unexpected hardware or software revisions.

A virus may be modified, either by the original author or someone else, so that a more harmful version of it appears. It is also possible that the modification produces a less harmful virus, but that has only rarely happened.

The damage caused by a virus may consist of the deletion of data or programs, maybe even reformatting of the hard disk, but more subtle damage is also possible. Some viruses may modify data or introduce typing errors into text. Other viruses may have no intentional effects other than just replicating.

Two different groups of viruses occur on PCs, boot sector viruses (BSV) and program viruses, although a few viruses belong to both groups.

A BSV infects the boot sector on a diskette. Normally the boot sector contains code to load the operating system files. The BSV replaces the original boot sector with itself and stores the original boot sector somewhere else on the diskette or simply replaces it totally. When a computer is then later booted from this diskette, the virus takes control and hides in RAM. It will then load and execute the original boot sector, and from then on everything will be as usual. Except, of course, that every diskette inserted in the computer will be infected with the virus, unless it is write-protected.

A BSV will usually hide at the top of memory, reducing the amount of memory that the DOS sees. For example, a computer with 640K might appear to have only 639K.

Many BSVs are also able to infect hard disks, where the process is similar to that described above, although they may infect the master boot record instead of the DOS boot record.

Program viruses, the second type of computer viruses, infect executable programs, usually .COM and .EXE files, but sometimes also overlay files.

An infected program will contain a copy of the virus, usually at the end, but in some cases at the beginning of the original program.

When an infected program is run, the virus may stay resident in memory and infect every program run. Viruses using this method to spread the infection are called "Resident Viruses."

Other viruses may search for a new file to infect, when an infected program is executed. The virus then transfers control to the original program. Viruses using this method to spread the infection are called "Direct Action Viruses." It is possible for a virus to use both methods of infection.

Most viruses try to recognize existing infections, so they do not infect what has already been infected. This makes it possible to inoculate against specific viruses, by making the "victim" appear to be infected. However, this method is useless as a general defense, as it is not possible to inoculate the same program against multiple viruses.

In general, viruses are rather unusual programs, rather simple, but written just like any other program. It does not take a genius to write one—any average assembly language programmer can easily do it. Fortunately, few of them do.

Now—to correct some common misconceptions, here are a few bits of information about what viruses cannot do.

A virus cannot spread from one type of computer to another. For example, a virus designed to infect Macintosh computers cannot infect PCs or vice versa.

A virus cannot appear all by itself, it has to be written, just like any other program.

Not all viruses are harmful; some may only cause minor damage as a side effect.

A virus cannot infect a computer unless it is booted from an infected diskette or an infected program is run on it. Reading data from an infected diskette cannot cause an infection.

A write protected diskette cannot become infected.

There is no way a virus can attach itself to data files, so viruses cannot be distributed with them. However, a BSV can be distributed on data diskettes.

The F-PROT package will provide protection against viruses, but there are other methods that also should be used. Before I list them, I want to warn you against three methods that are of very limited use.

One anti-virus measure consists of making every executable file read-only, by issuing commands like

ATTRIB +R *.EXE

This is actually not a bad idea, but it will not provide much protection against viruses. Most program viruses will remove this protection before they infect files, and restore it afterwards. Making files read-only will of course have no effect on BSVs. The main purpose of this method is actually to protect the user from his own mistakes, because this makes it harder to delete programs by mistake. However, some viruses are stopped by this method, "Lehigh" and "South African" in particular.

Another method is to hide the COMMAND.COM file, by giving the following sequence of instructions:

MKDIR C:\HIDDEN
COPY COMMAND.COM C:\HIDDEN
DEL COMMAND.COM

add SHELL=C:\HIDDEN\COMMAND.COM
/P to CONFIG.SYS

add SET COMSPEC=C:\HIDDEN\
COMMAND.COM to AUTOEXEC.BAT

This method is quite useless, to say the least. Few viruses infect COMMAND.COM, and some of them are able to do it, even if it has been hidden using this method.

A third useless method is to change the name of COMMAND.COM and patch other programs so they use the new name. Somebody who had only heard of the "Lehigh" virus got this "bright" idea. Apparently he thought that all other viruses operated like it, so he wrote and distributed a program to do this automatically. He thought it was a general cure for the virus problem, but he was wrong.

On the other hand, there are a number of ways to provide useful protection.

Rule #1 is: MAKE BACKUPS!!! Keep good backups (more than one) of everything you do not want to lose. This will not only protect you from serious damage caused by viruses, but is also necessary in the case of a serious hardware failure.

Never boot a computer with a hard disk from a diskette because that is the only way the hard disk could become infected with a BSV. (Well, strictly speaking, it can happen if you run a "dropper" program too, but that happens extremely rarely.)

Should you, by accident, have left a non-bootable diskette in drive A: when you turn the computer on, the message "Not a system disk" may appear. If the diskette was infected with a virus, it will now be active, but may not have infected the hard disk yet. If this happens, turn the computer off, or press the reset button. It is important to note that pressing Ctrl-Alt-Del will not be sufficient, as a few viruses can survive that.

If the computer has no hard disk, but is booted from a diskette, you should always use the same diskette, and keep it write-protected.

Keep all diskettes write-protected unless you need to write to them.

When you obtain new software on a diskette, write-protect the diskette before you make a backup copy of it. If it is not possible to make a backup of the diskette, because of some idiotic copy-protection, I do not recommend using the software.

Be really careful regarding your sources of software. In general, shrink-wrapped commercial software should be "clean," but there have been a few documented cases of infected commercial software.

Public-Domain, Freeware, and Shareware packages do not have to be any more dangerous—it all depends on the source. If you obtain software from a BBS, check what precautions the SysOp takes against viruses. If he does not screen the software made available for downloading, you should find another source.

Check all new software for infection before you run it for the first time. It is even advisable to use a couple of scanners from different manufacturers, as no single scanner is able to detect all viruses.

Obtain Shareware, Freeware, and Public-Domain software from the original author, if at all possible.

Look out for any "unusual" behavior on your computer, like:

Does it take longer than usual to load programs?
Do unusual error messages appear?
Does the memory size seem to have decreased?
Do the disk lights stay on longer than they used to?
Do files just disappear?
Anything like this might indicate a virus infection.

If your computer is infected with a virus—DON'T PANIC! Sometimes a badly thought out attempt to remove a virus will do much more damage than the virus could have done. If you are not sure what to do, leave your computer turned off until you find someone to remove the virus for you.

Finally, remember that some viruses may interfere with the disinfection operation if they are active in memory at that time, so before attempting to disinfect you MUST boot the computer from a CLEAN system diskette.

It is also a good idea to boot from a clean system diskette before scanning for viruses, as several "stealth" viruses are very difficult to detect if they are active in memory during virus scanning.

Questions

1. What choice of person is used in this case study? Is this person appropriate?
2. What is the chief tone of this document?
3. Could this subject have been discussed humorously? If so, how?
4. What level of formality does the author establish? Is the author personal or impersonal?
5. How does the author establish his or her authority, knowledge, or credibility?
6. Is the information provided here trustworthy? If so, how do you know? If not, how do you know?
7. In what ways would you check on the accuracy of the information provided here?
8. How would you rewrite several paragraphs in this document to make an audience doubt the reliability of the information provided?
9. If a computer virus destroyed all of the files on the hard drive of your home computer and you were told who created the virus, what points would you stress in an e-mail message to this person?
10. What other tones do you think you could use to treat the subject of computer viruses?

Questions/Topics for Discussion

1. In what ways are point of view and voice similar to tone?
2. How can you choose to be personal or impersonal with your audience? What are the advantages of both tones?
3. What are the advantages of a formal tone? Of an informal tone?
4. Are there important differences between attitudes and emotions? If so, what are these differences?
5. What other important attitudes or emotions not discussed in this chapter are sometimes used by writers of technical prose?
6. What is the difference between persona and ethos?
7. What are some of the ways an audience may detect the ethos of a writer?
8. Can a writer create a fake ethos? If so, how?
9. Are there any ways for an audience to detect or see through a fake ethos?
10. What strategies would you add to the list of ways to establish tone in a document?
11. Why is humor often effective for a novice audience? When can humor be inappropriate for an audience?

Exercises

1. In this excerpt, how does the writer establish tone and point of view? How would you characterize the tone and point of view?

   > With so many kinds of flies in nature's burgeoning storehouse of life, how does one choose a proper species for study? The answer is simple. Let the species choose you. This was how our laboratory came to work with the black blowfly fifteen years ago.
   >
   > As I recall, it was a steaming hot day. There was no air-conditioning in the laboratory because such nonsense was considered a luxury.... Later we did get air-conditioning because on hot, humid days the flies in our laboratory culture died like flies. While a prostrate stenographer evoked no compassion, a cage full of dead flies constituted a powerful argument for air-conditioning in the eyes of the administration.
   >
   > On this particular day, however, all the windows were open.... Somebody had left a liverwurst sandwich on the windowsill. It did not take it long to ripen in that heat. A gentle liver-like odor filled the laboratory and welled over to the outside. Somewhere in the wide acres beyond, an egg-burdened blowfly cruised urgently about seeking a suitable place to lay her cream-colored cylindrical eggs. In her wandering, she encountered the tantalizing odor and, unraveling the tangled skein that swirled and twisted in the lazy breeze, she came upon the liverwurst sandwich. Her busybody ovipositor sought out the crevices between the bread and the liverwurst and packed in row upon row of eggs. Twenty-four hours later we found the maggot-laden sandwich and transferred the squirming mass to a small cage.[34]

2. Find a brief technical memo or letter and discuss the persona of the document.
3. Read this excerpt from *The Declaration of Independence* and discuss the ethos of the passage.

   > Nor have we been wanting in attention to our British brethren. We have warned them from time to time of attempts by their legislature to extend an unwarrantable jurisdiction over us. We have reminded them of the circumstances of our emigration and settlement here. We have appealed to their native justice and magnanimity, and we have conjured them to the

ties of our common kindred to disavow these usurpations, which would inevitably interrupt our connections and correspondence. They too have been deaf to the voice of justice and of consanguinity. We must, therefore, acquiesce in the necessity, which denounces our separation, and hold them, as we hold the rest of mankind, Enemies in War, in Peace Friends.

4. Find a formal technical document (memo, letter, report, article, sample from a book) and discuss what makes the tone formal.
5. Find an informal technical document and discuss what makes the tone informal.
6. Consider the following passage. Discuss how the author achieves the tone (friendly, warm, gracious; empathetic; angry, accusative; familiar, cute) of the piece.

> When you look at multimedia tutorials, do you feel a pang of envy? Followed, perhaps, by a momentary feeling of exhaustion as you consider all the new technologies that you haven't yet conquered?
>
> Take heart. As a former technical writer who now develops computer-based training, I've found that many of the skills that I relied on as a writer are equally applicable to multimedia development. In this article I'll discuss what I learned in my transition and identify some of the strengths and skills you can transfer to multimedia development.
>
> As technical writers, we already know that in order to teach concepts and details, we must provide readers with task-oriented, step-by-step information. We appreciate technical writing as a learning and teaching tool.[35]

7. After reading the following passages, discuss how the author of each one establishes ethos, be it authority, credibility, trustworthiness, control, deference, awareness of a delicate situation, or something else.

Example 1

Since the summer of 1985, for an average of two hours a day, seven days a week, I've been plugging my personal computer into my telephone and making contact with the WELL (Whole Earth 'Lectronic Link)—a computer conferencing system that enables people around the world to carry on public conversations and exchange private electronic mail (e-mail). The idea of a community accessible only via my computer screen sounded cold to me at first, but I learned quickly that people can feel passionately about e-mail and computer conferences. I've become one of them. I care about these people I met through my computer, and I care deeply about the future of the medium that enables us to assemble.[36]

Example 2

Solutions built with Microsoft Office provide powerful tools for accessing, sharing and understanding information, allowing users to make better decisions and streamline business processes. A robust development platform for business solutions, Office for Windows® 95 offers a rich set of programmable objects, a common programming environment using Visual Basic® for Applications, and full support of OLE automation.

Today, more than 500,000 developers are creating solutions using Microsoft Office. These developers are helping companies of all sizes deploy line-of-business solutions more quickly, for less money, and using the familiar productivity tools with which their employees work every day. The result is better decisions, faster response, and a new ability to build business advantage through smart investment in technology.[37]

8. Write a letter of complaint about a product or a service. Establish a common ground in the opening of the later, explain the problem in the body of the letter, make a specific request in the closing of the letter. Maintain a professional and courteous tone throughout the letter.
9. Write a letter of complaint about a product or a service. Use an angry and accusatory tone throughout the letter.
10. For each of the following passages, discuss the use of humor in technical prose. How effectively does the writer use humor? Is the humor used appropriately?

Example 1

Retrieving Files from the Recycle Bin
One of the sad realities of computing life is that sometime, somewhere, you'll accidentally delete some crucial file that you'd give your eyeteeth to get back. Well, I'm happy to report that you can keep your teeth where they are because Windows 95's Recycle Bin is only too happy to restore the file for you.... Think about it: if you toss a piece of paper in the garbage, there's nothing to stop you from reaching in and pulling it back out. The Recycle Bin operates the same way: it's really just a special, hidden folder (called Recycled) on your hard disk. When you delete a file, Windows 95 actually moves the file into the Recycled folder. So restoring a file is a simple matter of "reaching into" the folder and "pulling out" the file. The Recycle Bin handles all this for you (and even returns your file *sans* wrinkles and coffee grounds).[38]

Example 2

Please try to be nice to the Technical Support staff. They're a fine bunch of hard-working, dedicated computer geeks...er, nerds...er, people. They didn't write the program or the manual, so any problems you have aren't really their fault. But they will do their best to help you in every way they can to get your games going. Yelling, screaming, swearing, death threats and bomb threats may result in a lower quality of technical support.[39]

11. Monitor a newsgroup on the Internet and find an example of several people "flaming" (verbally abusing) each other on one thread (one topic of discussion). Print the e-mail samples and turn them in with several paragraphs discussing the tones of the e-mail messages and how these tones are achieved.

Notes

1. John S. Fielden, "'What do you mean you don't like my style?'" *Harvard Business Review*, 60.3 (May–June 1982): 129.
2. Jacques Barzun, *Simple & Direct: A Rhetoric for Writers*, rev. ed. (New York: Harper & Row, 1985) 90.
3. David W. Ewing, *Writing for Results in Business, Government, the Sciences and the Professions*, 2nd ed. (New York: Wiley, 1979) 172.
4. Reginald L. Cook, *Passage to Walden*, 2nd ed. (New York: Russell & Russell, 1966) 216.
5. Katie Wales, *A Dictionary of Stylistics* (New York: Longman, 1989) 364.
6. William Zinsser, *On Writing Well: An Informal Guide to Writing Nonfiction*, 5th ed. (New York: HarperPerennial, 1994) 27.

7. Dona Hickey, *Developing a Written Voice* (Mountain View, CA: Mayfield 1993) 1.
8. Hickey, p. 1.
9. Lewis Thomas, *The Fragile Species* (New York: Collier, 1992) 54.
10. Lewis Thomas, "Warts," *The Medusa and the Snail: More Notes of a Biology Watcher* (New York: Bantam, 1980) 61.
11. Richard Lanham, *Revising Prose,* 3rd ed. (New York: Macmillan, 1992) 103.
12. Zinsser, p. 166.
13. Zinsser, p. 175.
14. Zinsser, p. 175.
15. *Merriam-Webster's Tenth Collegiate Dictionary and Thesaurus. Electronic Edition. CD-ROM. 1995.*
16. *Merriam-Webster's Tenth Collegiate Dictionary and Thesaurus. Electronic Edition. CD-ROM. 1995.*
17. Karen Lane, David Foster, and Heather Stuart. *LView Pro 1.6 User's Guide* (Merritt Island, Florida: DKH Productions, 1995) 1.
18. Ewing, p. 187.
19. Woody Leonhard, *The Underground Guide to Word for Windows™: Slightly Askew Advice from a WinWord Wizard* (Reading, MA: Addison-Wesley, 1994) 19.
20. *RoboHELP User's Guide* (La Jolla, CA: Electron Image, Inc., 1994) 290, 292, 294, 296.
21. Cook, p. 216.
22. *Merriam-Webster's Tenth Collegiate Dictionary and Thesaurus. Electronic Edition. CD-ROM. 1995.*
23. Richard Eastman, *Style: Writing and Reading as the Discovery of Outlook,* 3rd ed. (New York: Oxford UP, 1984) 101.
24. Zinsser, p. vii.
25. Howard Rheingold, *The Virtual Community: Homesteading on the Electronic Frontier* (Reading, MA: HarperCollins, 1994) 1.
26. Henry David Thoreau, *Walden and Civil Disobedience,* ed. Owen Thomas (New York: Norton, 1966) 1–2.
27. Michel de Montaigne, *The Complete Essays of Montaigne,* trans. Donald M. Frame (Stanford: Stanford UP, 1989) 240.
28. See Ewing, pp. 173–201, for a more detailed discussion of similar strategies.
29. Ewing, p. 197.
30. Zinsser, p. 230.
31. Stephen L. Nelson, *Field Guide to Microsoft Windows 95* (Edmond, WA: Microsoft, 1995) 48.
32. Tom Rugg, *LANtastic Made Easy* (Berkeley: Osborne McGraw-Hill, 1992) 15.
33. Document from files with F-Prot program, version 2.21.
34. Vincent G. Dethier, *To Know a Fly* (New York: McGraw-Hill, 1962) 8–9.
35. John Zuchero, "Computer-Based Training: Reinvent Yourself through Multimedia." *Intercom* (Mar. 1996): 18.
36. Rheingold, p. 1.
37. *Build Better Solutions for Your Business.* Microsoft, 1995. Part number 098–60680.
38. Paul McFedries, *The Complete Idiot's Guide to Windows® 95* (Indianapolis: Que, 1995) 134.
39. Tom Bentley, *SimFarm Addendum and Quick Start Guide* (Walnut Creek, CA: Maxis, 1993).

Chapter 10

Considering Style and Bias

Science is rooted in creative interpretation. Numbers suggest, constrain, and refute; they do not, by themselves, specify the content of scientific theories. Theories are built upon the interpretation of numbers, and interpreters are often trapped by their own rhetoric. They believe in their own objectivity, and fail to discern the prejudice that leads them to one interpretation among many consistent with their numbers.[1]
—STEPHEN JAY GOULD

The reason the practice of assigning masculine gender to neutral terms is so enshrined in English is that every language reflects the prejudices of the society in which it evolved, and English evolved through most of its history in a male-centered, patriarchal society.... What standard English usage says about males, for example, is that they are the species. What it says about females is that they are a subspecies.[2]
—CASEY MILLER AND KATE SWIFT

For all genre of documents, technical communication researchers will need to inquire into culturally determined differences of audience values and expectations. For example, North Americans have certain needs and expectations when they begin to read (or look at) instructions, but do audiences of North Africa or in other regions of the world have these same needs and expectations? Are there any universals of technical communication rhetoric?[3]
—TIMOTHY WEISS

Chapter Overview

Bias in Technical Prose
Kinds of Bias
Political Correctness
Chapter Summary

Case Study 7: Gender-Neutral Language
Questions/Topics for Discussion
Exercises
Notes

Bias in Technical Prose

> *Ideology informs the rhetorical choices that a technical writer makes. After all, the language of science and technology (for all its objective and predictive power) is not free from subjective bias; even the conventional symbols (circles, pyramids, arrows, Cartesian coordinates and axes, etc.) that underlie the pictographic depiction of cultural, biological, and mechanical systems often serve to misrepresent what they innocently represent.*[4]
>
> —W. JOHN COLETTA

We have discussed in several earlier chapters the paradigm shift in technical communication—the shift from the view that writing scientific and technical prose is an objective enterprise to the view that writing scientific and technical prose is fundamentally a rhetorical activity. (See Chapter 1 and Chapter 4.) This new view recognizes that *practicing* science and *creating* technology as well as *writing about* science and technology are human, social, and persuasive activities. This new view accepts the fact that, like it or not, writing technical prose is an inherently biased activity. Your technical prose is biased because you advocate a point of view, offer an interpretation, decide to convey information one way over many other possibilities.

One of the best studies of unconscious bias in science is Stephen Jay Gould's *The Mismeasure of Man*. Gould's major concern is "biological determinism, the notion that people at the bottom are constructed of intrinsically inferior material (poor brains, bad genes, or whatever)."[5] Among other things, Gould exposes the biases of craniometry (measurement of the skull and its contents) in the nineteenth century and intelligence testing in the twentieth century. Gould observes:

> The leaders of craniometry were not conscious political ideologues. They regarded themselves as servants of their numbers, apostles of objectivity. And they confirmed all the common prejudices of comfortable white males—that blacks, women, and poor people occupy their subordinate roles by the harsh dictates of nature.[6]

Gould shows repeatedly how the work of such craniometrists as Louis Agassiz, Samuel George Morton, and Paul Broca, and the intelligence testing research of Alfred Binet, H. H. Goddard, Lewis M. Terman, R. M. Yerkes, and Cyril Burt were unintentionally biased, "leading scientists to invalid conclusions from adequate data."[7]

Gould's examples of bias in science do not undermine science; instead, Gould illustrates how "facts are not pure and unsullied bits of information; culture also influences what we see and how we see it. Theories, moreover, are not inexorable inductions from facts. The most creative theories are often imaginative visions imposed upon facts; the source of imagination is also strongly cultural."[8]

Also, neither Gould's book nor this chapter is concerned with conscious fraud. As Gould observes, "fraud is not historically interesting except as gossip because the perpetrators know what they are doing and the *unconscious* biases that record subtle and inescapable constraints of culture are not illustrated."[9]

Of course, this chapter is concerned with more than the cultural biases of science or the fact that as a writer of technical prose you cannot escape the human or rhetorical point of view. Yes, as writers, we are all biased. Writers of technical prose are, like all other writers, products of their culture, and like other writers, they reflect, both intentionally and unintentionally, all levels and kinds of bias in their prose. Yet, as you will see in this chapter, some biases are worse than others. To write more competent and sophisticated technical prose, you must be aware of a whole range of biases—gender, corporate, philosophical, religious, political, racist, cultural, agist, biases against the physically and mentally disadvantaged—and how these biases may prevent you from communicating the message you want to communicate to your audience.

A Definition of Bias

What is bias? *Bias* is defined as "an inclination of temperament or outlook; especially: a personal and sometimes unreasoned judgment."[10] A synonym is prejudice. Prejudice is defined as a "preconceived judgment or opinion" or "an adverse opinion or leaning formed without just grounds or before sufficient knowledge," "an instance of such judgment or opinion," and "an irrational attitude of hostility directed against an individual, a group, a race, or their supposed characteristics."[11]

To understand what bias is more fully, let's also consider some contrasting terms: *fair, just, equitable, impartial, unbiased, dispassionate,* and *objective*. All of these terms "mean free from favor toward either or any side."[12] And there are important and subtle differences among these terms:

> FAIR implies an elimination of one's own feelings, prejudices, and desires so as to achieve a proper balance of conflicting interests (a *fair* decision). JUST implies an exact following of a standard of what is right and proper (a *just* settlement of territorial claims). EQUITABLE implies a less rigorous standard than JUST and usu. suggests equal treatment of all concerned (the *equitable* distribution of the property). IMPARTIAL stresses an absence of favor or prejudice, an impartial third party. UNBIASED implies even more strongly an absence of all prejudice (your *unbiased* opinion). DISPASSIONATE suggests freedom from the influence of strong feeling and often implies cool or even cold judgment (a *dispassionate* summation of the facts). OBJECTIVE stresses a tendency to view events or persons as apart from oneself and one's own interest or feelings (I can't be *objective* about my own child).[13]

To the degree people can eliminate their feelings, prejudices, or desires, these terms apply and are helpful to know. After all, every day many people in all kinds of activities and in most lines of work do their best to be fair, just, equitable, impartial, unbiased, dispassionate, or objective. But are we ever able to eliminate our feelings, prejudices, or desires completely? Are we ever able to be completely free of bias of one kind or another?

As the earlier definition of bias suggests, bias in its mildest form is simply an inclination, opinion, or premature judgment. All of us have preconceived notions about most matters. It's impossible for us to be free of opinions, feelings, emotions, or judgments. Stated simply, we are biased because we are human, not automatons. Of course, bias may be much more than an inclination. Bias in its worst form is an outright, unfounded, irrational prejudice. Many of us may pride ourselves on being free of such prejudice against minorities, the physically disadvantaged, or some other group, but, at the same time, many of us often take an unfounded dislike to certain individuals. In brief, all of us have biases of one kind or another.

Levels of Bias

As we discussed, there are many levels of bias. At its worst a bias may be a blatant prejudice. Another kind of bias is slanting. One meaning of slanting is "to interpret or present in line with a special interest."[14] Another meaning of slanting is "to maliciously or dishonestly distort or falsify."[15] Fraud in science is certainly one kind of slanting. In this chapter we will discuss some examples of blatant prejudice—whether it's racist, sexist, political, or some other kind—and we will discuss some examples of slanting. But the primary concern here is with the far more subtle ways we reflect our culture in our prose. We will look at some blatant examples, but the majority of the discussion will be on the less obvious ways we often convey bias even in technical prose.

Kinds of Bias

As we have discussed, many kinds of bias underlie technical prose, just as they do other kinds of prose. Some of the major kinds of bias—gender, corporate, philosophical, religious, political, racist, cultural, agist, and biases against the physically and mentally disadvantaged—are briefly discussed in this section.

Gender Bias

Gender bias is perhaps the most accurate term to reflect the whole range of stereotypes and other kinds of sexism in society, including sexist language. Tom McArthur observes that *gender bias* is "a term in sociology and women's studies for bias associated with sexual roles in society and gender terms in language. It extends the grammatical term *gender* to cover language-related differences in the behaviour of women and men and in perceptions of that behaviour. Such perceptions are expressed through casual stereotyping, as in: 'Well, she's supposed to be back by now but she's probably stopped off somewhere to gossip. You know how women are.'"[16]

Few today question the fact that both our culture and our language are biased against women. David Crystal comments, "There is now a widespread awareness, which was lacking a generation ago, of the way in which language covertly displays social attitudes towards men and women. The criticisms have been mainly directed at the biases built into English vocabulary and grammar which reflect a traditionally male-orientated view of the world, and which have been interpreted as reinforcing the low status of women in society."[17]

What is sexist language? Casey Miller and Kate Swift suggest that "sexist language is any language that expresses such stereotyped attitudes and expectations, or that assumes the inherent superiority of one sex over the other."[18]

In *The Handbook of Nonsexist Writing,* Miller and Swift discuss many examples of sexist language. They discuss the generic *man* in examples such as "A man who lies constantly needs a good memory," or "No man would be safe from nuclear fallout if...."[19] They discuss *man* in prefixes such as "Will mankind murder Mother Earth or will he redeem her?" and suffixes such as *chairman.*[20] They mention job titles such as *hat-check girl, stewardess, maid,* and *seamstress.* They give considerable attention to the pronoun problem, suggesting alternatives to the "fatally flawed" *he.*[21] They devote another chapter to generalizations such as "The average person finds it no problem at all to have three head colds, one sunburn, an attack of athlete's foot, 20 headaches, three hangovers and five temper tantrums with adolescent children, and still get in his 61 hours of shaving...."[22] They cover the difficulties for some in seeing women and girls as people ("I'll have my girl run off some copies right away"), and show how "sex-linked adjectives are not very useful in describing the human traits and qualities both sexes share."[23] And they comment on the commonplace nonparallel terminology in side-by-side references to the sexes, as in the example, "Three Stanford University students—two girls and a man—were abducted from a research station in Africa."[24]

Of course, Miller and Swift's definition of sexist language does not ignore the fact that gender bias may be anti-male as well as anti-female. Eugene August notes that "sex bias or sexism itself is widely held to be a male-only fault. When *sexism* is defined as 'contempt for women'—as if there were no such thing as contempt for men—the definition of *sexism* is itself sexist...."[25] He comments on three kinds of anti-male usage.

First, gender-exclusive language that omits males is common. August mentions *alma mater,* which means "nourishing mother" to refer to a university and *mammal,* which "clearly omits the male of the species" because it categorizes "animals according to the female's ability to suckle the young through her mammary glands."[26] Additionally, the male as victim is often ignored in many familiar expressions including as *innocent women and children* and *rape victim,* which August suggests, "means females only."[27]

Second, gender-restrictive language often limits males to a socially prescribed gender role. August discusses, for example, the word *coward.* He comments that it is a "word applied almost exclusively to males in our society, as are its numerous variants—*chicken, chickenshit, yellow, yellow-bellied, lily-livered, weak-kneed, spineless, squirrelly, fraidy cat, gutless wonder, weakling, butterfly, jellyfish,* and so on."[28] August also suggests that "if women in our society have been regarded as sex objects, men have been regarded as success objects, that is, judged by their ability to provide a standard of living."[29]

Third, anti-male bias occurs in the form of negative stereotyping. August suggests that crimes and vices of all kinds are typically attributed to males. August notes such expressions as *wife and child abuse.* Such a phrase "ignores battered husbands, but it does more: it suggests that males alone abuse children. In reality most child abuse is committed by mothers...."[30] As for vices, all kinds of words for inebriated are associated more with males than females: *boozer, drunkard, tippler, toper, swiller, tosspot, guzzler, barfly, frunk, lush, boozehound, souse, tank, stew, rummy,* and *bum.*[31]

Numerous changes have occurred over the past few decades to make English more sensitive to both sexes. Words such as *chairman* have become *chair* or *chairperson, miss* has been replaced by *ms.,* and *salesman* has become *sales assistant, sales clerk,* or *salesperson.* Many changes in job titles have become a legal requirement in most states. Of course, debate continues concerning how much change should occur, for example, whether *manhole cover, man in the street,* or *Neanderthal Man* should be changed.

Gender Stereotypes

Alleen Nilsen discusses three categories of stereotypes of women: 1) Women are sexy; Men are successful; 2) Women are passive; Men are active; and 3) Women are connected with negative connotations; Men with positive connotations.[32] Nilsen arrived at these conclusions by carefully reading a dictionary. She discovered pairs of words in which the feminine word has acquired sexual connotations while the masculine word has become respectable or businesslike. She cites the word *callboy* as an example, explaining that a callboy is the person who calls actors on stage. She then cites the word *callgirl,* which is obviously the feminine counterpart to *callboy* and which means prostitute. Also, compare the words *sir* and *madam.* Nilsen suggests that "sir is a term of respect, while madam has acquired the specialized meaning of a brothel manager."[33] Similarly, a man who is a *stud* (literally, a breeding stallion) is highly regarded, while a woman who is a *bitch* (a female dog) is thought of derogatively.

The category of "Women are passive; Men are active" is illustrated by our society's marriage ceremony. For example, the bride is "given away," almost always by a man. Also, the language of the ceremony shows the sexist nature of our society. Note that the phrase "man and wife" is not consistent. The phrase would make more sense as "man and woman" or "husband and wife." Even the fact that the woman takes on the man's name (and, therefore, his identity) shows the passivity of women.[34]

In the workplace, women are seen as unnaturally aggressive if they appear to be as ambitious as their male counterparts. Traditionally, women have occupied the passive careers (secretary, nurse, teacher) while men have established themselves as active professionals (lawyer, doctor, police officer). In fact, many publications still qualify females who enter traditional "male" professions. Newspapers often state "female lawyer" or "woman doctor." In the same way, they also qualify males in traditional "female" professions, such as referring to a "male nurse" or a "male secretary."

The third category of stereotypes—"Women are connected with negative connotations; Men with positive connotations"—is also well documented. Barbara Moely and Kimberly Kreicker asked students to rate a variety of words on several scales. For these students the word *gentleman* implies greater competence than the word *man,* while *lady* tends to suggest less competence than the word *woman.* In addition, the word *gentleman* is considered a "high status term" and is perceived as more flattering than its counterpart *lady.*[35]

Nilsen cites the example that when a young girl is told to "be a lady" she is being told to sit up straight with her legs together and to "be quiet and dainty."[36] But when a young boy is told to "be a man," he is being told to be "strong and virtuous—to have all the qualities that the speaker looks on as desirable."[37] In our society, masculine qualities are considered positive qualities.

In addition to stereotypes of women reinforced by our language, women and men have been stereotyped as using language in the following ways, among others:

1. Women tend to use such words as *adorable, cute, lovely, sweet* in describing people and objects and such vocatives as *my dear, darling, sweetie.*
2. Men tend to be more direct, less inclined to show their feelings, and more likely "to call a spade a spade." Tradition also requires them to be laconic: *men of few words; the strong, silent type.*
3. Women have often engaged in an "overflow" of adjectives and adverbs, found in an extreme form in the usage of society women between the world wars: My dear, it's just too simply wonderful to see you!
4. Women are often eager to talk about feelings and emotions in a way thought of as "gushing," while many men are almost tongue-tied in such matters.
5. Women frequently use *so, such, quite* as intensifiers (It's been so nice to see you again, and such a pleasure to meet the children—I'm really quite thrilled), or as qualifiers (Well, he's so, you know, so helpful, and it's such a shame he can't be here—I'm well, quite upset about it).
6. Women are considered to be more polite, using phrases such as *could you please,* and more concerned about "correct" and "proper" grammar and pronunciation.
7. In conversations, women are said to be by turn insecure and hedging (as shown by tags such as *do you?* or *isn't it?,* and qualifiers such as *I think*) and overbearing, talking and interrupting more than men do.
8. Women's "delicate sensibilities," especially in the middle classes, have traditionally kept them from using obscene or blasphemous language, and restricted its use by men in their presence.
9. Women are more likely to use polite euphemisms for topics such as death and sex.
10. Men typically talk about "important," "worldly" topics such as politics, sports, and war, whereas women's talk is "trivial" and usually "gossip."[38]

Ways to Make Writing Gender-Neutral

The debate continues whether English should be or can be gender-neutral. See, for example, Irving Younger's article in Case Study 7 at the end of this chapter for an argument that English is sex-neutral. Opinions on even the most basic issues vary widely. Compare, for example, William Strunk and E. B. White's advice on the generic *he* with the opinion of Casey Miller and Kate Swift. Strunk and White comment: "The use of *he* as pronoun for nouns embracing both genders is a simple, practical convention rooted in the beginnings of the English language. *He* has lost all suggestion of maleness in these circumstances. The word was unquestionably biased to begin with (the dominant male), but after hundreds of years it has become seemingly indispensable. It has no pejorative connotations; it is never incorrect."[39] In contrast Miller and Swift observe: "Use of the pronouns *he, his,* and *him* to refer to any unspecified or hypothetical person who may be either female or male is usually justified on two grounds. First, the practice is said to be an ancient rule of English grammar long and faithfully followed by educated speakers and writers. Second, it is asserted—

somewhat paradoxically, if the usage is thought to distinguish the educated from the uneducated—that everybody knows *he* includes *she* in generalizations. Historical and psychological research in the past few years has produced evidence to refute both claims."[40]

Whether you agree that English is inherently sexist or that a gender-neutral English is a good idea, you still must be sensitive to your audience. Words you may not think are sexist may be viewed as such by a majority of your audience. You can tell yourself that you're right and your audience is wrong, but this position will not help you to communicate with your audience. One simple way of summarizing the advice here is to suggest that the language you use is sexist if your audience perceives it to be sexist.

Specific advice on how to make your writing more gender-neutral can be found in many sources. David Crystal provides a good list below:

- Avoid so-called masculine generics such as the pronoun *he* with sex-indefinite antecedents or *man* and its compounds (except in unambiguous reference to males).
- Avoid using genuine generics as if they referred only to males (e.g., Americans use lots of obscenities but not around women).
- Avoid adding modifiers or suffixes to nouns to mark sex of referents unnecessarily. Such usage promotes continued sexual stereotyping in one of two ways: by highlighting referents to sex, modification can signal a general presupposition that referents will be of the other sex *lady professor, male secretary*), and thus that these referents are aberrant; and conventionalized gender-marking "naturalizes" the presumptive or unmarked sex of the noun's referents (*stewardess, cleaning lady*).
- Use parallel forms of reference for women and men, e.g., do not cite a male scholar by surname and a female scholar by the first name plus surname.
- Avoid gender stereotyped or demeaning characterizations, e.g., presenting men as actors and women as passive recipients of others' actions. Men are frequently the agents, women the recipients, of violent acts. We recommend that the portrayal of violent acts be avoided altogether, regardless of the sex or species of the participants.
- Avoid peopling your examples exclusively with one sex.
- Avoid consistently putting reference to males before reference to females. Not only does this order convey male precedence, in English it will put males in subject position and women in object position.
- Avoid sexist (or otherwise derogatory) content in examples (e.g., *The man who beats his mistress will regret it sooner than the man who beats his wife*).
- Ask yourself whether you have remembered to cite or acknowledge women as well as men whose own research is relevant or whose comments may have helped you. Given traditional views of men's and women's place in intellectual endeavour, there is a danger that ideas advanced by women and adopted by men will be remembered as having originated with men, and that more generally women's intellectual contributions will tend to be underestimated.[41]

Corporate Bias

The corporate world consists of many different discourse communities even within one company. As an employee you often write representing the views of one discourse community before another. At the simplest level, you may be writing to another employee in a different department who also happens to be a friend of yours. You might be writing to remind your co-worker about an upcoming social event. In this case, you are more than likely representing only yourself and not projecting any kind of corporate image.

More frequently, however, your communications are corporate communications of one kind or another. You might be writing to a co-worker and representing your department's or your team's position on an issue. You could be writing to your supervisor or to your supervisor's supervisor on a matter of vital interest to your job security. Or you could be writing as a representative of the company to someone outside of the company, for example, a vendor with whom the company does frequent business.

In all of these examples and many other possible scenarios, you are writing within a corporate context. You are representing an interest within the company to another party in the company or you are representing the company to the world outside. In all of these cases, you are reflecting a corporate bias. The ethos you project in your documents is most often not just your ethos, but an ethos that will gain you an advantage, that will aid your point of view.

A typical example of corporate bias may be found in a letter from the Commissioner of the Internal Revenue Service:

Dear Taxpayer:

Last year we at the IRS made a commitment to serve you better in several areas. I'm pleased to report that we've made significant progress.

We said we'd make it easier to file a tax return. Last year more than 11 million taxpayers filed electronically, nearly 3 million taxpayers used a short machine-scored answer sheet and 700,000 filed by telephone. This year telephone filing will be available to about 20 million 1040EZ filers throughout the United States.

We made a commitment to issue refunds within 21 days to taxpayers who filed electronically and within 40 days to those who filed on paper. We came close to achieving this goal, but did have to slow the process sometimes. This extra effort kept us from paying out $400 million in improper refunds, but it also forced us to miss our deadline at times. I apologize to those taxpayers who were inconvenienced.

Providing information about our tax laws or your account status when you want it is another of our priorities. Last year we responded to 118 million taxpayers, an increase of nearly 60% from the year before. Automated information was always available. Information on refunds was available 16 hours each day, and IRS personnel could be reached by phone for 10 hours each business day.

We've made real progress, and we remain committed to doing even better. We appreciate your suggestions about how we can do that.

The writer is representing the biases of the Internal Revenue Service (IRS), her employer. Some of these biases are the belief that taxpayers should cheerfully comply with the laws for paying taxes and that taxpayers should consider the IRS personnel their helpful friends who are eager to assist them in preparing their tax forms. Another professed bias of the government agency is that it exists for the benefit of citizens (or taxpayers, in this case).

The writer expresses her awareness that she is representing the agency when she repeats the IRS commitment to better service. She justifies the slowness of refund processing by appealing to another, unstated but still evident, bias of the IRS: unwillingness to allow "improper refunds." IRS personnel believe they have to be viewed as responsive to taxpayers to overcome the negative view many taxpayers have of them. By touting the agency's availability by phone and ignoring the incessant busy signal the average taxpayer encounters, the writer projects the proper corporate image of a caring, helpful executive, representing a responsible, concerned organization.

Philosophical or Religious Bias

Philosophical means "of or relating to philosophers or philosophy" or "based on philosophy."[42] Religious means "relating to or manifesting faithful devotion to an acknowledged ultimate reality or deity," or "of, relating to, or devoted to religious beliefs or observances."[43] Of course, many writers write with a philosophical or religious bias. Indeed, for countless writers a philosophy or a religion is their principal subject. Marcus Aurelius wrote advocating his stoicism, Ralph Waldo Emerson spoke out for his transcendentalism, William James expounded upon his pragmatism, and Jean-Paul Sartre defended his brand of existentialism. Thomas Merton wrote about Catholicism, and Philip Kapleau writes about Zen Buddhism.

Many writers, of course, write against other philosophies or religions in favor of one philosophy or religion. And many writers reveal one blatant kind of religious prejudice or another. One of H. L. Mencken's biographers, Carl Bode, for example, comments on Mencken's diaries in which "he displays a general anti-Semitism ("The Jews of Wall Street, The Jews of Hollywood") that would have distressed his friends and gladdened his enemies."[44]

Religious bias can, of course, be far more subtle than that of Mencken's diaries. David Crystal comments, for example, on "the language of religion, where a male-dominated conception of God has been handed down from patriarchal times. Along with it has come the attributes which are stereotypically associated with men, such as toughness, coolness, and authority. Missing are attributes such as caring and weeping. God, it seems, could not possibly cry for a lost creation."[45] Some would suggest that the very language we use to discuss God begins with biases of one kind or another. Casey Miller and Kate Swift comment: "What a growing number of theologians and scholars are saying is that the myths of the Judeo-Christian tradition, being the products of patriarchy, must be reexamined, and that the concept of an exclusively male ministry and the image of a male god have become idolatrous."[46]

In general, religious bias is not a common problem in technical prose (except, again, in the area of international technical communication, as we discuss under **Cultural Bias** later in this chapter). Of course, some companies cater to followers of one religion, and the biases in communications of these companies are self evident.

Philosophical bias also occurs in many kinds of ways. Within a company a philosophical bias may be something as simple as a company slogan, such as "The customer always comes first" or "Quality products are our highest priority." In many companies where such slogans are adopted, employees view them cynically as yet another ploy by the company. Yet there are many companies where such slogans are taken seriously. In these companies these "philosophies" can affect all kinds of internal correspondence as well as documentation made available to the world outside. Companies that will not allow substandard documentation to be shipped with computer hardware or software (there are some companies like this, aren't there?) show how a philosophy may have a direct effect on how technical prose is written.

Political Bias

Political bias may be as simple as creating internal e-mail, memos, and letters that show a sensitivity toward office politics (the corporate world and academe are both very political) or as complex as interpreting the human condition from a political point of view, as in Marxism, for example. As with the other kinds of biases discussed here, political bias may be low key or extreme.

Propaganda is one type of writing in which the political biases are usually blatant. Donna Cross warns us about some common types of propaganda that most of us are unaware of. She suggests, "People are bamboozled mainly because they don't recognize propaganda when they see it."[47] Some common techniques she refers to are name-calling ("labeling people or ideas with words of bad connotation"[48]), glittering generalities (using words with good connotations "to get us to *accept* and *agree* without examining the evidence"[49]), plain folks appeal ("a speaker tries to win our confidence and support by appearing to be a person like ourselves"[50]), *argumentum ad hominem* (distracting "our attention from the issue under consideration with personal attacks on the people involved"[51]), bandwagon (urging "us to support an action or an opinion because it is popular"[52]), and testimonial ("having some loved or respected person give a statement of support [testimonial] for a given product or idea"[53]).

Of course, much propaganda is used to deceive, and the language of doublespeak is filled with all kinds of clever deception. Chapter 11 will discuss this deceptive aspect of style in detail.

Racial Bias

Racial bias is concerned with far more than perceptions of blacks among whites or whites among blacks. There are, after all, many races, ethnic groups, and nationalities, and there are many kinds of racial bias even within one race, ethnic group, or nation. Note, for example, the prejudice of colorism, a bias held by lighter skinned people against darker skinned or vice versa. This prejudice exists not only in the black community but in Hispanic, Asian, and white communities as well.

Racial bias is expressed in many subtle and not so subtle ways. Of course, racial prejudice is commonly expressed as an insult. Linguists Keith Allan and Kate Burridge comment in their study of racist vocabulary:

> So far as we know, all human groups have a derogatory term available in their language for at least one other group with which they have contact.... Among the racist dysphemisms of English, are: *mick* for Irish person (or in Australia, a Roman Catholic), *frog* (Cockney *jiggle and jog*) for a French person, *kraut* and *hun* for a German, *chink* (Cockney *widow's wink*) for a Chinese, *jap* or *nip* (Cockney *orange pip*) for a Japanese,

paki for a Pakistani, *polaks* for a Pole, *wop* (Cockney *grocer's shop*) and *eyetie* for an Italian, *ayrab, dune coons,* and *camel jockeys* for an Arab, *kike, yid* (Cockney *dustbin lid* and *four-by-two*) for a Jew, *chief, Hiawatha,* and *Geronimo* for male North American Indians and *squaw* for their womanfolk, and so forth. English whites may use *black, nigger, nignog, wog* (Cockney *spotty dog*), *coon* (Cockney *silvery moon*), and so on for people of African ethnicity and for other people with similar skin color to Africans, such as Australian Aborigines and south Indians.[54]

This kind of dysphemistic language (substituting offensive expressions for inoffensive ones) will probably always be with us. People fear what they do not understand and what is different from themselves.

More intriguing, however, are the more subtle types of racial bias. William Raspberry comments on the very effects of the words *white* and *black* in our culture. He argues that our perceptions of black in our culture are far too narrow. According to Raspberry, black youngsters associate black with success in music and athletics, for example, but they don't associate black with success in business, in academia, and the nonentertainment professions. Raspberry believes the results of this type of association "are devastating."[55] He comments, "One reason black youngsters tend to do better at basketball...is that they assume they can learn to do it well, and so they practice constantly to prove themselves right."[56] Raspberry asks, "Wouldn't it be wonderful if we could infect black children with the notion that excellence in math is 'black' rather than white, or possibly Chinese?"[57] He believes, "There is no doubt in my mind that most black youngsters could develop their mathematical reasoning, their elocution and their attitudes the way they develop their jump shots and their dance steps: by the combination of sustained, enthusiastic practice and the unquestioned belief that they can do it."[58]

Cultural Bias

The culture in which you write influences the way you write your technical documents. Cultural norms vary from country to country. What is permissible in one country may be frowned upon in another. As Mohsen Mirshafiei observes, "In the United States (and in many industrialized countries), technical communication requires simple, concise expression and clear thinking. Conflict with these requirements can arise in persons whose culture values detailed, subjective analyses and excessively philosophical argumentation."[59] Mirshafiei discusses the example of Middle Eastern students: "Middle Eastern students are highly rhetorical, and they use a highly complex and decorative language in their communication."[60] Stated another way, "Arabs, Indians, Pakistanis, and Persians are encouraged by their culture to use a highly decorative and often hyperbolic language to express themselves. Americans, however, prefer to be more explicit in communicating with one another."[61] He suggests that "some of the culturally related devices that a number of our foreign students often use are exaggeration, generality, circular structures, and digression from the central idea in their writing."[62]

Mirshafiei refers to a reference letter he received from Iran to illustrate some of the differences:

> Now that Afshin wants to continue his studies it is impossible here in Iran due to the political turmoil, unrest, war, religious persecutions and the discrimination against the religious minorities. Therefore he wants to continue his studies in a country like the United States where they draw tremendous power from 'Faith and Freedom.' So then any help done to him will be a service done to the reward and enrichment of the humanity at large. Hence I earnestly request you to give him a seat in your esteemed University. The Good God shall bless you a hundredfold for all your endeavors. Thanking you in anticipation for all the services you render....[63]

This letter clearly shows the influence of a writer's culture. As Mirshafiei observes, "Facts in this letter are given in a highly exaggerated language, and generality has replaced specificity."[64]

Another example concerns the Japanese. Consider the following description of native Japanese in technical communication:

> Japan is highly homogeneous, and social cohesion is prized. By speaking in a vague manner, conflicting opinions can be implied, and so smooth social discourse is attained. Outright statements are considered abrupt and impolite; because of this, even obvious facts are often couched in probable terms. The phrase "ishin tenshin" refers to the "telepathic communication" that has sprung up because of Japanese homogeneity. When hearing a vague remark, the Japanese listener usually knows what was meant, through "ishin tenshin." However, Westerners have great difficulty in these contexts in determining whether an implication or a fact has been offered.[65]

Speakers of English have difficulty understanding Japanese "telepathic communication." Native speakers of English expect primarily a linear development of thought in their communications with others. In the same fashion, Japanese have much difficulty in understanding communications written in English.

Carolyn Boiarsky discusses other rhetorical conventions in international communication. She observes, "While cultural influences are less overt in written communication, they nonetheless inform the rhetorical conventions of a text, often regulating the content, organizational pattern and sequence, presentation of argument, style, and format."[66] She discusses, for example, the Asian custom of developing trust before engaging in business. This custom not only affects person-to-person contact but also the conventions of a letter: "Even a business letter begins with an inquiry into the reader's health, the state of the family, or some topic related to a reader's personal affairs. A letter providing information on a valve may begin, "We hope this letter is finding you in good health." The information about the valve may not appear until page two."[67]

Of course, writers of technical prose are presented with special challenges when it comes to translating technical documents into another language. As Timothy Weiss suggests, "Translation demands not only the analytic and interpretive ability required to understand the pattern of ideas of a source text but also the ability to re-express those ideas analogously within the world view of the target audience."[68]

Age Bias

Federal law prohibits companies from discriminating against individuals on the basis of their age. However, this fact does not prevent our culture as a whole, and companies in particular, from encouraging or practicing age discrimination. Just a casual glance at the target audiences for most television programs and advertising in all kinds of media, to mention only two examples, shows how we are a nation obsessed with youth or staying youthful. Of course, senior citizens are a large target audience for advertisers, but the focus of most advertising is for much younger groups. Advertisers have every right to advertise to any group they choose, but when certain groups of people are singled out with potentially harmful consequences, then the problem becomes one not only of bias but also of ethics. The advertising campaign concerning Joe Camel is one current example of a major tobacco company allegedly singling out teenagers in the hope of making cigarette smoking appealing. Whatever else this campaign is, it is at least biased toward the youth of our culture.

Bias against the elderly is even more common. Companies may not refuse to hire a prospective employee on the basis of age, nor can companies fire anyone on the basis of age either. Still companies have ways of forcing early retirements, transferring people, and in many other ways making them ineffectual on their jobs, so they may be let go for what appear to be other, more legitimate reasons. It's almost impossible to find examples of age bias in a company's policies or other internal communication because such documents are so sensitive and even subject to litigation. As a writer, you need to be sensitive to the age or ages of your audience, and you need to be careful to avoid using any language that calls attention to one age group over another unnecessarily or to avoid using language that reflects any age discrimination.

Bias against the Physically and Mentally Challenged

Two other large groups frequently victimized by language bias are the physically and mentally challenged. In our culture we often refer to people in both groups as *handicapped* or *disadvantaged* while most of the people in these groups resent such labels. Many feel they are neither handicapped nor disadvantaged; they are just different. Instead of *handicapped* or *disadvantaged, challenged* has become a more acceptable label. As in the case of age, Federal law prohibits discrimination against those who are physically or mentally challenged, but such discrimination continues nonetheless. As with other kinds of potential language discrimination, you have to be sensitive to perceptions within our culture and within physically and mentally disadvantaged groups. Avoid using language that unnecessarily calls attention to one group over another, and be sensitive to the labels groups prefer to be called.

Political Correctness

It's hard to conclude a discussion about levels and kinds of bias without also discussing political correctness (also referred to as PC or P.C.). Political correctness is a pejorative phrase referring to attempts by proponents to police words that are "overtly or covertly *sexist* (especially as used by men against women), *racist* (especially as used by whites against blacks), *ableist* (used against the physically or mentally impaired), *ageist* (used against any specific age group), *heightist* (especially as used against short people), etc."[69] In the 1980s many people became increasingly committed to decreasing prejudice, particularly language prejudice in these areas. By 1990, suggests David Crystal,

> Anyone who used vocabulary held to be "politically incorrect" risked severe condemnation by PC activists. Organizations, fearful of public criticism and litigation, went out of their way to avoid using language which might be construed as offensive.... The generic use of *man* was widely attacked.... *Mentally handicapped* people were to become people *with learning difficulties. Disabled* people were to become *differently abled. Third World* countries were to be *developing nations.* All but the most beautiful or handsome were *aesthetically challenged.* And, in the academic literary world there would need to be safeguards against the unhealthy influence wielded by such *DWEMs*('Dead White European Males') as Shakespeare, Goethe, and Moliere.[70]

People reacted to this legislation of terminology in expected ways. Some equated the "inflexible condemnation of 'incorrect' vocabulary" with "the 'thought police' of futuristic novels."[71] For some, political correctness became a weapon for silencing people. The whole concept remains controversial even today, with proponents of political correctness taking a strong stand against opponents. Crystal summarizes the current situation:

> Critics of PC believe that the search for a "caring" lexicon is pointless, as long as the inequalities which the language reflects do not change. Proponents of PC argue that the use of language itself helps to perpetuate these inequalities. At present, the speed at which fashions change in the use of PC terms suggest that it is not so easy to manipulate language as the reformers think. Dissatisfaction over one term tends to spread to its replacement, as has been seen with such sequences as *negro* to *black* to *Afro-American* to *African-American.*"[72]

Crystal and others believe that recent humor directed at political correctness may help serve to ease tensions. (See the discussion of humor in Chapter 9.) At the very least, the positions of both sides reveal how easy it is to associate biases with language.

Chapter Summary

Writing prose, whether it's nontechnical or technical, is an inherently biased activity. Your prose is biased because you advocate a point of view, offer an interpretation, decide to convey information one way over many other possibilities.

Yes, as writers, we are all biased. Writers of technical prose are, like all other writers, products of their culture, and like other writers, they reflect, both intentionally and unintentionally, all levels and kinds of bias in their prose. Yet some biases are worse than others. To write more competent and sophisticated technical prose, you must be aware of a whole range of biases—gender, corporate, cultural, philosophical, religious, political, and racist—

and how these biases may prevent you from communicating the message you want to communicate to your audience.

Case Study 7: Gender-Neutral Language

The issue of gender-neutral language is controversial. The following brief article, by Irving Younger, the Marvin J. Sonosky Professor at the University of Minnesota Law School, is one of many arguing that English is already a sex-neutral language.[73] Read the article and answer the questions that follow.

The English Language is Sex-neutral

Because the higher animals come by twos, their names tend to occur in couples: hen and rooster, cow and bull, woman and man. Should we wish to speak of both sexes together, we sometimes find both-sex terms at the ready such as chickens, cattle or people.

For some word-couples, though, no both-sex term exists. There are goose and gander, for example, but what shall we call geese and ganders mixed? Or ducks and drakes? Or lionesses and lions?

The rules of language tell us. One word of every word-couple serves as a both-sex term, including not only its own sex but, where the context requires, the other as well. "This goose lays eggs" refers only to a female of the species. "This is a goose farm" says that here are raised both geese and ganders.

The Unmarked Usage

The employment of one word of a word-couple as a both-sex term is called "the unmarked usage." The other word of the word-couple is "marked"—it means only the sex it denotes. Which of the members of a word-couple will be marked and which unmarked is a matter of language, not of social status or gender preference. For example, goose and duck, both female, and lion and man, both male, are all unmarked.

These considerations also apply to personal pronouns. He (or his or him) means a male, she (or her) a female. One of the couple is marked (that's she), the other unmarked (that's he). When Jesus said, "He that is without sin among you, let him first cast a stone at her," everyone knows that "he" and "him" meant any man or woman while "her" meant a particular woman.

For reasons of tone, rhythm or variety, writers frequently use the unmarked member of a word-couple. Pope entitles his poem "An Essay on Man" because "An Essay on Human Beings" falls flat. Darwin writes "The Descent of Man" because "The Descent of Men and Women" is redundant. Lincoln says, "Our fathers brought forth...a new nation," because "our parents" strikes a sour note. Pope, Darwin and Lincoln understood themselves to be speaking of women no less than of men and were confident that others would understand them exactly.

I am saying that sexism and the English language are unrelated. Between male and female the language is neutral. The marked–unmarked usage is an aspect of the internal economy of English, nothing more. Only to the uninformed do fisherman, fireman and handyman suggest that women may not wield hook, hose or hammer.

Five Defenses

So stick with the marked–unmarked usage. When you're challenged as a sexist, make a five-fold response.

1. The marked–unmarked usage is a quirk of language, not a tool of oppression. Virginia Woolf's *A Room of One's Own* is among the classics of feminism. In it, after stating that both men and women can be great writers, the author goes on to say, "The whole of the mind must be wide open if we are to get the sense that the writer is communicating his experience with perfect fullness." Virginia Woolf was no enemy to women. The words "his experience" hardly refer to the experience of men only. They refer to the experience of both men and women, as the context makes clear.
2. English has no adequate substitute for the marked–unmarked usage. "He or she" and "him or her" are clinkers. "S/he" is grotesque. Watchperson, draftsperson and fireperson sound ridiculous because they play false with the language. Nor is person the neutral word its proponents think it is. Person comes from the Latin *perona* and is grammatically feminine. Cicero wrote, "peronam in republica

teuri principis"—in English, "to be a leading man in the state"—using *persona* to mean man, not man or woman. The marked–unmarked usage avoids these tangles.
3. Because the rules on marked–unmarked usage are well-known and definite, they make for clarity. Violate them and the result is confusion. When a law review writer says, "A judge can be trusted not to abuse her discretion in these circumstances," you think first that only female judges can be trusted, then that the writer means "his or her," and finally that you're unsure.
4. Our language is alive. That's why Shakespeare's English is different from Tennyson's and Tennyson's from Auden's. The English of the future may turn out to handle gender better than our English does, but we must let it happen slowly. How can we cast out the old without first developing the new? To jettison the marked–unmarked usage overnight, with nothing to take its place, would cripple the language. We lawyers must not permit that to happen, for it is we who assert a professional competence to think, speak and write clearly about the complicated things. To do it, we need a language undiminished, flexible and precise, qualities preserved only when we obey the rules by which the language works.
5. Equality for women is a great cause. It demands of its champions determination, intelligence and energy. Then how foolish of them to waste effort on attacking the marked–unmarked usage, a linguistic, not a political phenomenon. Erase it from our language and you've neither enhanced the status of women nor increased the esteem in which they are held by the other half of mankind.

Questions

1. Irving Younger writes, "I am saying that sexism and the English language are unrelated. Between male and female the language is neutral." Do you agree or disagree? Why or why not?
2. Which of Younger's five defenses (if any) do you find convincing? Why? Which (if any) do find unconvincing? Why?
3. Younger believes marked and unmarked usage is "a linguistic, not a political phenomenon." What points made earlier in this chapter may be used to argue against Younger's position?
4. Do you detect any biases in Younger's article? If so, what are they?
5. What important points about sexist language and culture can be used to counter some of Younger's points?

Questions/Topics for Discussion

1. What is bias? What is bias in technical prose?
2. What is political correctness? Why is it so controversial? Do you think political correctness has gone too far? Why or why not?
3. What is gender bias? What is sexist language?
4. What are the moderate and extreme positions on sexist language?
5. Review the discussion of gender stereotypes. Which ones do you agree with and which ones do you disagree with? Why? Can you think of other gender stereotypes that were not mentioned?
6. Why do writers of technical prose need to be concerned about cultural bias?
7. Is corporate bias a big problem in technical prose? Why or why not?
8. How much should writers of technical prose be concerned about philosophical, religious, political, or racist bias? Are these biases big concerns for technical documents?
9. What are the kinds and levels of biases writers of technical prose should be most concerned about?
10. If writers are well informed about biases can they be free of biases? Why or why not?

Exercises

1. Find an example of fraud in science. Discuss how this fraud is different from the kind of unconscious bias Stephen Jay Gould comments on.
2. Look up the words *inclination, opinion, judgment, bias, prejudice, slanting, propaganda,* and *fraud* in a good dictionary. Discuss what the definitions of these words have in common and how they are different.
3. In the following report, how does the writer organize the information in order to increase its psychological effect on the audience? Are there any biases in the document? Why or why not?

   > We Norte-Americanos, who have become somewhat resistant to the "everything man-made causes cancer" cries of the environmentalists and the EPA, have taken thus far a sane attitude towards this. Berkeley biologist Bruce Ames has noted that if there is any risk of cancer from chlorinated water at all, it is a thirtieth that of a serving of peanut butter. But Peruvian bureaucrats bought the rhetoric wholesale and greatly reduced the chlorine pumped into the country's water supply.
   >
   > This set the stage for horror when, Pan American Health Organization officials suspect, a Chinese freighter released its cholera-contaminated bilge water into the harbor in Lima. Eventually the bacteria made its way into open wells which hadn't been chlorinated at all and in other fresh water supplies in which chlorine levels had fallen too low to kill the germ.
   >
   > As is usual with epidemics, the rich have fared considerably better than the poor. Peru's rich can afford private water supplies and bottled water. It is the poor, says Ghersi, who "have been sacrificed for the EPA and the environmentalists."
   >
   > Undaunted by all this, the EPA is now taking action that may well endanger much of the U.S. water supply.
   >
   > Citing the hypothetical cancer risks from chlorination, the agency in 1994 proposed a rule that would require water systems across the country to eliminate the process known as pre-disinfection as a means of controlling "disinfection by-products."[74]

4. Look up ten words in a good dictionary that appear to have definite feminine or masculine associations. Discuss why these words have these associations.
5. Read the following passage. Do you consider it sexist? What choices did the writer make that might be construed as sexist?

   > The models' day jobs may be a drag, but L. A. after dark more than makes up for the disappointment. Models are in such social demand that trendy nightspots regularly host evenings in honor of the girls and their agencies. The agencies get a free party, the clubs a throng of camera-primed beauties.
   >
   > Which only serves to draw men. Instinctual as migrating caribou, the fifty-year-old millionaire and twenty-year-old model find each other quickly. For her, any subsequent dates represent a kind of blue-chip food stamp—that is, a free meal at Drai's, Cicada, The Ivy. For him, the ego boost of being seen with such a pretty young thing is reward enough. But only for a while.
   >
   > It's no surprise that this fragile understanding tends to crumble after a few weeks. Not that our professional-girlfriend-in-the-making doesn't try to extend the flirtation as long as she can. "These girls work the men," says the fashion executive. "They're pros." So when he starts to want more, she just gets a new phone number. Some L. A. models, adds the fashion executive, change numbers every two or three months.[75]

6. Find an example of a letter or memo written in English by a U.S. corporation to a corporation in another country. Identify, if you can, some of the nuances in the letter showing an awareness of various cultural biases.
7. Many customers develop a strong loyalty to certain products by certain manufacturers. For example, many people who ride motorcycles are usually loyal fans of Harley Davidson motorcycles. Others prefer Japanese, British, or German motorcycles. Owners of Harley Davidson motorcycles are not usually fans of Japanese motorcycles and vice versa. Owners of Ford trucks are not usually fans of Chevrolet trucks, and owners of Macintosh computers are not usually fans of IBM-compatible computers. What are some of the biases commonly held by owners of one kind of product? Are all of their biases reasonable? Why or why not? If you own or use a computer, do you prefer a Macintosh or an IBM-compatible? Why? Are all of your preferences based on logic or good reasons alone? What do you think some of your biases are?
8. Talk to a member of another race and ask this person about various stereotypes and cultural biases this person has encountered. Summarize this person's views in several paragraphs.
9. Find a recent example of political correctness on campus or in the news. Discuss what your views are concerning this political correctness.
10. Provide several paragraphs discussing what your likes and dislikes are. Discuss how some of these biases may affect your writing.

Notes

1. Stephen Jay Gould, *The Mismeasure of Man* (New York: Norton, 1981) 74.
2. Casey Miller and Kate Swift, *The Handbook of Nonsexist Writing,* 2nd ed. (New York: Harper & Row, 1988) 4.
3. Timothy Weiss, "Translation in a Borderless World," *Technical Communication Quarterly* 4.4 (1995): 417.
4. W. John Coletta, "The Ideologically Based Use of Language in Scientific and Technical Writing," *Technical Communication Quarterly* 1.1 (1992): 60.
5. Gould, p. 31.
6. Gould, p. 74.
7. Gould, p. 27.
8. Gould, p. 22.
9. Gould, p. 27.
10. *Merriam-Webster's Tenth Collegiate Dictionary and Thesaurus.* Electronic Edition. CD-ROM, 1995.
11. *Merriam-Webster's Tenth Collegiate Dictionary and Thesaurus.* Electronic Edition. CD-ROM, 1995.
12. *Merriam-Webster's Tenth Collegiate Dictionary and Thesaurus.* Electronic Edition. CD-ROM, 1995.
13. *Merriam-Webster's Tenth Collegiate Dictionary and Thesaurus.* Electronic Edition. CD-ROM, 1995.
14. *Merriam-Webster's Tenth Collegiate Dictionary and Thesaurus.* Electronic Edition. CD-ROM, 1995.
15. *Merriam-Webster's Tenth Collegiate Dictionary and Thesaurus.* Electronic Edition. CD-ROM, 1995.
16. Tom McArthur, ed. *The Oxford Companion to the English Language* (New York: Oxford UP, 1992) 430.

17. David Crystal, *The Cambridge Encyclopedia of the English Language* (New York: Cambridge UP, 1995) 368.
18. Casey Miller and Kate Swift, "One Small Step for Genkind," *The Gender Reader,* eds. Ashton-Jones and Olson (Needham Heights, MA: Allyn & Bacon, 1991) 248.
19. Miller and Swift, *The Handbook of Nonsexist Writing,* p. 18.
20. Miller and Swift, *The Handbook of Nonsexist Writing,* p. 27.
21. Miller and Swift, *The Handbook of Nonsexist Writing,* p. 45.
22. Miller and Swift, *The Handbook of Nonsexist Writing,* p. 61.
23. Miller and Swift, *The Handbook of Nonsexist Writing,* p. 84.
24. Miller and Swift, *The Handbook of Nonsexist Writing,* p. 101.
25. Eugene August, "Real Men Don't: Anti-Male Bias in English," *Language Awareness,* 5th ed., eds. Paul Eschholz, Alfred Rosa, and Virginia Clark, (New York: St. Martin's, 1990) 297.
26. August, p. 291.
27. August, p. 291.
28. August, p. 294.
29. August, p. 295.
30. August, p. 296.
31. August, p. 296.
32. Alleen Pace Nilsen, "Sexism in English: A 1990s Update," *The Gender Reader,* eds. Ashton-Jones and Olson, (Boston: Allyn & Bacon, 1991) 260.
33. Nilsen, p. 262.
34. Nilsen, p. 264.
35. Nilsen, p. 350.
36. Nilsen, p. 267.
37. Nilsen, p. 267.
38. McArthur, pp. 430–31.
39. William Strunk and E. B. White, *The Elements of Style,* 3rd ed. (New York: Macmillan, 1979) 60.
40. Miller and Swift, *The Handbook of Non-Sexist Writing,* pp. 43–44.
41. Crystal, p. 369.
42. *Merriam-Webster's Tenth Collegiate Dictionary and Thesaurus.* Electronic Edition. CD-ROM, 1995.
43. *Merriam-Webster's Tenth Collegiate Dictionary and Thesaurus.* Electronic Edition. CD-ROM, 1995.
44. Carl Bode, *Mencken* (Baltimore: Johns Hopkins UP, 1986) vii.
45. Crystal, p. 368.
46. Casey Miller and Kate Swift, "One Small Step for Genkind," *Language Awareness,* 5th ed., eds. Paul Eschholz, Alfred Rosa, and Virginia Clark, (New York: St. Martin's, 1990) 313.
47. Donna Cross, "Propaganda: How Not to be Bamboozled," *Language Awareness,* 5th ed., eds. Paul Eschholz, Alfred Rosa, and Virginia Clark, (New York: St. Martin's, 1990) 149.
48. Cross, p. 150.
49. Cross, p. 150.
50. Cross, p. 151.
51. Cross, p. 152.
52. Cross, p. 154.
53. Cross, p. 158.
54. Keith Allan and Kate Burridge, *Euphemism and Dysphemism: Language Used as Shield and Weapon* (New York: Oxford UP, 1991) 121–22.

55. William Raspberry, "What It Means to Be Black," *Language Awareness,* 5th ed., eds. Paul Escholz, Alfred Rosa, and Virginia Clark, (New York: St. Martin's, 1990) 270.

56. Raspberry, p. 270.

57. Raspberry, p. 270.

58. Raspberry, p. 270.

59. Mohsen Mirshafiei, "Culture as an Element in Teaching Technical Writing," *Technical Communication* 41.2 (1994): 281.

60. Mirshafiei, p. 281.

61. Mirshafiei, p. 277.

62. Mirshafiei, p. 282.

63. Mirshafiei, p. 280.

64. Mirshafiei, p. 280.

65. Mirshafiei, p. 281.

66. Carolyn Boiarsky, "The Relationship between Cultural and Rhetorical Conventions: Engaging in International Communication," *Technical Communication Quarterly* 4.3 (1995): 247.

67. Boiarsky, p. 247.

68. Weiss, p. 415.

69. McArthur, p. 794.

70. Crystal, p. 177.

71. Crystal, p. 177.

72. Crystal, p. 177.

73. Irving Younger, *Persuasive Writing* (Minnetonica, MN: The Professional Education Group, 1990) 85–88.

74. Michael Fumento, "Will the U.S. Repeat Peru's Deadly Chlorine Folly?" Online. http://www.townhall.com/townhall/columnists/fumento/fume021496.html

75. Sandra Tsing Loh, "In L. A. They're Called Professional Girlfriends," *Cosmopolitan* Feb. 1996: 182–187.

Chapter 11

Determining the Ethics of Style

Doublespeak is not the product of carelessness or sloppy thinking. Indeed, most doublespeak is the product of clear thinking and is carefully designed and constructed to appear to communicate when in fact it doesn't. It is language designed not to lead but mislead. It is language designed to distort reality and corrupt thought.[1]
—WILLIAM LUTZ

Sometimes we want to be unclear. We don't know what we're talking about, and we don't want anyone else to know that. Or we do know what we're talking about, and we don't want anyone else to know what we know. On those occasions, we write unclearly deliberately, and if what we write gets the job done, then we say the writing is "good." But "good" has two meanings. Assassins can be "good" at their jobs but not be "good" people. In the same way, writing can be "good" if it gets the job done, but if the job is ethically questionable, then the writing may be bad just because it is so good.[2]
—JOSEPH WILLIAMS

What kind of behavior is "prose behavior"? Prose is usually described in a moral vocabulary—"sincere," "open" or "devious," and "hypocritical"—but is this vocabulary justified? Why, for that matter, has it been so moralistic? Why do so many people feel that bad prose threatens the foundations of civilization? And why, in fact, do we think "bad" the right word to use for it?[3]
—RICHARD LANHAM

Chapter Overview

What Is Ethics?
Ethics and Technical Prose
Ethics and the Professions
Criteria for Judging Ethical Actions
Unethical Language
The Ethics of Style
Chapter Summary
Case Study 8: Deception and the Exxon Valdez
Questions/Topics for Discussion
Exercises
Notes

What Is Ethics?

Simply defined, "Ethics is the study of right and wrong conduct."[4] More broadly, ethics is "the discipline dealing with what is good and bad and with moral duty and obligation."[5] Ethics also means "a set of moral principles or values" or "a theory or system of moral values."[6] More broadly still, ethics may be defined as a guiding philosophy.

Vincent Ruggiero observes that "the focus of ethics is moral situations—that is, *those situations in which there is a choice of behavior involving human values* (those qualities that are regarded as good and desirable)."[7]

An ethical person, then, is someone who is governed by right conduct. But what is right conduct? What is moral? What is good? You'll have to go elsewhere for detailed answers to these questions. A detailed discussion of ethics is beyond the scope of this textbook. After all, ethics is a major branch of philosophy with over a 2500 year history in the West. And this textbook is about how to make your technical prose style more effective, not how to live more ethically. Still, many writing choices are ethical choices, so a chapter outlining these ethical concerns is necessary.

This chapter is primarily concerned with the ethical and unethical uses of language. But language cannot be separated from the people who use it. You may recall from Chapter 1 that part of the definition of *style* is the following: "Style is who you are and how you reflect who you are, intentionally or unintentionally, in what you write." In Chapter 4 ethos was defined as "the character of the writer, the writer's integrity, competency, credibility." We learned in that chapter how it is an essential element in helping to change the mind of any audience. And, as we saw in Chapter 9, ethos is a dominant concern of tone, and tone is a major element of style. The ethos or character of a writer may be crucial in determining the effectiveness of that writer's style. The ethics of a writer then is tied to the ethics of style.

Ethics and Technical Prose

Just as people must make many ethical decisions throughout their lives, you must make many ethical decisions concerning what you write throughout your career. For example, are you doing your best to document a product honestly and accurately? Are you knowingly omitting any essential information? If you are unclear or imprecise, and if your poor instructions cause injury to someone, are you morally responsible? If you are in a company's marketing department, and you are asked to exaggerate a product's features, are you guilty of lying? If you know you are promising more than you can deliver in a proposal written in response to a request for a proposal (RFP), are you unethical or just keenly competitive? If you fail to point out numerous known bugs in a program in your software manual, are you a good software documentation writer or are you unethical? If you exaggerate your qualifications on your resume or in your cover letter to gain an advantage over the competition, are you unethical? These are just some of many possible scenarios faced by writers of technical prose everywhere.

Why isn't it possible just to make a list of ethical language choices that everyone could agree on and everyone could abide by? Unfortunately, it's not that easy. Suppose that you wrote the clearest instructions you could write, but someone neglected to provide some essential information and a customer is injured following your instructions. Are you unethical in this instance? Suppose because of unreasonable deadlines, you have to cut corners and you just don't have time to document some important features of a software program. Are you a bad person? Suppose you have been told by your boss in marketing that you're expected to hype the product just to keep up with the competition. Are you unethical for trying to keep your job and doing what you are told to do? As these possible scenarios and many others show, it's not always easy to determine which writing decisions are right or wrong, moral or immoral, ethical or unethical.

Ethics and the Professions

Of course, ethical concerns are not new to those who must write technical prose. Most professions and professional organizations have published ethical guidelines. The Computer Ethics Institute, for example, published the following guidelines:

The Ten Commandments of Computer Ethics

1. Thou shalt not use a computer to harm other people.
2. Thou shalt not interfere with other people's computer work.
3. Thou shalt not snoop around in other people's computer files.
4. Thou shalt not use a computer to steal.
5. Thou shalt not use a computer to bear false witness.
6. Thou shalt not copy or use proprietary software for which you have not paid.
7. Thou shalt not use other people's computer resources without authorization or proper compensation.

8. Thou shalt not appropriate other people's intellectual output.
9. Thou shalt think about the social consequences of the program you are writing or the system you are designing.
10. Thou shalt always use a computer in ways that insure consideration and respect for your fellow humans.[8]

Search for information on *ethics* using any search engine on the World Wide Web, and you'll see all kinds of databases offering other codes of ethics and all kinds of information on ethics. You'll see medical ethics, business ethics, computer ethics, military ethics, media ethics, journalism ethics, and so on. You'll see links to many professional societies and their published codes of ethics. You'll see course syllabi, papers, online journals, and much more.

It seems as though almost everyone is concerned about ethics in one way or another. Developing ethical guidelines for technical communicators parallels the challenges of doing so for computer professionals and engineers. All three groups have obligations to society, to their employers, to their clients, and to co-professionals and even professional organizations.

In *Computer Ethics* Tom Forester and Perry Morrison list many ethical questions faced by computer professionals:

- Is copying software really a form of stealing? What sort of intellectual property rights should software developers have?
- Are so-called "victimless" crimes (against, e.g., banks) more acceptable than crimes with human victims? Should computer professionals be sued for lax computer security?
- Is hacking merely a bit of harmless fun or is it a crime equivalent to burglary, forgery and/or theft? Or are hackers to be seen as guardians of our civil liberties?
- Should the creation of viruses be considered deliberate sabotage and be punished accordingly?
- Does information on individuals stored in a computer constitute an intolerable invasion of privacy? How much protection are individuals entitled to?
- Who is responsible for computer malfunctions or errors in computer programs? Should computer companies be made to provide a warranty on software?
- Is "artificial intelligence" a realistic and a proper goal for computer science? Should we trust our lives to allegedly artificially intelligent "expert" systems?
- Should we allow the workplace to be computerized if it de-skills the workforce and/or increases depersonalization, fatigue and boredom?
- Is it OK for computer professionals to make false claims about the capabilities of computers when selling systems or representing computers to the general public? Is it ethical for computer companies to "lock-in" customers to their products?
- Should, indeed, computer professionals be bound by a Code of Conduct and if so, what should it include?[9]

As you would expect, the Society for Technical Communication also has ethical guidelines:

STC Ethical Guidelines for Technical Communicators

Introduction. As technical communicators, we observe the following ethical guidelines in our professional activities. Their purpose is to help us maintain ethical practices.

Legality. We observe the laws and regulations governing our professional activities in the workplace. We meet the terms and obligations of contracts that we undertake. We ensure that all terms of our contractual agreements are consistent with STC Ethical Guidelines.

Honesty. We seek to promote the public good in our activities. To the best of our ability, we provide truthful and accurate communications. We dedicate ourselves to conciseness, clarity, and creativity, striving to address the needs of those who use our products. We alert our clients and employers when we believe material is ambiguous. Before using another person's work, we obtain permission. In cases where individuals are credited, we attribute authorship only to those who have made an original, substantive contribution. We do not perform work outside our job scope during hours compensated by clients or employers, except with their permission; nor do we use their facilities, equipment or supplies without their approval. When we advertize our services, we do so truthfully.

Confidentiality. Respecting the confidentiality of our clients, employers, and professional organizations, we release business-sensitive information only with their consent or when legally required. We acquire releases from clients and employers before including their business-sensitive information in our portfolios or before using such material for a different client or employer or for demo purposes.

Quality. With the goal of producing high-quality work, we negotiate realistic, candid agreement on the schedule, budget, and deliverables with clients and employers in the initial project planning stage. When working on the project, we fulfill our negotiated roles in a timely and responsible manner and meet the stated expectations.

Fairness. We respect cultural variety and other aspects of diversity in our clients, employers, development teams, and audiences. We serve the business interest of our clients and employers, as long as such loyalty does not require us to violate the public good. We avoid conflicts of interest in the fulfillment of our responsibilities and activities. If we are aware of a conflict of interest, we disclose it to those concerned and obtain their approval before proceeding.

Professionalism. We seek candid evaluations of our professional performance from clients and employers. We also provide candid evaluations of communication products and services. We advance the technical communication profession through our integrity, standards, and performance.[10]

Codes of conduct are valuable because they establish ideals and help define the character of a profession. These codes help to establish an atmosphere of professionalism, and they help to encourage members of a profession to act ethically even in the most difficult of circumstances. However, guidelines on paper may have little meaning if people do not value the guidelines. And, of course, no matter how thorough your written guidelines are, there will always be a situation not covered by the guidelines. One practical approach is to have a strategy for determining whether an action is ethical or unethical. The next section offers such a strategy or framework.

Criteria for Judging Ethical Actions

The purpose here is to outline briefly one clear approach (of many possible approaches) in order for you to apply some criteria to language issues.

One of the most difficult aspects of ethics is how to evaluate moral issues, how to determine what is and is not ethical. Is there a strategy we can use to help us evaluate actions? Of course, there are many. (In fact, the number is probably countless.) Some of us are guided entirely by our religious beliefs. Others believe that many ethical questions cannot be completely answered by religious beliefs or by one institutional religion and its doctrine. The distinction between a religion and a system of ethics is quite complex, but, it should suffice to say, Protestants, Catholics, Buddhists, Hindus, Muslims, and Jews (to mention only most of the major religious followers) face daily ethical choices just as do sophists, stoics, existentialists, agnostics, and atheists.

In *Thinking Critically about Ethical Issues,* Vincent Ruggiero comments: "To develop a sound and meaningful basis for discussing moral issues, we must find a standard for judging the morality of actions, a standard that is acceptable to men and women of various moral perspectives. In other words, we must find a standard that reflects the principles that most ethical systems have in common."[11]

Ruggiero points to one important principle that underlies almost all ethical systems: the principle of *respect for persons.*[12] This principle has three separate requirements:

> First, that each and every person should be regarded as worthy of sympathetic consideration, and should be so treated; secondly, that no person should be regarded by another as a mere possession, or used as a mere instrument, or treated as a mere obstacle, to another's satisfaction; and thirdly, that persons are not and ought never to be treated in any undertaking as mere expendables.[13]

This principle of respect for persons is either explicit or implicit in most ethical systems.

Ruggiero suggests that three basic criteria emerge from this principle of respect for persons—*obligations, ideals,* and *consequences.* In one way or another, these three criteria are used in most ethical systems to determine what actions are ethical or unethical.[14]

Concerning obligations, "Every significant human action occurs, directly or indirectly, in a context of relationships with others. And relationships usually imply *obligations;* that is, restrictions on our behavior, demands to do something or avoid doing it."[15] One kind of obligation is a formal agreement. Other kinds include those of friendship, citizenship, and business. Obligations, of course, have to be valued: "When we say obligations bind morally, we mean they exist to be honored. To honor them is right; to dishonor them is wrong."[16]

The criterion of ideals is also essential: "In the general sense *ideals* are notions of excellence, goals that bring greater harmony in one's self and between self and others. In ethics, ideals are also specific concepts that assist us in applying the principle of respect for persons in our moral judgments. Some common ideals that figure prominently in ethical reasoning are fairness, tolerance, compassion, loyalty, forgiveness, amity, and peace."[17]

Finally, there is the criterion of consequences:

> *Consequences* are the beneficial or harmful effects that result from the action and affect the people involved, including, of course, the person performing the action. Some consequences are physical; others are emotional. Some occur immediately; others occur only with the passing of time. Some are intended by the person performing the act; others are unintended. Finally, some consequences may be obvious, and others may be very subtle and hidden by appearances. The ethicist is concerned with all significant consequences occurring in a moral context.[18]

As you can imagine, applying these three criteria to any action to determine whether or not the action is ethical is not easy. Obligations may conflict with each other. Ideals may compete with each other and many conflict with obligations, and many actions may produce many consequences, some good and some harmful. To complicate matters further, Ruggiero reminds us that we are also often confronted by moral dilemmas. A moral dilemma "is defined as any situation or predicament that arises from the impossibility of honoring all the moral values that deserve honoring. A moral dilemma exists whenever the conflicting obligations, ideals, and consequences are so very nearly equal in their importance that we feel we cannot choose among them, even though we must."[19]

Ruggiero advises us to use the following approach for determining whether an action (or action in a hypothetical case) is ethical or unethical.

First, he suggests that you study the details of the action. Are there any extenuating circumstances that make this action different from similar actions?

Second, you should identify how the three basic criteria may be relevant to the action: "Are there any *obligations?* What *ideals* are involved? What are the *consequences* of the case? Whom will they affect? In what way?"[20] Then decide if the emphasis is on one of the three criteria or all three almost equally.

Third, you should list all of the possible actions that are available.

Fourth, after reviewing what you now know from the first three steps, decide which action is most ethical.

Ruggiero believes this approach helps people analyze complex moral issues more systematically. You can use this approach to decide what is or is not ethical behavior and to explain your decisions to others.[21]

Unethical Language

If you lie to someone in a memo, letter, or manual, are you being unethical? What if you intentionally state something ambiguously? Pretentiously? Deceptively? Of course, the answers to these questions depend on a great deal. You need much more information before you can judge whether or not the action is ethical or unethical. We can use the criteria briefly explained in the preceding section to help us evaluate language choices.

Doublespeak

William Lutz's *Doublespeak* may be viewed as a valuable collection of unethical language used in everyday living, advertising, business communication, around the world, the military, the bureaucracy of government, and for nuclear weapons.

Lutz defines doublespeak as the following:

> Doublespeak is language that pretends to communicate but really doesn't. It is language that makes the bad seem good, the negative appear positive, the unpleasant appear attractive or at least tolerable. Doublespeak is language that avoids or shifts responsibility, language that is at variance with its real or purported meaning. It is language that conceals or prevents thought; rather than extending thought, doublespeak limits it.
>
> Doublespeak is not a matter of subjects and verbs agreeing; it is a matter of words and facts agreeing. Basic to doublespeak is incongruity, the incongruity between what is said or left unsaid, and what really is. It is incongruity between the word and the referent, between seem and be, between the essential function of language—communication—and what doublespeak does—mislead, distort, deceive, inflate, circumvent, obfuscate.[22]

As we discussed in Chapter 6 for Lutz there are four kinds of doublespeak: euphemisms, jargon, gobbledygook or bureaucratese, and inflated language.

A euphemism becomes doublespeak when it "is used to mislead or deceive."[23] He cites the example of the U.S. State Department announcing it would no longer use the word *killing* in annual reports on the condition of human rights in various countries: "Instead, it would use the phrase 'unlawful or arbitrary deprivation of life,' which the department claimed was more accurate. Its real purpose for using this phrase was simply to avoid discussing the embarrassing situation of government-sanctioned killings in countries that are supported by the United States and have been certified by the United States as respecting the human rights of their citizens."[24]

Jargon can also be doublespeak when it is "pretentious, obscure, and esoteric terminology used to give an air of profundity, authority, and prestige to speakers and their subject matter. Jargon as doublespeak often makes the simple appear complex, the ordinary profound, the obvious insightful. In this sense it is used not to express but impress."[25] Lutz offers the example of National Airlines inserting a footnote to its annual report to stockholders. Because of an airplane crash, National Airlines made an after-tax insurance benefit of $1.7 million dollars. In its footnote in the annual report National Airlines refers to "the involuntary conversion of a 727."[26] In this way National Airlines recognizes the crash of the airplane and the profit, but avoids referring to the accident or the deaths.

Gobbledygook or bureaucratese is doublespeak because this kind of language "is simply a matter of piling on words, of overwhelming the audience with words, the bigger the words and the longer the sentences the better."[27] Lutz cites many examples including the words of Jesse Moore, NASA's associate administrator, when asked during an investigation of the *Challenger* disaster in 1986 if the performance of the shuttle program was improving with each launch or remaining the same. Moore replied:

> I think our performance in terms of the liftoff performance and in terms of the orbital performance, we knew more about the envelope we were operating under, and we have been

pretty accurately staying in that. And so I would say the performance has not by design drastically improved. I think we have been able to characterize the performance more as a function of our launch experience as opposed to it improving as a function of time.[28]

Inflated language is also a kind of doublespeak because it "is designed to make the ordinary seem extraordinary; to make everyday things seem impressive; to give an air of importance to people, situations, or things that would not normally be considered important; to make the simple seem complex."[29] Examples of inflated language include Chrysler stating that it "initiates a career alternative enhancement program" instead of admitting it is letting five thousand employees go. Another example is a hospital bill sent to a family for "negative patient care outcome" instead of admitting the patient died.[30]

In all of these examples and many others that Lutz provides, you can see prose that is deliberately ambiguous, pretentious, or deceptive.

According to Lutz, one of "the most chilling and terrifying" uses of doublespeak occurred in 1981 when then Secretary of State Alexander Haig testified before congressional committees about the murder of three American nuns and a Catholic lay worker in El Salvador. The four women were raped and shot at close range. And there was ample evidence that soldiers of the Salvadoran government committed the crime. Secretary Haig made the following comment before the House Foreign Affairs Committee:

> I'd like to suggest to you that some of the investigations would lead one to believe that perhaps the vehicle the nuns were riding in may have tried to run a roadblock, or may accidentally have been perceived to have been doing so, and there'd been an exchange of fire and then perhaps those who inflicted the casualties sought to cover it up. And this could have been at a very low level of both competence and motivation in the context of the issue itself. But the facts on this are not clear enough for anyone to draw a definitive conclusion.[31]

In follow-up questioning the next day, Haig was asked to clarify his comment. He finally said, "What I meant was that if one fellow starts shooting, then the next thing you know they all panic."[32] Lutz suggests that Haig was deliberately vague to imply "that the women were in some way responsible for their own fate."[33] He uses vague expressions such as "would lead one to believe" and "may accidentally have been perceived to have been doing so." He avoids the word "kill" and instead says "inflicted the casualties." In effect, Lutz argues, "the secretary of state has become an apologist for rape and murder."[34]

How can you apply the three criteria of obligation, ideals, and consequences to this action? Haig had an obligation to his employer, the State Department and U.S. government, to diffuse a volatile situation. As a representative of the government, he also had an obligation to tell the public the truth. Ideals involved here range from the public's right to know, the government's need to maintain a good relationship with other governments, to individual commitment to government service. The consequences of Haig's testimony include increased cynicism on the part of the public and increased distrust of public officials, unbelievable and ineffective testimony, and damage to the credibility of the State Department, to mention only a few. Certainly, the bad consequences outweigh any good consequences in this particular action. Therefore, Haig's testimony was clearly unethical.

As for other examples of potentially unethical communication, consider warranties and disclaimers from software developers. As Forester and Morrison note, "Given the incidence of faulty software and system failure in general, it is perhaps not surprising that software developers rarely provide their clients or purchasers with warranties of any substance. Indeed, if one compared this situation with that of almost any other form of purchased goods, then it is difficult not to be amused and perhaps even incredulous."[35]

Consider the following example:

> We don't claim Interactive EasyFlow is good for anything—if you think it is, great, but it's up to you to decide. If Interactive EasyFlow doesn't work: tough. If you lose a million because Interactive EasyFlow messes up, it's you that's out of the million, not us. If you don't like this disclaimer: tough. We reserve the right to do the absolute minimum provided by law, up to and including nothing.
>
> This is basically the same disclaimer that comes with all software packages, but ours is in plain English and theirs is in legalese.
>
> We didn't really want to include a disclaimer at all, but our lawyers insisted. We tried to ignore them, but they threatened us with the shark attack at which point we relented.[36]

Is such a disclaimer fair? Ethical?

Other Strategies for Lying with Language

Of course, companies, governments, and people have other ways of using the language unethically besides deliberately choosing to be ambiguous, deceptive, or pretentious. Essential or potentially damaging information may be omitted in a document. What you are told may be true, but you may not be told the whole story. Also, information supporting your prose or verbal comments—statistics, graphics, and photographs, for example—may be distorted in numerous subtle ways. The ways of lying through language are almost countless.

The Ethics of Style

In Chapter 3 we discussed Richard Lanham's "The Official Style" as an example of a genuine style. In the last chapter of his book *Revising Prose,* Lanham addresses the ethics of style. In this chapter Lanham is concerned with the question of why we should be concerned about revising "The Official Style" into a plain style. Lanham believes there are two kinds of answers to this question.

First, as Lanham explains, "Generally, it helps if you write better prose. It makes for a better statement of purpose when you apply for law school or a job; later on, it will help you write a better legal brief or progress report or memo."[37] Yet Lanham also realizes that writing in a plain style is not always appropriate: "Where The Official Style is mandated by rule or custom (as in some bureaucratic situations it is), plain prose may sound simpleminded or even flip. The sensible procedure here: learn both languages, the plain and The Official Style."[38] Lanham is a pragmatist when it comes to effective writing. He knows, as anyone who writes frequently knows, *"Stylistic* judgment is, last as well as first, always *political* judgment."[39] Different situations or contexts determine different styles. We must adjust the style to suit the occasion.

Second, Lanham offers a more complex answer

> The second kind of answer to "Why bother?" is simply, "Are you willing to sign your name to what you have written? To present yourself in public—whether it matters to anyone else or not—as this kind of person?" In a sense, it is a simple question: "Whatever the advantage—or disadvantage—ought I do this?" The primary moral question: If everyone else is committing murder, ought I do the same? Do you choose to encounter the world on its terms or on your own? A simple question but one we must all answer for ourselves.[40]

For Lanham, the ethics of style is a matter of personal ethics. Lanham speaks of the "primary ground for morality, the self."[41] He refers to this self as both a dynamic and static entity: "It is static when we think of ourselves as having central, fixed selves independent of our surroundings, an 'I' we can remove from society without damage, a central self inside our heads. But it becomes dynamic when we think of ourselves as actors playing social roles, a series of roles that vary with the social situation in which we find ourselves."[42] Lanham observes, "Our complex identity comes from the constant interplay of these two kinds of self. Our final 'self' is usually a mixed one, few of us being completely the same in all situations.... If we were completely sincere we would always say exactly what we think—and cause social chaos. If we were always acting an appropriate role, we would be certifiably insane."[43]

Lanham's static and dynamic selves correspond closely to ethos and persona as they are discussed in Chapter 9. Ethos is the character of the writer, the writer's integrity, competency, objectivity. Persona is the mask or assumed identity of a writer. In the process of writing, ethos and persona are in constant interplay. Whenever a writer tries to be sincere, at least part of his or her ethos is usually apparent. At the same time, the act of writing requires playing various social roles, requires assumed selves or identities or personas. But where does ethos end and persona begin in the act of writing? Where does ethos end and persona begin in the self? As we mentioned earlier, Lanham suggests our identity lies in the interplay between the static and dynamic selves. And this is the case for ethos and persona.

Chapter Summary

Simply defined, ethics is the study of right and wrong conduct. Ethics concerns all kinds of behaviors, not just decisions concerning language. But when decisions concerning language are involved, both ethical and stylistic decisions are involved. Writers of technical prose are faced daily with all kinds of ethical choices, and it's not always easy to make the most ethical decision. Professional or organizational ethical guidelines may provide some help, but, usually, a more systematic approach is more helpful.

Knowing how to determine obligations, ideals, and consequences for a particular action can be particularly helpful in judging the morality of an action involving language. Also, you can use a four-step approach to judge the morality of a language decision—studying the details of an action, identifying how the three basic criteria are relevant to the action, listing all possible actions that are available, and reviewing what you know after these first three steps and deciding which action is most ethical.

Often a style may be ambiguous, deceptive, or pretentious and not be unethical. More often, however, your style may be unethical when you deliberately use ambiguous, deceptive, or pretentious language for questionable obligations or ideals and if there are damaging consequences. The four kinds of doublespeak—euphemisms, jargon, gobbledygook or bureaucratese, and inflated language—are often good examples of unethical language. Of course, there are other strategies for lying with language, by omitting essential or damaging information, for example.

The ethics of style cannot be separated from the ethics of the self. As Lanham suggests, the ethics of style is a matter of personal ethics. Who you are as a writer is a matter of your static and dynamic selves, your ethos and persona.

Ultimately, a bureaucratic style is often an unethical style because it undermines both the ethos of the self and the necessary respect for others that underlies most ethical systems. It depersonalizes the self and society.

Case Study 8: Deception and the Exxon Valdez

The *Exxon Valdez* oil spill affected 1244 miles of the Alaskan coastline and resulted in devastating effects on the ecosystem and the Alaskan people. The disaster also provided proof of the devastating effects of deceptive writing. The Exxon Corporation and the *Anchorage Daily News* were two of the agencies that intentionally offered false information. To avoid as much publicity and liability as possible, Exxon not only deliberately withheld or offered vague information, but also minimized the extent of damage and exaggerated the ecological recovery. The *Anchorage Daily News* exaggerated the extent of damage, deliberately misquoted sources, and twisted information to produce inflammatory news stories to increase sales of the publication and to protect the ecology.

Exxon Corporation

The Exxon Corporation went in with big guns and "hired Burson Murstellar as its public relation firm after the spill. Previous clients include the military dictatorship in Argentina, the Rumanian dictator Nicolae Ceausecescu and Union Carbide after the Bhopal disaster."[44] As the recovery progressed, the people and agencies dealing with Exxon came to realize the corporation could not be trusted with the truth. As an industry analyst observed, "big oil had once again raped the public trust."[45]

People

For many of the Alaskan people, the coastline and water provided their livelihood. Even before Exxon offered to pay for their help with the cleanup, many Alaskans responded. However, the health warning Exxon issued to the people about exposure to the oil and dispersant used was not made available until long after the work started, and the warning minimized the effects of exposure.

An Exxon Material Safety Data Sheet for Inipol Eap22, issued on 28 July 1989, in the middle of the cleanup operations, states:

> In case of skin contact, remove any contaminated clothing and wash skin thoroughly with soap and water...As a precaution, exposure to liquids, vapors, mists or fumes should be minimized...inhalation of high vapor concentrations may have results ranging from dizziness, headache, and respiratory irritation to unconsciousness and possibly death.... Prolonged or repeated skin contact may cause skin irritation ...components of the product (2-butoxyethanol) may be absorbed through the skin and could produce blood and kidney damage. Symptoms of overexposure include paleness and red discoloration of the urine...Exxon shall not be liable for any loss or damage arising out of the use thereof.[46]

Although the statement does suggest that exposure can cause death, that phrase is buried in the middle. Phrases like "as a precaution" and "prolonged or repeated skin contact may cause skin irritation" mislead the reader. Facing a possible loss of their means of support, the people could easily read the statement as the lesser of two evils.

Because oil dispersants add another chemical, harmful to humans, minimizing the available information had harmful affects on many Alaskans. In a data sheet released by the Arco Oil and Gas Company, the dangers of exposure to oil are more clearly expressed:

> Prolonged exposure to crude oil vapors can cause central nervous system depression. Hydrogen sulfide can be absorbed through the skin and can produce toxic effects…. Aspiration into lungs may produce chemical pneumonia…May contain benzene which in humans is associated with blood disorders such as anemia and leukemia…Signs and symptoms of central nervous depression produced by prolonged inhalation of crude oil vapors may include headache, drowsiness, nausea, incoordination, convulsions, unconsciousness, and death.[47]

Even after reports started appearing of workers being directly exposed to the oil and mixture of oil and dispersants, "an Exxon hygienist reported that, after preliminary analysis, crews could continue working."[48] The report by the Exxon hygienist deliberately misled the workers and assured them there was no serious damage in the exposure. Hundreds of workers suffered health related problems due to the exposure to oil and oil dispersants. According to Randall Scarlett, an attorney for many of the workers affected by the exposure, "There is no doubt that some of the individuals are going to die."[49]

Ecosystem

The extent of damage to the ecosystem was so wide spread it is still being investigated and is not completely determined. In *The* Exxon Valdez: *A Case of Virtual Reality,* Andrew Rowell recorded a statement in 1993: Rick Steiner, an Associate Professor at the University of Alaska, comments: "Four years after the spill, oil still remains trapped in mussel mats in the inter-tidal zone, being picked up into the food chain…."[50] Steiner adds, "There really is no such thing as oil spill restoration. You cannot fix a broken ecosystem like you fix a broken car."[51]

Despite the vast amounts of data available on the ecosystem damage in 1991 and still being reported four years after the spill, Exxon released statements such as this in 1991: "It is extremely unlikely that hydrocarbon concentrations resulting from the spilled oil have had or will have any adverse effects on plants and animals living in the water column of Prince William Sound and the Western Gulf of Alaska, including commercial fishery species."[52]

Exxon deliberately reported false information: "An Exxon spokesperson said on National Public Radio in the USA in June 1989 that the company had found only 330 dead birds and 70 otters, despite the fact that tens of thousands of corpses had already been recovered. The official later said he had misunderstood the question."[53]

Because of the lack of concern shown by Exxon for the Alaskan people and the ecosystem, another statement in 1991 also rang false: "Concern for the environment and for the well-being of our employees is a long-standing Exxon tradition which remains a guiding principle today…."[54]

Anchorage Daily News

The *Anchorage Daily News* produced inflammatory and misleading reports that kept the Alaskan people unsettled and angry. For example, *Anchorage Daily News* staff reporter Charles Wohlforthe reported that NOAA Scientific Support Coordinator John Robinson had proclaimed that "if [subsistence users] don't like it (oil on the shoreline), they can go somewhere else."[55] The Alaska Department of Environmental Conservation increased the problem when "an information employee faxed copies of the article to native villages and corporations throughout the state."[56] Later, Wohlforthe admitted that Robinson really hadn't actually said that "subsistence users can go somewhere else," and that perhaps he had been "too glib" in his summary of what he had observed.[57]

Although the reporters and editors of the *Anchorage Daily News* were at times intentionally deceptive in their writing, the deception did help to create opposition to Exxon. The *Anchorage Daily News* came to function as the opposition or thorn in the side of the big oil company. And employees of the *Anchorage Daily News* had another big stake in the story. They, unlike the Exxon officials involved in the cleanup, lived in Alaska. While Exxon was trying to divert the world's view in another direction, the *Anchorage Daily News* kept the world's view open to the disaster and caused Exxon to try continually to fix the problem.

Questions

1. How would you apply the three basic criteria of obligations, ideals, and consequences to Exxon's deception in this case study?
2. How would you apply the three basic criteria to evaluate the Exxon hygienist's actions?
3. In what ways was Exxon wrong to deceive the public about the extent of the oil disaster? Justify your answer based upon some of the strategies discussed in this chapter.
4. In what ways was the *Anchorage Daily News* wrong to deceive the public?
5. What other examples can you identify of corporations deceiving the public on such a grand scale?

Questions/Topics for Discussion

1. How would you define ethics?
2. Do you think having published ethical guidelines influences behavior within a profession? Why or why not?
3. How can line graphs and photographs be used to distort information?
4. What ethical questions for computer professionals mentioned by Tom Forester and Perry Morrison overlap with ethical concerns of writers of technical prose?
5. What is an obligation?
6. What is an ideal?
7. What is a consequence?
8. What is an example of a deception in a communication that is not unethical? Why?
9. What is an example of ambiguity in a communication that is not unethical? Why?
10. What is an example of pretentiousness in a communication that is not unethical? Why?
11. Do you agree with Lanham that a bureaucratic style is ultimately a bad style? Why or why not? Is a bad style necessarily an unethical style?

Exercises

1. Look at the following ethical guidelines. For each one, discuss the quality of ethics established in the document.

    ### Example 1

    An Engineer's Hippocratic Oath

    I solemnly pledge myself to consecrate my life to the service of humanity. I will give to my teachers the respect and gratitude which is their due; I will be loyal to the profession of engineering and just and generous to its members; I will lead my life and practice my profession in uprightness and honor; whatever project I shall undertake, it shall be for the good of mankind to the utmost of my power; I will keep far away from wrong, from corruption, and from tempting others to vicious practice; I will exercise my profession solely for the benefit of humanity and perform no act for a criminal purpose, even if solicited, far less suggest it;

I will speak out against evil and unjust practice wheresoever I encounter it; I will not permit considerations of religion, nationality, race, party politics, or social standing to intervene between my duty and my work; even under threat, I will not use my professional knowledge contrary to the laws of humanity; I will endeavor to avoid waste and the consumption of non-renewable resources. I make these promises solemnly, freely, and upon my honor.[58]

Example 2

Oath of an Informatician

Daniel Loeffler

Student in Informatics, University of Hamburg

On the dignity of my humanity, I freely engage myself to fulfill this oath by my best knowledge and conscience:

I want to live and work honestly and sincerely and always do my best to follow my vocation (in the sense of this oath).

I am aware of my responsibility as a computer expert to process and safeguard all information in a correct way. I will work in an ecological consciousness of information to prevent the "information-sphere" from pollution by faulty programs and wrong data.

I will act in a holistic and interdisciplinary way. I will be cautious not to separate subject and object, environment and interior system, or developers, users and usees, as all consequences of my actions in turn affect myself. I take responsibility for myself, for my thinking, feeling, speaking and acting, and I will only promise what I can keep.

I do not serve any egoistic interests, I will not allow misuse of information technics nor falsification of information, and I will give nobody the possibility to influence me adversely.

I will not use power to control others, and I shall not develop or distribute any destructive software. I will make public any unjust practice and any problem. I will do my best to find reasons of my own faults, and I will forgive others and correct mistakes. I will be open for suggestions and any constructive criticism.

I respect the human rights, the privacy of individuals and the democratic freedom of information. I accept myself and all human beings as they are, regardless of sex, religion, nationality and birth. I will stand up for keeping nature alive.

I respect my teachers and continue their work, in an evolutionary sense. I will work to increasing my knowledge and to a further global development.

If I follow this oath, I am allowed to do all things successfully, otherwise I disqualify myself.[59]

Example 3

While the importance of informed consent is unquestioned, controversy prevails over the nature and possibility of an informed consent. Nonetheless, there is widespread agreement that the consent process can be analyzed as containing three elements: information, comprehension, and voluntariness.

Information

Most codes of research establish specific items for disclosure intended to assure that subjects are given sufficient information. These items generally include: the research procedure, their purposes, risks and anticipated benefits, alternative procedures (where therapy is involved), and a statement offering the subject the opportunity to ask questions and to withdraw at any time from the research.

Comprehension

Another way to implement the Belmont Report principles is by assuring that research subjects comprehend the information and understand what they are consenting to:

The manner and context in which information is conveyed is as important as the information itself. For example, presenting information in a disorganized and rapid fashion, allowing too little time for consideration or curtailing opportunities for questioning, all may adversely affect a subject's ability to make an informed choice.

Voluntariness

Finally, the Belmont Report principles can be implemented only if the subjects give consent voluntarily:

This element of informed consent requires conditions free of coercion and undue influence. Coercion occurs when an overt threat of harm is intentionally presented by one person to another in order to obtain compliance. Undue influence, by contrast, occurs through an offer of an excessive, unwarranted, inappropriate or improper reward or other overture in order to obtain compliance.[60]

2. Use a search engine on the World Wide Web to search for links to the word *ethics*. Identify ten useful sites you found with this type of search.
3. Create a one- to two-paragraph fictional case study involving a writer of technical prose and an ethical choice. Make sure the case study involves the three criteria discussed by Vincent Ruggiero.
4. Identify and discuss three films that center around difficult ethical choices for the major actors.
5. Consider this warning from a bottle of drain opener:

Misuse may cause injury. Not for use in the presence of other chemicals.

Sounds like a typical chemical warning found on many household products, right? Suppose in fact that the product contains sulfuric acid, a violent reagent and deadly chemical. And further suppose that when this product is mixed with certain other chemicals commonly found in home-use products, it will explode.

    In your opinion, is the warning sufficient? Is it unambiguous? Is its wording ethical? Could a consumer reading the warning reasonably conclude that the product requires no special handling? Discuss the ethical considerations of labeling dangerous household substances.
6. Find an example of a pretentious style that is unethical. Discuss why it is unethical.
7. Find an example of a deceptive style that is unethical. Discuss why it is unethical.
8. Find an example of prose that is not ambiguous, deceptive, or pretentious, but that is still unethical. Discuss why it's unethical.
9. Defend The Official Style against Lanham's charge that it's a bad style.
10. Discuss how Lanham's concepts of the static and dynamic selves are related to the concepts of ethos and persona.

Notes

1. William Lutz, *Doublespeak* (New York: Harper & Row, 1989) 18–19.
2. Joseph Williams, *Style,* 4th ed. (New York: HarperCollins, 1994) 134.
3. Richard Lanham, *Revising Prose,* 3rd ed. (New York: Macmillan, 1992) 96–97.
4. Vincent Ryan Ruggiero, *Thinking Critically about Ethical Issues,* 3rd ed. (Mountain View, CA: Mayfield, 1992) 4.
5. *Merriam-Webster's Tenth Collegiate Dictionary and Thesaurus.* Electronic Edition. CD-ROM, 1995.
6. *Merriam-Webster's Tenth Collegiate Dictionary and Thesaurus.* Electronic Edition. CD-ROM, 1995.
7. Ruggiero, p. 5.
8. Computer Ethics Institute Home Page, http://www.cpsr.org:80/dox/cei.html
9. Tom Forester and Perry Morrison, *Computer Ethics: Cautionary Tales and Ethical Dilemmas in Computing* (Cambridge, MA: MIT P, 1992) 4–5.
10. *STC Membership Directory 1995–96* 42:3A (September 1995) xi.
11. Ruggiero, p. 53.
12. Ruggiero, p. 54.
13. Ruggiero quoting Errol E. Harris, p. 54.
14. Ruggiero, p. 54.
15. Ruggiero, p. 55.
16. Ruggiero, p. 55.
17. Ruggiero, pp. 55–56.
18. Ruggiero, p. 56.
19. Ruggiero, p. 91.
20. Ruggiero, p. 57.
21. Ruggiero, p. 58.
22. Lutz, pp. 1–2.
23. Lutz, p. 3.
24. Lutz, p. 3.
25. Lutz, p. 4.
26. Lutz, p. 4.
27. Lutz, p. 5.
28. Lutz, pp. 5–6.
29. Lutz, p. 6.
30. Lutz, p. 6.
31. Lutz, pp. 17–18.
32. Lutz, p. 18.
33. Lutz, p. 18.
34. Lutz, p. 18.
35. Forester and Morrison, p. 76.
36. Cited in Forester and Morrison, p. 77.
37. Lanham, p. 96.
38. Lanham, p. 96.
39. Lanham, p. 96.
40. Lanham, p. 96.
41. Lanham, p. 97.
42. Lanham, p. 97.

43. Lanham, p. 97.

44. C. Deal, *The Greenpeace Guide to Anti-Environmental Organizations* (Berkeley: Odonian, 1993) 15.

45. B. William, *United States Petroleum Strategies in the Decade of the Environment* (Tulsa, OK: Penn Well, 1991) 7.

46. Exxon Company USA, 1989, Material Safety Data Sheet—INIPOL, EAP 22, Houston, 28 July, quoted in Andrew Rowell's *The* Exxon Valdez: *A Case of Virtual Reality,* ed. Andrea Goodall (Amsterdam: Greenpeace International, 1994).

47. Arco Oil and Gas Company, 1987, Crude Oil—Material Safety Data Sheet, Texas, 27 May, quoted in Rowell.

48. Alaskan Oil Spill Commission, 1990, *Spill—The Wreck of the* Exxon Valdez, State of Alaska, Final Report, February, Appendix N, T/V, *Exxon Valdez* Oil Spill Chronology, Day Twenty-Four, pp. 1–2.

49. W. Coughlin, 1992, "Illness Tied to the Exxon Cleanup is Cited in Spate of Lawsuits," *Boston Globe,* 4 April, quoted in Rowell.

50. R. Steiner, 1993, Lessons from Alaska for Shetland—Lessons from Both for the World, Associate Professor of the Marine Advisory Program, Cordova, Alaska, quoted in Rowell.

51. R. Pagano, 1993, "After the Spill, Congress Asks, Where's the Oil?" *Anchorage Daily News,* 24 May, quoted in Rowell.

52. Reuter News Service, 1991, Valdez Mayor Touts Recovery in Exxon Commercial, Anchorage, Alaska, 29 April, quoted in Rowell.

53. Rowell, p. 25.

54. I. G. Rawl and I. R. Raymond, 1991, "Dear Shareholder, Environment, Health and Safety, A Progress Report," Texas p. 3, quoted in Rowell.

55. C. Wohlforth, "State Adopts High Standard for Oil Cleanup," *Anchorage Daily News,* 19 July 1990, quoted in *T/V* Exxon Valdez *Oil Spill: Federal on Scene Coordinator's Report.* Vol. 1, U.S. Coast Guard, Department of Transportation, 1993.

56. E. McMullen (Port Graham Village Council), letter to J. Robinson (NOAA), 20 July 1990, no. W1695; J. Larsen, Jr. (The Aleut Corp.), letter to C. Ehler (NOAA), 25 July 1990, no. W1704; and J. Fall (ADF&G), letter to J. Robinson (NOAA), quoted in *T/V* Exxon Valdez *Oil Spill: Federal on Scene Coordinator's Report.* Vol. 1, U.S. Coast Guard, Department of Transportation, 1993.

57. A. Smith, Esq. (North Pacific Rim Assn.), letter to Tribal Councils Association boards, 31 July 1990, no. W1449, FOSC Exxon Valdez Archive, quoted in *T/V* Exxon Valdez *Oil Spill: Federal on Scene Coordinator's Report.* Vol. 1, U.S. Coast Guard, Department of Transportation, 1993.

58. Charles Susskind, *Understanding Technology* (Baltimore: Johns Hopkins UP, 1973) 188. Online http://ei.cs.vt.edu:80/~cs3604/spring.95/Hippocr.Oath.html

59. http://ei.cs.vt.edu:80/~cs3604/spring.95/Loeffler.Oath.html

60. From *The Belmont Report,* 1979; Federal Register (Part II): Federal Policy for the Protection of Human Subjects; Notices and Rules, 1991. http://sun.soci.niu.edu/~jthomas/ethics.cmu

Chapter 12

Editing for Style

I got frustrated because people came to interviews for editing jobs who didn't have a clue what the job was. They didn't understand that an editor has to be more than a proofreader. An editor has to be an advocate for the reader and a partner for the writer.[1]
—JUDITH TARUTZ

Technical editing involves a wide-ranging, deeply probing, thorough review of a technical manuscript and is performed for the purpose of improving the communication of scientific and engineering concepts. In fact, many authors consider the technical editor to be one who can be relied upon to transform a mass of rough draft material into a polished and publishable report.[2]
—ROBERT VAN BUREN AND MARY FRAN BUEHLER

Technical editing is a fast-paced, demanding job that requires good judgment, the ability to manage long-term projects, a capacity to collaborate with others yet maintain independence—and mastery of punctuation.[3]
—CAROLYN RUDE

Chapter Overview

Misconceptions about Editing
Approaches to Editing
The Challenges of Editing for Style
Strategies to Help You Edit for Style
Copyediting for Style
Chapter Summary
Questions/Topics for Discussion
Exercises
Notes

Misconceptions about Editing

Many people have misconceptions about editing and editors. Some people think editing is merely light proofreading, catching, for example, typos, transposed letters, problems in punctuation and spelling. Some people think the primary role of an editor is to take a technical document and make it "look pretty." Others think the writer–editor relationship is an adversarial relationship in which the editor is always trying to change the writer's style or meaning. In this closing chapter of this textbook, it's important to address these misconceptions and others before discussing how writers of technical prose can edit for style.

Let's start with the misconceptions for some that editing is merely proofreading or that the primary task of an editor is to make a document "look pretty." Proofreading is, of course, an important part of editing, but editing consists of far more than proofreading. To some, proofreading is merely a light edit, checking for basic spelling and punctuation errors.[4] More accurately, proofreading is the last stage of the editing process, the three-part process of substantive editing, copyediting, and proofreading. And this complicated process requires people who are highly skilled, well-educated, remarkably patient, and extremely tolerant. Far from merely proofreading or making a document "look pretty," serious editing makes certain that a technical document is as effective for its intended purpose and audience as it possibly can be.

Let's take a brief look at this editing process. In *substantive editing* the editor thoroughly checks the substance or the content and meaning of the technical document. In this phase the editor checks, for example, the purpose, scope, accuracy of the information, organization, design and layout, style, and logic of the document. Substantive editing may occur with each successive draft of a document.

The *copyediting* of a text occurs later in the process. The copyeditor checks for punctuation and spelling errors, correctness, consistency, accuracy, and completeness.

Proofreading, ideally, occurs at the end of the process. Many often confuse copyediting with proofreading, but the two have some distinct differences. As Carolyn Rude points out, "Copyediting prepares the text for printing; proofreading verifies that the text has been printed according to specifications. Copyediting takes place early in the production; proofreading is done later. Copyediting marks are interlinear; proofreading marks are mostly marginal."[5]

Okay, so editing is a complicated process, but what about the adversarial relationship between writers and editors? Of course, the answer is that many writer–editor relationships are adversarial and many are not. Many editors are too overbearing, too meddlesome, and too inflexible. Many others are fair, open-minded about choices in style, and flexible about many of the writer's choices.

In the last chapter of *On Writing Well* William Zinsser comments on editors and their relationships with writers. Zinsser admits he is grateful for the good editors who have edited his work. He says they have helped him sharpen his writing in many ways, from changing the focus or tone to finding weaknesses in the logic. As an experienced writer, Zinsser understands the value of an experienced editor:

> What a good editor brings to a piece of writing is an objective eye that the writer has long since lost, and there is no end of ways in which an editor can improve a

manuscript: pruning, shaping, clarifying, tidying a hundred inconsistencies of tense and pronoun and location and tone, noticing all the sentences that could be read in two different ways, dividing awkward long sentences into short ones, putting the writer back on the main road if he has strayed down a side path, building bridges where the writer has lost the reader by not paying attention to his transitions, questioning matters of judgment and taste. An editor's hand must also be invisible. Whatever he adds in his own words shouldn't sound like his own words; they should sound like the writer's words.[6]

Zinsser tells us his relationships with editors have been both good and bad. He objects to editors who significantly alter a point a writer is trying to make in a sentence, or who go beyond changes in style or structure and significantly alter content. For Zinsser, "Ideally the relationship between a writer and an editor should be one of negotiation and trust.... The process...is one in which the writer and the editor proceed through the manuscript together, finding for every problem the solution that best serves the finished article."[7]

Approaches to Editing

The focus of this chapter is editing for style. As you will see, good editors edit a document for many other elements in addition to elements of style. Still, to understand how to edit for style more effectively, you need a better understanding of the broader concerns of editing and of some of the many major approaches to editing.

Whether you are editing your own documents or the documents of others, the problem is how to edit effectively. How do you go about the process of editing? Where do you begin? What can you do to make your editing more systematic?

There are probably as many approaches to editing as there are editors. After all, how you edit your own work or edit someone else's work is very much a matter of individual preference. Many people prefer to start with basic sentence, spelling, punctuation, and other errors on the first reading. Then they read the document a second time looking for larger errors in content, organization, and format. And then they will read the document a few more times to catch other errors not caught in the first few attempts. Others use no system at all. They merely go from page to page looking for any errors, both large and small, they can find.

Experienced editors, however, use some kind of system. One useful approach is to use *top-down* editing instead of *bottom-up* editing. As Carolyn Rude explains, in *top-down* editing, "You start with the most comprehensive document features, the ones that will affect others. These features include content, organization, and format, all of which need to be consistent throughout a document."[8] *Bottom-up* editing has some disadvantages compared to *top-down* editing. Rude observes that "*bottom-up* editing, beginning with sentences, can waste time and divert your attention from the comprehensive goals. You could start by correcting sentences for grammar and editing them for style, and then proceed to format and content; however, you might ultimately decide to delete many of those same sentences altogether."[9]

Sam Dragga and Gwendolyn Gong provide a helpful model for editing by showing how the canons of rhetoric—invention, arrangement, style, and delivery—correspond with the "rhetorical objectives of editing: accuracy, clarity, propriety, and artistry."[10] Editing for accuracy requires determining "the correctness and appropriateness of the writer's communicative aim, analysis of audience, and collected information."[11] Editing for clarity "requires the editor to examine the effectiveness of a writer's rhetorical choices regarding the arrangement of text."[12] Here the editor evaluates visual–verbal orientation, organization, and coherence and cohesion. Editing for propriety requires creating a comprehensive list or a style guide that lists numerous style decisions. The propriety of a document's visual style may also be evaluated. Finally, editing for artistry requires determining the delivery specifications, for example, "the printing process, paper, folding and binding, typography, illustrations, and page design."[13]

Some editors have proposed various levels to editing. Joseph Mancuso discusses the Radford levels approach in *Technical Editing*:

1. *Minimum Edit.* An editor proofreads the document and checks grammar, spelling, and mechanics.
2. *Light Copy Edit.* An editor checks for consistency among manuscript elements, for instance, between the table of contents and text headings.
3. *Full Copy Edit.* An editor checks style, usage, and parallelism.
4. *Supportive Edit.* An editor prepares part of the document, including table of contents, index, and cover sheet. The editor also specifies the physical appearance of the document.
5. *Substantive Edit.* The editor ensures the completeness and coherence of the document.[14]

Robert Van Buren and Mary Fran Buehler developed five types of editing for their work at the Jet Propulsion Laboratory at the California Institute of Technology in Pasadena, California. These types vary depending on the levels of editing each contains. The nine levels of edit are: coordination, policy, integrity, screening, copy clarification, format, mechanical style, language, and substantive. A type 1 edit requires applying all nine levels; a type 2 requires using all but substantive; a type 3 requires using all but mechanical style, language, and substantive; a type 4 involves only coordination, policy, integrity, and screening; and a type 5 requires checking only coordination and policy.

A brief discussion of the nine levels of edit will help clarify the challenges of editing for style over editing for other elements. For our purposes here, it is helpful to discuss these levels of edit into two broad groups—one group consisting of coordination, policy, integrity, screening, copy clarification, and format edits; and one group consisting of mechanical style, language, and substantive edits.

The first group of edits involves only minimal editing for style. A *coordination edit* consists of planning and estimating a document project, record maintenance, scheduling and schedule follow-up, manuscript markup, and monitoring and liaison. A *policy edit* makes certain that a publication represents the policy of the Jet Propulsion Laboratory. Required report elements are checked to see if they are present, headings are checked, and so on, but, most important, the document is checked for any inappropriate comments about companies or government agencies. And the document is checked to make sure no statements are included that endorse any products or services of a company. In an *integrity edit* the elements of a document are checked to make certain that the parts of the publication match. The table of contents is matched with headings in the document, page numbers and other numbers are checked, figure captions and table titles are checked, references are checked, and so on. A *screening edit* is a minimum editorial standard at the Jet Propulsion Laboratory. Spelling, subject–verb agreement, sentence completeness, incomprehensible statements, figures, and titles, among other elements, are checked. A *copy clarification edit* focuses on cleaning up illegible material or unproducible artwork. Finally, a *format edit* assures conformity in typography, layout, and figures and visual aids.

The second group of edits—a mechanical style edit, a language edit, and a substantive edit—all involve varying degrees of editing for style. A *mechanical style edit* requires checking many elements of a document for an appropriate and consistent style. This basic kind of style edit involves using the *U.S. Government Printing Office Style Manual* to check for consistency in capitalization, spelling, word compounding, acronyms and abbreviations, bibliographic reference style, use of special fonts, callouts, and other elements.

A *language edit* is "an in-depth review concerned with the way in which the ideas in a report are expressed."[15] This kind of edit covers spelling; grammar and syntax; punctuation; usage; fluency; parallelism; conciseness; proper use of description, exposition, narrative, and argument; identification of inconsistent or erroneous terminology; and abbreviations, acronyms, and symbols.

A *substantive edit* "deals with the meaningful content of the publication."[16] Many elements, overlapping with elements checked with the other kinds of edits, are covered in reviewing the overall publication, tables, and figures. This level of edit, however, does not review for technical accuracy.

Although some of these levels are specific to the kinds of writing done at the Jet Propulsion Laboratory, the nine levels of edit have become a widely used model for technical editors elsewhere. This kind of model can be particularly useful for freelance technical writers and editors who are asked to estimate the cost of editing a technical documentation set. Using this kind of model, a freelancer can clearly spell out types and level of edits to be done and charge accordingly.

The Challenges of Editing for Style

This chapter doesn't focus on the entire editing process because this is a textbook about style, not editing. There are some excellent books on editing and technical editing available, and, if you are serious about your writing, these texts should be part of your library. In addition to Carolyn Rude's *Technical Editing,* there is Judith Tarutz's *Technical Editing,* Sam Dragga and Gwendolyn Gong's *Editing: The Design of Rhetoric,* Donald Bush and Charles Campbell's *How to Edit Technical Documents,* and Joseph Mancuso's *Technical Editing.*[17] There are also numerous technical communication books with excellent chapters on editing the work of others, such as Gerald Alred's, Walter Oliu's, and Charles Brusaw's *The Professional Writer: A Guide for Advanced Technical Writing.*[18]

Before we begin discussing the specifics of editing for style, let's discuss a few more points about the challenges of editing.

First, editing your own style is often harder than editing someone else's style. The reason is simple. You are too familiar with your own material to look at it objectively. It's difficult to look at your writing from someone else's point of view. You know what you are trying to say, and you think you have stated your meaning clearly. Because you are too familiar with your own work, you will help yourself considerably if you set aside a draft for a day or two before editing it. This delay will help you look at the work with fresh eyes. You will often be surprised that what you wrote just a day or two ago sometimes misses the mark. A sentence you thought was clear is now not clear at all. The technical terms you thought were appropriate are now not suitable at all for either your purpose or your audience.

Second, editing someone else's style is often easier because you are unfamiliar with the document and you have the objectivity you need, but editing someone else's style also presents special challenges. The constant temptation is to alter the writer's style so that it's more like your own. You don't think the writer has made the point clearly in a particular sentence, and you believe that if only the writer phrased the sentence the way you would phrase it, then the sentence would be better. You have to be careful about imposing your habitual language choices on the work of others. There is no easy solution to this problem. You can begin, however, by being aware of stylistic choices and knowing that the same thought can often be phrased many different ways.

Strategies to Help You Edit for Style

In this textbook we have covered many style choices. Keeping these choices in mind and applying them systematically to your writing and the writing of others can be an effective way to edit for style. As you may recall, we covered ten broad topics, from being aware of your discourse community to being aware of your ethical choices. You should think of these topics as both strategies for helping you to improve your style and strategies to help you edit your style and the style of others. What follows is a kind of informal checklist (instead of another editing process) to help you examine your style and the style of others more carefully. Many of the questions or suggestions are part of a substantive edit, others are more appropriately part of a copyedit, and some are part of the final process of proofreading. Use the questions that follow to help you create your own approach to editing for style.

Checking the Discourse Community

As we have discussed, no single factor is more important for determining your stylistic choices than your audience. All of your stylistic choices should be carefully selected for their effectiveness on your audience. As Chapter 2 illustrates, discourse communities consist of audiences, and the expectations of audiences can vary widely within a discourse community as well as differ greatly in different discourse communities. Different discourse communities have different styles, and to write successfully in these discourse communities you must know what these different styles are and how to use them to your advantage.

Chapter 2 already provides a checklist of questions for you to ask concerning your discourse community. It's helpful to revise that list slightly here to aid you in editing for an awareness of a discourse community or communities in a technical document:

- What is your relationship to the audience? Peer? Subordinate? Supervisor?
- Is the audience expecting your document? Motivated to read it?
- What will your audience use the document for?
- How knowledgeable is your audience about the subject?
- Are you assuming any basic knowledge (theories, assumptions, ideas) you should not assume the audience knows?
- Are there special preferences in this discourse community you need to give particular attention to?
- Is your audience comfortable with the level of formality in your document?
- Is your audience comfortable with your other stylistic preferences?
- Are there particular political elements (tensions, relationships among people, histories, backgrounds) you need to pay special attention to?
- Is this the best time to present this document or would a later time be better?
- What will the results be if your document fails? Succeeds?
- Have you considered all of the other important elements necessary for an effective document in this particular discourse community?

Checking the Appropriateness of the Style

In Chapter 3 we discussed a variety of writing styles: a primer style, a telegraphic style, a passive style, a verb style, a nominal style, an affected style, a plain style, a complex style, and a literary style. We discussed how some of these styles often overlap, how, for example, writers of technical prose will often combine a verb style and a noun style. As you check your document for appropriateness in style, you need to ask some basic questions:

- Is this the most effective style for this particular subject, purpose, audience, and context?
- What kind of style does my audience expect in this particular situation?
- Is my audience familiar and comfortable with this style?

Checking the Persuasiveness

In Chapter 4 we discussed the fundamental persuasive nature of technical documents. Of course, many technical documents are not overly persuasive, but many are. If your document or the document of another aims at being persuasive, you need to check carefully its persuasive elements.

- Have you carefully considered your readers' position concerning your subject and purpose?
- Have you considered the impact your purpose will have on your audience? What is that expected impact? What other reactions could also result?

- Do you use formal persuasive strategies such as Aristotelian argument or Rogerian argument?
- Is the logic of your argument sound? Do you contradict or undermine your points in any way?
- Have you presented your argument in the best possible order for your audience? Should you change the organization in any way to achieve a greater psychological impact?
- Does any part of your document alienate or offend your audience? If so, what kinds of changes can you make to the content?

Checking the Diction

As you know, Chapter 5 covers many levels of diction and many diction topics. It's not practical to offer a comprehensive checklist here, but the following questions offer a good start:

- Have you chosen the appropriate level of diction? Why is it formal, informal, colloquial, or slang?
- Have you avoided problems with connotations? Are any words unnecessarily emotionally charged?
- Have you checked to make sure that you are using the most accurate technical terms?
- Have you used active verbs and active voice wherever appropriate and possible?
- Have you checked your nontechnical diction to make sure you are using the most appropriate words for the intended effect or tone?
- Are any of your words unclear or could any be more clear?
- Have you examined each sentence carefully to cut out wordiness and to achieve conciseness?
- Are you using concrete and specific words and avoiding, wherever it's appropriate, unnecessary abstract words?
- Are you consistent in your terminology throughout the document?
- Are you avoiding pretentious words and using simpler or less complex words wherever possible?
- Are you using words that best express your sincerity and seriousness?
- Are you avoiding common diction faults, from archaic words to wooliness?
- If appropriate, have you used style elements common to a literary style yet avoided an ornate or embellished literary style?

Checking the Technical Terms and Jargon

Of course, technical terms and jargon present special editing challenges. Chapter 6 reminds you that what technical terms you use very much depends on your audience. If you are writing for an audience of peers within your discourse community, there is little need for definition, clarification, or elaboration. Often, however, you must write for a lay audience or a combined audience, so the challenges of handling technical terms are more difficult.

- Have you defined any technical terms you use for an audience that is unfamiliar with these terms? How do you define these terms?
- Can you simplify any of your definitions, offer additional examples, or provide any illustrations that may give your audience a better understanding of some of the terms you use?
- Are you avoiding pretentious words? Have you checked to see if a simpler word can be substituted for any of the technical terms you use?
- Have you carefully considered how familiar your audience is with your subject? Do you have a good idea of the knowledge level, education level, and other important background details concerning your audience?
- If appropriate, have you provided a helpful glossary?
- Have you prepared your audience in every way for the technical terms you use?
- If you use pretentious language, does this use suit your purpose?
- After reading your document, will your readers have a good understanding of your topic?
- Have you checked every sentence to make sure you have avoided creating gobbledygook for your readers?
- Have you handled all abbreviations, acronyms, and initialisms carefully and clearly? Should you also include a list of abbreviations, for example, with your document?

Checking the Sentences

Good editing requires you to check every sentence to make sure that it is as clear, concise, and accurate as it can be.

- Have you appropriately varied the forms (simple, compound, complex, compound-complex) of your sentences?
- Have you avoided unnecessary compound-complex sentences in your technical prose?
- How would you characterize your sentence style? Chiefly loose sentences with some periodic and balanced sentences? Could more sentences be balanced?
- Are you using lists effectively wherever possible? Are these lists parallel?
- Have you effectively used sentence combining techniques (coordination, subordination, relative clauses, participial phrases, appositives, absolutes, noun substitutes, rearrangement, addition/deletion, and emphasis wherever possible?
- Have you ended your sentences effectively by, for example, trimming the end, shifting less important ideas to the front of the sentence, or shifting important ideas toward the end of the sentence?
- When referring to technical terms for the first time, do you make every effort to place them at the end of the sentence?
- Have you paid attention to the rhythm of your sentences?
- Have you avoided putting too many short sentences together?
- Have you varied the length of your sentences? Are any of your sentences too long?
- Are some of your sentences too artful or too elegant?

- Have you checked carefully for common sentence faults? Comma splices? Fragments? Fused sentences? Run-on sentences? Dangling modifiers? Awkward sentences?

Checking the Paragraphs and Other Segments

Technical prose must be carefully edited at the word, phrase, clause, sentence, paragraph, and larger segment level. Editing technical paragraphs can be as painstaking as editing at any other level.

- Is each of your paragraphs on one topic only?
- Is the length of each paragraph appropriate for its purpose?
- Do you demonstrate an awareness of the principles of effective page layout and design in your paragraphing? Are the paragraphs short and introduced by headings where appropriate? Do your headings clearly indicate the topic to be discussed? Is your heading hierarchy clear? Are your headings parallel? Have you provided enough headings to help guide your reader through your document?
- Have you used the most appropriate paragraph pattern to develop each of your paragraphs?
- Do you achieve an effective cohesion within your paragraphs?
- Do you achieve an effective coherence within your paragraphs and from paragraph to paragraph? Have you used effective transitional words wherever appropriate? Have you repeated key words where necessary? Repeated key ideas where necessary?
- Are your paragraphs complete? Are any of the paragraphs undeveloped or lacking in detail? If you have decided to use very short paragraphs throughout your technical document, do you still provide your reader enough information?
- Does your document reveal careful attention to segments larger than paragraphs? Are these larger segments unified, ordered, cohesive, coherent, and complete?
- Have you avoided common paragraph faults? Are there any inconsistencies in verb tense, mood, voice, person and number, tone, and point of view?

Checking the Tone

Checking for an appropriate tone may be one of the most difficult editing tasks because often it is difficult to establish an appropriate tone.

- Have you been appropriately informal or formal with your audience?
- Have you been appropriately impersonal or personal?
- Have you successfully conveyed the attitude or emotion you want to convey? Have you checked carefully for any words, phrases, clauses, sentences, or paragraphs that may undermine the attitude or emotion you want to convey?
- Have you successfully established the ethos you want to establish? Have you established the authority or credibility you want to establish? Have you established trust if necessary? Have you appropriately established control, deference, or an awareness of a delicate situation if necessary?

- If you use humor, is it appropriate for your subject, purpose, and audience? If you don't use humor, would using humor help make your document more effective?

Checking for Bias

As Chapter 10 shows, bias may be unconscious or blatant. It may be subtle or overbearing. To edit your work well, you must also check for bias.

- Do you reveal a particular bias or biases in your technical document? If so, what is this bias? Is this bias reasonable or is it an unreasonable prejudice?
- Are you guilty of any slanting in your document?
- Have you checked your document carefully for any problems in gender bias? Have you made every effort to make your writing gender-neutral? Have you fully considered the gender biases of your audience?
- Have you checked for any corporate, philosophical or religious, political, racial, cultural, and age bias, and avoided any bias against the physically or mentally challenged?
- Have you checked your document to see if it is politically correct on the matters that most concern your audience?

Checking for Ethics

Of course, editing for ethics is not something you save for last or do at the end of the editing process. Ethics is something you must be aware of from the first word you write (if not even before you begin to write).

- Are you deliberately ambiguous, deceptive, or pretentious in any part of your technical document? Are your reasons ethical?
- Have you measured your language decisions within the framework of obligations, ideals, and consequences? Are you satisfied that this framework shows your decisions to be ethical?
- Do you have any doublespeak (euphemisms, jargon, gobbledygook or bureaucratese, or inflated language) in your document? If so, why?
- Have you deliberately used other strategies (besides choosing to be ambiguous, deceptive, or pretentious) to lie with language in your document? Have you omitted any essential or potentially damaging information? Have you distorted any significant data?

Copyediting for Style

As we discussed earlier, many of the style questions listed concern substantive editing. And, as we also discussed earlier, copyediting occurs after a thorough substantive editing. The goal is to check carefully for errors in correctness, consistency, accuracy, and completeness.[19] Document correctness focuses on spelling, punctuation, grammar, and style. Document consistency assures that the text or visuals are treated the same way throughout. Hyphenation, heading hierarchies, and fonts are just some of the many elements checked for consistency. Document accuracy focuses on content, for example, possible errors in dates, names, numbers, and visuals. Finally, document completeness assures that all of the necessary front and back matter are included, that all of the necessary pages are provided, and that all other items essential for making the document complete are included.

As you can see, when you copyedit a document for correctness you have your last chance to catch errors in style. The other kinds of copyediting and the final part of the editing process, proofreading, are concerned with other issues.

Chapter Summary

Editing technical documents, like writing technical documents, is a craft and takes years of practice to do well. While it may be true that most writers are not good editors, it is also true that writers can learn how to become better editors of their own work and the work of others.

You can become a better editor by understanding that editing is a complicated process, a process involving substantive editing, copyediting, and proofreading. You need to know how these kinds of editing differ from each other and why substantive editing is done first and proofreading last. You can also become a better editor by editing more systematically, preferably using a *top-down* approach instead of a *bottom-up* approach. And you also need to be familiar with other editing approaches such as the one involving the four canons and editorial objectives or the five types and nine levels used at the Jet Propulsion Laboratory.

You need to be aware that editing for style presents special challenges whether you are editing your own work or the work of another. When you are editing your own style, it can be especially difficult, for example, to notice when your sentences are not as clear as they should be or when certain words are the most appropriate words. When you are editing the work of others, it is especially difficult to resist imposing your own stylistic preferences.

Ten major style topics should be checked in every technical document that you write or edit. You should check every technical document for how effectively it conveys information to a discourse community, the appropriateness of the style, its persuasive elements, the diction, the technical terms and jargon, the sentences, the paragraphs, the tone, the bias, and the ethics. Editing carefully for style will help make your document or the documents of others much more effective.

Questions/Topics for Discussion

1. What are some other common misconceptions about editing and editors in addition to those mentioned here?
2. How do you feel about someone else editing your work? What good experiences have you had? What bad experiences have you had?
3. Do you think it makes sense to refer to editing as a three-part process of substantive editing, copyediting, and proofreading? Why or why not?
4. What are some of the important differences between substantive editing and copyediting?
5. What are some of the important differences between copyediting and proofreading?

6. What qualities and skills do you think a good technical editor must have?
7. What do you think are the special challenges of editing someone else's style?
8. Editing for style has been presented in ten broad sections in this chapter beginning with checking the discourse community to checking the ethics. Have any important topics or questions been left out? If so, what are they?
9. What do you think is the best way to handle an editor who has substantially changed the style and meaning of your technical document?
10. What's the difference between a top-down and a bottom-up approach to editing? Which approach works best for you? Why?
11. What are some advantages for freelance editors of using a systematic approach to editing?

Exercises

1. For the following examples, discuss what each one needs in terms of further editing. What are some of the changes you would recommend in the style of the passages? Discuss why you think your recommendations would improve the passages.

Example 1

The problem of tennis elbow affects many athletes. A regime of massage therapy can help to eliminate the pain and stiffness that accompanies this condition. If the symptoms of tennis elbow are present, the elbow should be used as little as possible. At the onset of the symptoms of tennis elbow, 24–48 hours of rest should be given to the joint. Apply ice to the elbow for no more than ten minutes each day to reduce soreness. A program of therapeutic massage should then be started. Therapeutic massage will prevent the muscles in the area from shortening and bring blood to the injured area to promote healing.

Example 2

Every language is made up of three basic elements: phonemes, allophones, and morphemes. A phoneme is a basic unit or category of sound such as /p/ or /t/ (the phoneme is the sound made when /p/ or /t/ is pronounced). Allophones are the various sounds, or manifestations, of a particular phoneme in any given language environment. Morphemes are basic units of meaning, and are made up of allophones. Morphemes can be either "free," or "bound." Free morphemes can stand alone as words. For example, the words "hat" and "auto" are free morphemes. Bound morphemes can only function as part of a word, such as an "s" added onto a word to make it plural. Bound morphemes can be derivational morphemes which change the category or meaning of a word; or, bound morphemes can be inflectional in which case the meaning of a word is changed only slightly as when a word becomes plural or past tense.

Example 3

How to use the self-service pump at the gas station without having to go inside to pay:

Park your car next to the gas pumps, so that it is parallel to the pump containing the type of gas you want. Be sure your gas tank is within reach of the pump hose. Now turn off the motor and step out by the pumps. Locate the button pad with the message window beside it on your selected pump. The window message says "Select payment type." Press the green square button that says "Pay Outside." The window message tells you to insert your card. Use one of the credit cards listed on the pumps, such as Mastercard or Visa. The vertical card slot is located to the left of the selection buttons and below the window. Face the magnetic stripe on the card to the left and toward the top of the card. (The diagram on the pump shows you which way to insert the card.) Insert the card and removed it quickly. If you don't withdraw the card quickly enough, the window tells you to reinsert the card. Once you withdraw the card successfully, the window says "Authorizing..." If your card is accepted, the window tells you that the pump is ready and to lift the lever.

Note: If your credit card is unacceptable for some reason, the window gives you a message to that effect. In this instance, select the "Pay Inside" button, and pump your gas as described below. Then go in the store and pay the clerk with cash or a valid credit card.

If your card is authorized, open the cover to your gas tank and remove the cap. Next lift the hose off the pump. You will see a lever where the hose was resting. Lift it up. Now you are ready to pump your gas. Insert the pump and squeeze the inside portion of the handle against the stationary part of the handle. Watch the center window on the pump to see how much gas you have pumped into your tank. Continue to squeeze the handle until you have the amount of gas in your tank that you want. Then release your grip on the handle to stop the flow of gasoline. Remove the hose from your tank, push the lever on the gas pump down, and set the hose back in its original position. Be sure you put the cap back on your gas tank and close the cover. The window on the gas pump tells you it is printing your receipt. Remove your paper receipt when it is through.

Example 4

The First Century

People ride bicycles for different reasons. They may commute, ride leisurely, cycle for sport, or perhaps live out all three. However, for the commuters and the leisure cyclers the adventure of the 100 mile obstacle does not loom as great as it does above the helmet of the sport cycler. A cycler's first century requires much more preparation than the average bike trek, and being safe will allow you to have a faster and more enjoyable 100 miles.

Most sport cyclers have experienced the twenty or fifty mile ride, but imagine being fifty miles into your trip—without the proper clothing, fuel supply, or tools—and the only thing you have left is the thought of the next fifty miles.

Thinking is the most important aspect of the century, whether you are on the road or training the week before. First, take in consideration your clothing. Bike shorts are a must for long trips, as those rashes from previous trips without them have preached to you. A tight shirt is also recommended to increase your aerodynamics which will increase your speed proportionally. A strong pair of gloves should be worn in order to remove wire or glass from your tires or perhaps cushion a fall. In addition, a pair of sport glasses will provide safety from flying debris and shade your eyes from the sun's ultra-violet rays, and speaking of which, make sure you wear a coat of sunscreen. And finally to protect your thinker—always wear a helmet.

Secondly, a proper diet before and during that first century is crucial. Your body needs adequate fuel to build endurance, to free you from cramps, and most of all to provide energy and strength to those leg muscles that will be pedaling for close to five hours. You don't want to pack heavy with fuel because it can weigh you down, but you do want to pack smart. Liquids are severely important, whether they are fluids high in carbohydrates or just regular water, you need a good supply. A good tip in fuel planning is to map parts of your century close to several locations such as stores or rest areas to use as refueling pit stops.

2. Consider the following short document. Discuss how it takes into account the concerns of the discourse community of novice photographers. Suppose instead that it was written for a photography magazine read by professional photographers or a magazine for both novice and experienced photographers or the discourse community of weekend birdwatchers. What changes would have to be made to make the document more effective for each of these discourse communities?

> When enlarging a picture from a 35 mm negative, you would follow these steps.
>
> 1. Place the negative strip into the negative holder, shiny side facing up.
> 2. Place the negative holder into the slot on the enlarger just under the light mechanism.
> 3. Place the paper into the easel at the bottom of the enlarger. The shiny side of the paper should face up.
> 4. Turn on the lamp on the enlarger to expose the paper. Desired exposure time is usually between 3–7 seconds. A test strip (a small sample of paper) should be done first to acquire appropriate time. (Follow these steps for the test strip too.)
> 5. After exposure is complete, place the paper into a small tub of developer chemical. The picture will begin to appear.
> 6. Remove picture from developer when picture has completed developing. Place picture in a tub of stop bath or tap water. This will stop the developing process.
> 7. Remove picture from stop bath and place in a tub of fixer for a minimum of 5 minutes. This makes the picture permanent.
> 8. Remove picture from fixer and rinse chemicals off with water. Hang up or place on a flat wire rack to dry. (Usually overnight)

3. Make your own list of the twenty most important style elements to edit in a technical document. Are any of the elements in your list not in this chapter?
4. Discuss the advantages of a systematic approach to editing such as the one offered by the Jet Propulsion Laboratory. Are there any disadvantages to this process? If so, what are they?
5. Write a memo or letter discussing why you think you are a better writer than editor or a better editor than writer.
6. Find a brief technical document available on the Internet, print it, and edit it for weaknesses in style. Discuss your reasons for your suggestions.
7. Why do page layout and design, as well as technical illustrations, often require editing for style?
8. Make a list of your stylistic preferences and pet-peeves. Discuss whether you would call attention to any errors in these preferences if you found them in someone else's work.
9. Consider the following excerpt from an installation manual. What are its weaknesses in the use of technical terms and jargon? What kinds of editing suggestions would you offer for improving this document?

> Step 5   Initialize the ProDrive hard disk for use in your system.
>
> Some ProDrive configurations come with an automated installation program designed for DOS applications. This program will set the drive type as well as partition and format the drive. Use this program if it was provided with your drive.
>
> a. For programmable or custom drive type tables:
>
> Under many computer SETUP utilities, a programmable or custom drive type entry can be configured for the exact specifications of the hard disk (i.e. cylinders, heads, sectors/track). If your system has this capability, Quantum recommends that you use this option. This op-

tion will give you use of the entire storage capacity of the hard disk under most operating systems and does not require a device driver for the hard disk. Refer to the computer operation manual for more information.[20]

10. Find a document that is clearly unethical. In what ways is it practical to suggest that the writer make changes? In what ways is it impractical?

Notes

1. Sally Smith, "Getting There from Here: Total Immersion Editing," *The Editorial Eye: Focusing on Publications Standards and Practices,* URL: http://www.eei-alex.com/eye/tarutz.html.

2. Robert Van Buren and Mary Fran Buehler, *The Levels of Edit,* 2nd ed. (Pasadena, California: Jet Propulsion Laboratory, 1980) 1.

3. Carolyn Rude, *Technical Editing* (Belmont, CA: Wadsworth, 1991) 2.

4. The professional editors of *Editorial Eye* have three general levels: proofreading, copyediting, and substantive editing, which they refer to as "the levels *light, medium,* and *heavy.*" See Robin A. Cormier, "Estimating Editorial Tasks: A Five-Step Method," *The Editorial Eye: Focusing on Publications Standards and Practices.* Online, http://www.eei-alex.com/eye/estimate.html

5. Rude, p. 181.

6. William Zinsser, *On Writing Well: An Informal Guide to Writing Nonfiction,* 5th ed. (New York: HarperPerennial, 1994) 278–79.

7. Zinsser, p. 280.

8. Rude, p. 208.

9. Rude, p. 208.

10. Sam Dragga and Gwendolyn Gong, *Editing: The Design of Rhetoric* (Amityville: Baywood, 1989) 15.

11. Dragga and Gong, p. 29.

12. Dragga and Gong, p. 56.

13. Dragga and Gong, p. 183.

14. Joseph Mancuso, *Technical Editing* (Englewood Cliffs, NJ: Prentice Hall, 1992) 44.

15. Van Buren and Buehler, p. 21.

16. Van Buren and Buehler, p. 23.

17. Judith Tarutz, *Technical Editing* (Boston: Addison-Wesley, 1992); Sam Dragga and Gwendolyn Gong, *Editing: The Design of Rhetoric* (Amityville, NY: Baywood, 1989); Donald Bush and Charles Campbell, *How to Edit Technical Documents* (Phoenix, AZ: Oryx, 1995); Joseph Mancuso, *Technical Editing* (Englewood Cliffs, NJ: Prentice Hall, 1992).

18. Gerald Alred, Charles Brusaw, and Walter Oliu, *The Professional Writer: A Guide for Advanced Technical Writing* (New York: St. Martin's, 1992).

19. See Rude's Chapter Three, "Copyediting: An Introduction," for an excellent overview of this editing process, pp. 34–67. The discussion here is indebted to Rude's overview.

20. Quantum ProDrive®, IDE-AT Interface, Hardware Installation Manual.

Glossary

abbreviation   a shortened version of a word, formed from its first letter or letters

absolute   a simple sentence subordinated to an independent clause by removing or altering its verb form to establish a closer connection between clauses

abstract word   a word that expresses an idea or quality that cannot be perceived by the five senses

academic style   a complex style characteristic of much academic prose, using literary or elevated vocabulary

accuracy   freedom from mistake or error; exactness in technical prose

acronym   a pronounceable abbreviation formed from the first letter or letters of several words

action verb   a verb that describes what is, was, or will be done

addition/deletion   the building up of sentence elements by deleting material from shorter sentences and adding them together to form longer sentences

affected style   a pretentious style that strives for complexity; an unnecessarily complex style

alliteration   the repetition of initial consonant sounds in two or more neighboring words or syllables

ambiguity   the quality or state of being unclear, resulting from using words subject to more than one interpretation

analogy paragraph   a type of comparison paragraph that applies a familiar relationship to an unfamiliar instance

anthropomorphizing   the attribution of human form or personality to non-human things (*see* personification; mechanism)

antithesis   a technique to achieve balance by pitting one half of a sentence against the other

appositive   a phrase that identifies or defines something

appropriate style   a style suitable for its intended purpose

archaic word   a word having the characteristics of the language of the past

argot   criminal slang (*see* cant)

assonance   the occurrence of identical vowel sounds preceded and followed by different consonant sounds

attitude   one of the three elements of tone (*see* level of formality, ethos); a state of mind with regard to a fact or situation (*also* emotion) *audience* those for whom a product, production, written work, or speech is intended

bad style a style exhibiting poor control of the elements that produce well-crafted writing

balanced sentence one of three types of sentences classified according to the arrangement of their material (*see* periodic sentence, loose sentence); a sentence that uses parallel structure, antithesis, or symmetry to achieve its rhythm

barbarism an idea, act, or expression that offends contemporary standards of good taste or acceptability

bias a premature judgment or inclination; a prejudice

blooper an embarrassing public blunder

bottom-up editing editing a work by beginning at the sentence level for grammar and style, and proceeding to higher-level features of form and content

bureaucratic style a style common in business, academe, and industry; characterized by dense, complex, and often pompous phrasings (*also* The Official Style)

buzzwords fashionable, trendy words or phrases

cant language of the underworld or other group, meant to be obscure to outsiders

catchword/catchphrase a memorable word or phrase, repeated by many people

cause-and-effect paragraph a paragraph that describes a process as a chain of causes and effects

chiasmus a technique to achieve balance by using two expressions, the second of which reverses the word order of the first (*also* symmetry)

classification paragraph a paragraph that performs an analysis by dividing its subject into parts and showing how these parts are related

clause a group of words that contains a subject and a predicate

cliché a trite phrase or expression

clipped word a shortened word created by dropping off part of a longer word

coherence a property of a smooth-flowing paragraph that promotes its topic through the use of effective transitions and the repetition of key words and ideas

cohesion a property of a smooth-flowing paragraph that promotes its topic by the joining of old and new elements in each sentence

colloquial diction one of the four levels of diction (*see* formal diction, informal diction, slang); the language of conversation, used in informal, everyday speech

colloquialism a local or regional dialect expression

combined audience an audience composed of readers with various degrees of familiarity or expertise in a subject (*also* complex audience, mixed audience)

comma splice a sentence fault resulting from joining together two independent clauses using a comma instead of a coordinating conjunction, a period, or a semi-colon

comparison paragraph a paragraph that points up similarities between two or more things

complex audience an audience composed of readers with various degrees of familiarity or expertise in a subject (*also* combined audience, mixed audience)

complex sentence one of the four sentence forms (*see* simple sentence, compound sentence, compound-complex sentence); a sentence composed of one independent clause and one or more dependent clauses

complex style a style characterized by technical, abstract, theoretical, or difficult-to-understand material; an intricate or involved style

compound sentence one of the four sentence forms (*see* simple sentence, complex sentence, compound-complex sentence); a sentence composed of two or more simple sentences connected by coordinating conjunctions or by punctuation

compound-complex sentence one of the four sentence forms (*see* simple sentence, compound sentence, complex sentence); a sentence composed of two or more independent clauses and at least one dependent clause (*also* complex-compound sentence)

computerese computer jargon applied to non-computer applications

concrete word a word that expresses specific realities

confusibles words easily confused with one another

connective a word that joins sentence parts (*also* connector)

connector a word that joins sentence parts (*also* connective)

connotation the affective meaning of a word, which suggests a meaning apart from the thing it explicitly names or describes

consistency the parallel treatment of similar or identical terms and expressions each time they appear

consonance the recurrence or repetition of consonant sounds

contraction the shortening of a word, syllable, or word group by the omission of a sound or letters

contrast paragraph a paragraph that points out differences between two or more things

coordination the use of conjunctions to combine simple sentences

coordination edit one of the nine levels of edit; an edit covering the planning and estimating of a documentation project, record maintenance, scheduling and monitoring, and liaison

copy-clarification edit one of the nine levels of edit; an edit to ensure that artwork and text are legible

copyediting an intermediate stage in the editing process in which the editor checks for errors in punctuation, correct spelling, consistency, accuracy, and completeness (*see* substantive editing, proofreading)

corporate bias the tendency of a corporation to reflect its values and mission in its communications

cultural bias the tendency to promote the views or beliefs of one particular racial, religious, or national group as superior to that of other groups

dangling modifier a modifying word or phrase that does not modify the apparent subject of the sentence

dead metaphor a word or phrase whose metaphoric meaning has been weakened through overuse

deadwood an unneeded or redundant word or phrase

declarative sentence one of four kinds of sentences classified according to function (*see* imperative sentence, interrogative sentence, exclamatory sentence); a sentence that makes a statement or asserts a fact

deduction the affirmation of a particular instance from a generalization

definition paragraph a paragraph that clarifies new terms, concepts, or ideas using definitions

denotation the cognitive meaning of a word, which relates a word directly to an object, action, or idea *dependent clause*   a clause that is used as a part of speech in a sentence but does not make sense when standing alone (*also* subordinate clause)

descriptive approach   an empirical, context-based approach to language that focuses on how people actually use language

descriptive paragraph   a paragraph that provides a vivid impression of a person, place, or object

detailed paragraph   a paragraph that uses many details to develop its topic

dialect   a regional variety of language

diction   a writer's or speaker's choice of words

discourse community   a group of people who commonly associate with each other and share a special language, experiences, basic knowledge, expectations, values, methodologies, idiosyncrasies, and goals

double meaning   the interpretation of a word or phrase in more than one way

double negative   the use of two negatives for emphasis within a sentence

doublespeak   deliberately obscure or deceptive language

dysphemism   an expression used for its offensive connotation

editing for accuracy   editing to ensure the correctness of factual information in a text

editing for artistry   editing to ensure the attractive presentation of material in a text

editing for clarity   editing to ensure that presented information communicates effectively

editing for propriety   editing to ensure compliance with a style guide

elegance   the effect achieved when creating successful artful prose

elegancies   words or phrases that writers believe to be more elegant than the words they replace; genteelisms

emotion   one of the three elements of tone (*see* level of formality, ethos); a state of mind with regard to a fact or situation (*also* attitude)

emphasis   drawing attention to important sentence elements by positioning them advantageously within the sentence

enumeration paragraph   a paragraph that presents its supporting points as a numerical list

equivocation   intentional ambiguity

ethics   the study of right and wrong conduct

ethos   one of Aristotle's three modes of persuasion (*see* logos, pathos); one of the three elements of tone (*see* level of formality, attitude); focuses on the speaker or writer's character, integrity, competence, and objectivity

etymology   the study of word origins

euphemism   a polite word or expression that makes an unpleasant subject seem less offensive

example paragraph   a paragraph that uses examples to illustrate its topic

exclamatory sentence   one of four kinds of sentences classified according to function (*see* declarative sentence, imperative sentence, interrogative sentence); a sentence that expresses surprise or strong emotion

executive   someone, usually in business, who leads a division or a company and who is responsible for making decisions

expert   someone who has extensive knowledge about a subject

expletive   a syllable, word, or phrase used as a placeholder, with no meaning of its own; an exclamatory word or phrase, usually obscene or profane *expository prose*   prose whose primary purpose is to inform

expressive prose   prose whose primary purpose is to convey emotion or feeling, which may in the process reveal the writer in the prose; emotive prose

figure of speech   an expression that conveys meaning or heightens effect, often through a comparison with more familiar images

formal diction   one of the four levels of diction (*see* informal diction, colloquial diction, slang); language used in formal contexts for serious occasions: speeches, legal documents, academic journal articles

format edit   one of the nine levels of edit; an edit to ensure uniformity in typography, layout, figures, and visual aids

fragment   an incomplete sentence

fused sentence   a sentence composed of two independent clauses with no punctuation or coordinating conjunctions in between

gender bias   an expression of a stereotyped attitude toward a person or group based on sex or sexual role

gender-neutral language   language that includes men and women, without explicitly or implicitly excluding either

gender stereotype   a judgment on the usual or desirable traits or behaviors characteristic of either men or women

general-to-specific paragraph   a paragraph that begins with a topic sentence and provides specific supporting details to illustrate the topic sentence

generic "he"   a masculine pronoun used as if it were gender-neutral

gobbledygook   the unintelligibility resulting from using jargon and long sentences inappropriately

good style   a mastery of the elements that produce well-crafted writing

grammar   a summary of the rules of language that govern syntax and morphology

high (grand) style   writing elevated in tone, subject, and expression; associated with epic poetry or ceremonial occasions

homographs   words that are spelled the same but pronounced differently

homonyms   words that are pronounced the same but spelled differently

homophones   words that are pronounced the same but have different meanings

hyperbole   extravagant exaggeration

hypernym   a general word used in preference to a specific one

hyponym   a specific word used in preference to a general one

idiom   a fixed expression whose meaning cannot be deduced from the meanings of its constituent parts

immeasurable word   a word representing a concept not generally quantified

imperative sentence   one of four kinds of sentences classified according to function (*see* declarative sentence, interrogative sentence, exclamatory sentence); a sentence that expresses a command, request, or entreaty

improprieties   improper or indecorous acts or remarks

inappropriate style   writing unsuited for its intended use

independent clause   a clause that makes sense by itself (*also* principal clause, main clause)

induction   the process of reaching a generalization from particulars *informal diction*   one of the four levels of diction (*see* formal diction, colloquial diction, slang); language used in ordinary, casual, everyday settings

initialism   an abbreviation formed from the first letters of the important words in an expression and pronounced as separate letters

integrity edit   one of the nine levels of edit; an edit to ensure that internal references of a publication match

interrogative sentence   one of four kinds of sentences classified according to function (*see* declarative sentence, imperative sentence, exclamatory sentence); a sentence that asks a question

"izing"   the formation of a verb by adding "ize" to a noun or adjective

jabber   incoherent, unintelligible speech

jargon   the specialized vocabulary of a trade, profession, hobby, or science; technical terminology used inappropriately for an audience; pretentious or unnecessarily complex words used to impress others

language edit   one of the nine levels of edit; an edit to ensure correct grammar and syntax, punctuation, usage, fluency, parallelism, conciseness, correct terminology, abbreviations, and symbols

layperson   someone who does not belong to a specific discourse community and who is therefore unfamiliar with the usual terms and expressive devices of that group

level of formality   one of the three elements of tone (*see* attitude or emotion, ethos); the distance between writer, subject, and audience; may range from highly formal to very informal, from very personal to very impersonal

levels of diction   a hierarchy of formality in word choice

levels of edit   the editing process broken down into categories according to the facet of the document being examined (*see* nine levels of edit)

literarisms   words used characteristically by the literary or learned

literary style   a style of writing that relies heavily on literary devices to carry and embellish its message

logos   one of Aristotle's three modes of persuasion (*see* ethos, pathos); focuses on the nature of the message or the quality of the data; the facts

loose sentence   one of three types of sentences classified according to the arrangement of their material (*see* periodic sentence, balanced sentence); a sentence that is composed of add-on phrases and clauses, without a set pattern (*also* running style)

low style   writing suitable for entertainment, popular forums, folk language

main clause   a clause that makes sense by itself (*also* independent clause, principal clause)

malapropism   humorous misuse or distortion of a word or phrase

meaningless modifiers   words that neither restrict nor clarify the words they modify

mechanical style edit   one of the nine levels of edit; an edit to ensure appropriate and consistent style

mechanism   the attribution of technological attributes to humans (*see* anthropomorphizing, personification)

metaphor   a figure of speech that implicitly compares unlike things; an implied comparison

metathesis   transposition of the letters or syllables of a word

metonym   a figure of speech in which a word that represents a part is used in place of the word for the whole

middle style   writing suitable for educational or edifying purposes

mixaphor   a term coined by Theodore Bernstein for a mixed metaphor

mixed audience   an audience composed of readers with various degrees of familiarity or expertise in a subject (*also* combined audience, complex audience)

mixed metaphor   a figure of speech combining inconsistent or incongruous metaphors

modifier   a word or group of words that restricts, limits, or makes more exact the meaning of other words; a word or words that qualify another word

modifier strings   groups of modifiers used together without a noun or phrase in between

moral dilemma   a situation in which obligations, ideals, or consequences conflict, making it impossible to satisfy them simultaneously

morphology   the structure of words and how they are formed from smaller units

narrative paragraph   a paragraph that reports on something that happened

neologism   a newly invented word or expression

new paradigm of technical writing   the view that technical writing is a rhetorical discipline and that technical writers interpret technical material for their audience

new rhetoric of science   the view that science writing is a rhetorical activity and that hypotheses, theories, and laws of science are interpretations

nine levels of edit   a system developed at the Jet Propulsion Laboratory at the California Institute of Technology to delimit elements covered in various types of edits (*see* coordination edit, policy edit, integrity edit, screening edit, copy-clarification edit, format edit, mechanical style edit, language edit, and substantive edit)

nominal style   a style that relies heavily on nouns and their derivatives (nominalizations, noun strings, noun adjectives) to carry meaning

nominalization   the formation of a noun from a verb or adjective

nonce word   a word coined for a single occasion

nonsense word   a word conveying no intelligible idea

noun adjective   a noun used to modify another noun

noun phrase   a noun with its attached modifier

noun string   two or more nouns placed together; an expression created by combining noun phrases

noun substitute   a clause beginning with "that" or with an altered form of the verb, functioning in a sentence as a noun

objective style   a style of writing undistorted by personal feelings or prejudices

old paradigm of technical writing   the view that technical writing is wholly objective and that writers of technical material serve as passive bridges between subject matter and audience

oxymoron   a combination of contradictory or incongruous words; a self-contradictory expression

paragraph   a block of text composed of a variable number of sentences that generally treats one topic or one aspect of one topic

parallel structure   a technique to achieve balance by using a series of nouns, verbs, phrases, clauses, or sentences

paraprofessional   someone in the process of learning about a specialized community

participial phrase   a dependent clause introduced by a participle

participle   a verb ending in *-ed* or *-ing*

passive style   a style that uses a preponderance of verbs in the passive voice
pathos   one of Aristotle's three modes of persuasion (*see* ethos, logos); focuses on convincing the audience through appeals to emotion
periodic sentence   one of three types of sentences classified according to the arrangement of their material (*see* balanced sentence, loose sentence); a tightly structured sentence that makes its point in the last word or phrase
person   the personal pronouns that refer to the writer, the reader, and the subject
persona   the role constructed by the narrator of a text
personification   a figure of speech that attributes human characteristics to non-human things or abstract ideas (*see* anthropomorphism, mechanism)
philosophical bias   the tendency to promote the general beliefs and attitudes of a particular group, institution, or philosophy as superior to others
phrase   a group of related words used as part of a sentence but not having a subject or predicate
plain style   a straightforward, easy-to-understand style (*also* reader-friendly style)
point of view   the relationship of the writer to the information, expressed in the use of person
policy edit   one of the nine levels of edit; an edit to ensure that a document conforms with company policy for reports
political bias   the tendency to promote the interests of one political group or subgroup over other groups
political correctness   an exaggerated tendency to render language inoffensive by removing supposedly inequitable references to various groups of people
precision   the degree of refinement with which an operation is performed or a measurement stated
predicate   that which is said about the subject in a sentence
prescriptive approach   a traditional, authoritarian, rule-based approach to language
primer style   a style characterized by short, simple sentences, suitable for children or military training manuals and useful for those with poor reading skills and limited education
principal clause   a clause that makes sense by itself (*also* independent clause, main clause)
proofreading   the last stage of the editing process, correcting errors in typeset material (*see* substantive editing, copyediting)
propaganda   slanted writing that promotes the agenda of an institution, cause, or person
provincialism   a word or expression common to a specific geographic area (*also* regionalism)
purpose   the writer's intended goal or aim; what writers want their readers to learn from their writings
question-to-answer paragraph   a paragraph that asks a question at the beginning and then provides an answer
racial bias   the tendency to view one group as inherently superior to another based on race or its supposed characteristics
rearrangement   the positioning of sentence elements to improve a sentence's effectiveness
redundant words   words in excess of those needed to express an idea
regionalism   a word or expression common to a specific geographic area (*also* provincialism)

relative clause   a dependent clause related to a noun in an independent clause through a relative pronoun
relative word   a word that depends on the location or position of the reader for its meaning
religious bias   the tendency to promote the religious beliefs and attitudes of a particular group as superior to that of other groups
rhetoric   the art of persuasive writing or speaking
rhythm   a recurring pattern of strong and weak elements or stresses
Rogerian argument   a form of communication that presents information from the other person's point of view, emphasizing cooperation over conflict
run-on sentence   an unnecessarily long sentence
scope   the boundaries or limits of a subject and its depth of detail
screening edit   one of the nine levels of edit; an edit to ensure correct spelling, subject-verb agreement, sentence completeness
sentence   a word or group of words that expresses a complete thought
sentence combining   a technique to create effective sentences by reducing complex ideas to their simplest parts and then blending the simple parts into sentences
sentence variety   the use of a mixture of different sentence forms, functions, styles, combining techniques, and lengths to produce more effective prose
sexist language   language that expresses the gender bias of a writer or speaker
shop talk   jargon used within only one organization or occupation subgroup
simile   a figure of speech that explicitly compares unlike things
simple sentence   one of the four sentence forms (*see* compound sentence, complex sentence, compound-complex sentence); a sentence with one independent clause and no dependent clauses
slang   one of the four levels of diction (*see* formal diction, informal diction, colloquial diction); a specialized, colloquial language; language peculiar to a particular group; the most informal level of language, often used for its ability to express an attitude rather than for the information it conveys
slanting   a distorted presentation
solecism   a violation of the rules of grammar
specific-to-general paragraph   a paragraph that begins with supporting details and ends with the topic sentence
style   the cumulative effect of the words, phrases, clauses, sentences, and paragraphs chosen and arranged by a writer
stylistics   the analysis of style and its elements
subject   the topic of a sentence
subordinate clause   a clause that is used as a part of speech in a sentence but does not make sense when standing alone (*also* dependent clause)
subordination   the use of conjunctions to transform independent clauses into dependent clauses
substantive editing   one of the nine levels of edit; an early stage of the editing process in which the editor checks the content and meaning of the technical document, verifies the accuracy, organization, design, style, and logic of the document (*see* copyediting, proofreading)
suggestible   a word invoked by the use of another that it resembles, however superficially

symmetry   a technique to achieve balance by using two expressions, the second of which reverses the word order of the first (*also* chiasmus)
synonyms   words that have the same or nearly the same meaning
syntax   the structure of sentences and the arrangements of words and word clusters
taboo words   language unacceptable in polite society
tautology   the unnecessary repetition of a word, phrase, or idea
tech speak   a parody of jargon; language that is equally mystifying to the expert and the layperson
technical terminology   the vocabulary used in the technical trades and professions
technical writing   the art of writing about a discipline for a specific audience
technician   someone whose job involves working with technology or machinery
technobabble   an affected style commonly used in the computer industry (coined by John Barry)
telegraphic style   a style that omits articles, conjunctions, and transitional phrases; suitable for taking notes or making lists
term   a word having a particular meaning in a specialized field
terminology   the vocabulary of a specialized field
The Official Style   *see* bureaucratic style
tone   the attitude you take toward your subject, your audience, and yourself
top-down editing   editing a work beginning with the most comprehensive features and proceeding to the lowest-level features
topic sentence   one sentence in a paragraph that captures the essential message of that paragraph
totality word   a generalization that expresses an all-or-nothing attribute
traditional rhetoric of science   the view that science writing is a transparent medium for recording scientific fact without distortion
transitional words   words that link sentences or parts of sentences to show addition, contrast, comparison, or result
understatement   the underplaying of an idea to create an effect
unmarked usage   the gender-inclusive use of a single-gender noun
unnecessary word   a word that can be omitted without affecting the clarity of the message
unusual word   a word with several meanings, some of which may be unfamiliar to the reader
vagueness   the quality of language not clearly expressed
verb style   a style that relies extensively on verbs
verbiage   a profusion of words
verbizing   the formation of verbs from other parts of speech
vogue word   a fashionable or popular word
voice   the relationship of subject and object in a sentence or clause; a narrative point of view
vulgarism   idiomatic or informal language not generally considered respectable
-wise word   a word created by adding the suffix "-wise" to another word
wooliness   inexact language
wordiness   an abundance or overabundance of words
zeugma   the use of one word to modify or govern two words or phrases, even though it has a different sense when applied to each

Bibliography

Ahearne, John F. "Telling the Public About Risks." *Bulletin of the Atomic Scientists* (Sep. 1990): 37–39.

Allan, Keith, and Kate Burridge. *Euphemism and Dysphemism: Language Used as Shield and Weapon.* New York: Oxford UP, 1991.

Allen, Jo. "The Case Against Defining Technical Writing." *Journal of Business and Technical Communication* 4.2 (1990): 68–77.

Alred, Gerald, Charles Brusaw, and Walter Oliu. *The Professional Writer: A Guide for Advanced Technical Writing.* New York: St. Martin's, 1992.

Altman, Rebecca Bridges, and Rick Altman. *Mastering PageMaker 5.0® for Windows™.* San Francisco: Sybex, 1993.

Anderson, Paul. *Technical Writing: A Reader-Centered Approach.* 3rd ed. New York: Harcourt Brace Jovanovich, 1995.

August, Eugene. "Real Men Don't: Anti-Male Bias in English." *Language Awareness.* Ed. Paul Eschholz, Alfred Rosa, and Virginia Clark. 5th ed. New York: St. Martin's Press, 1990.

Baker, Russell. "Little Red Riding Hood Revisited." *New York Times Magazine,* 13 Jan. 1980, 10.

Barklund, Jonas. "Bounded Quantifications for Iteration and Concurrency in Logic Programming." *New Generation Computing* 12.2 (1994): 161–182.

Barnes, J., ed. *The Complete Works of Aristotle: The Revised Oxford Edition.* Vol. 2. Trans. W. R. Roberts. Princeton, NJ: Princeton UP, 1984.

Barry, John A. *Technobabble.* Cambridge, MA: MIT P, 1991.

Barth, John. "Teacher." *The Best American Essays 1987.* Ed. Gay Talese. New York: Ticknor & Fields, 1987.

Barzun, Jacques. *Simple and Direct: A Rhetoric for Writers.* Rev. ed. New York: Harper and Row, 1985.

The Belmont Report 1979. Federal Register II: Federal Policy for the Protection of Human Subjects. Notices and Rules. 1991. Online. Available at URL: http://sun.soci.niu.edu/~jthomas/ethics.cmu

Bentley, Tom. *SimFarm Addendum and Quick Start Guide.* Walnut Creek, CA: Maxis, 1993.

Bernstein, Theodore M. *The Careful Writer: A Modern Guide to English Usage.* New York: Atheneum Verlag, 1977.

Bode, Carl. *Mencken.* Baltimore: Johns Hopkins UP, 1986.

Boiarsky, Carolyn. "The Relationship Between Cultural and Rhetorical Conventions: Engaging in International Communication," *Technical Communication Quarterly* 4.3 (1995): 245–259.

Bolinger, Dwight L. *Language, The Loaded Weapon: The Use and Abuse of Language Today.* New York: Longman, 1980.

Build Better Solutions for Your Business. Redmond, WA: Microsoft, 1995.

Burgess, Anthony. *A Clockwork Orange.* New York: Ballantine, 1986.

Burkett, Eva M., and Joyce S. Steward. *Thoreau on Writing.* Conway, AR: University of Central Arkansas P, 1989.

Bush, Donald, and Charles Campbell. *How to Edit Technical Documents.* Phoenix, AZ: Oryx, 1995.

Chapman, Robert L., ed. *American Slang.* New York: Harper and Row, 1987.

Charrow, Veda. "What Is Plain English, Anyway?" Document Design Center, Technical Report. Washington: American Institutes for Research, Dec. 1979.

Ciardi, John. *A Browser's Dictionary and Native Guide to the Unknown American Language.* New York: Harper and Row, 1980.

Coletta, W. John. "The Ideologically Based Use of Language in Scientific and Technical Writing." *Technical Communication Quarterly* 1.1 (1992): 59–70.

Cook, Reginald. *Passage to Walden.* 2nd ed. New York: Russell, 1966.

Cooper, Henry S. F., Jr. *Thirteen: The Flight That Failed.* New York: Dial, 1973.

Cormier, Robin A. "Estimating Editorial Tasks: A Five-Step Method." *The Editorial Eye: Focusing on Publications Standards and Practices.* Online. Available at URL: http://www.eei-alex.com/eye/estimate.html

Cowart, Robert. *Mastering Windows® 95: The Windows 95 Bible.* San Francisco: Sybex, 1995.

Cross, Donna. "Propaganda: How Not to be Bamboozled." *Language Awareness.* Ed. Paul Eschholz, Alfred Rosa, and Virginia Clark. 5th ed. New York: St. Martin's, 1990.

Crystal, David. *The Cambridge Encyclopedia of the English Language.* New York: Cambridge UP, 1995.

Davey, Peter. "An Intemperate Argument." *Architectural Review* July 1994: 4–5.

Day, Rob. *Designer Photoshop: From Monitor to Printed Page.* New York: Random House, 1995.

Day, Robert. *How to Write and Publish a Scientific Paper.* 3rd ed. Phoenix, AZ: Oryx, 1988.

———. *Scientific English: A Guide for Scientists and Other Professionals.* Phoenix, AZ: Oryx, 1992.

Deal, C. *The Greenpeace Guide to Anti-Environmental Organizations.* Berkeley: Odonian, 1993.

DeGeorge, James, Gary A. Olson, and Richard Ray. *Style and Readability in Technical Writing: A Sentence-Combining Approach.* New York: Random House, 1984.

DeNesti, Linn Susanne. "A Practical Box of Fun." *Adobe Magazine* Mar-Apr. 1996: 31.

Dery, Mark. *Escape Velocity: Cyberculture at the End of the Century.* New York: Grove, 1996.

Dethier, Vincent G. *To Know a Fly.* New York: McGraw-Hill, 1962.

Dombrowski, Paul. "*Challenger* and the Social Contingency of Meaning: Two Lessons for the Technical Communication Classroom." *Technical Communication Quarterly* 1.3 (1992): 73–86.

Donnellan, LaRae. "Technical Writing Style: Preferences of Scientists, Editors, and Students." Paper. Conference on College Composition and Communication. 1985. ERIC ED 258 262.

Dragga, Sam, and Gwendolyn Gong. *Editing: The Design of Rhetoric.* Amityville, NY: Baywood, 1989.

Duncan, Ray. "The Windows File-Oriented Common Dialog Functions." *PC Magazine* 26 May 1992: 379–386.

Eastman, Richard. *Style: Writing and Reading as the Discovery of Outlook.* 3rd ed. New York: Oxford UP, 1984.

Eiseley, Loren. "More Thoughts on Wilderness." *A Writer's Reader.* 7th ed. New York: HarperCollins, 1994.

The Even More Incredible Machine: Journal Entries June Thru August. Coarsegold, CA: Sierra On-Line, n.d.

Ewing, David W. *Writing for Results in Business, Government, the Sciences, and the Professions.* 2nd ed. New York: Wiley, 1979.

Fahey, Tom. *The Joys of Jargon.* New York: Barron's, 1990.

Fielden, John. "'What Do You Mean You Don't Like My Style?'" *Harvard Business Review* 60.3 (1982): 128–138.

Fine, Michael J., Melanie A. Smith, et al. "Prognosis and Outcomes of Patients with Community-Acquired Pneumonia." *Journal of the American Medical Association* 10 Jan. 1996: 135.

Finegan, Edward. *Language: Its Structure and Use.* Fort Worth, TX: Harcourt Brace Jovanovich, 1994.

Flanagan, David. *Motif Tools: Streamlined GUI Design and Programming with the XMT Library.* CD-ROM. O'Reilly, 1994.

Forester, Tom, and Perry Morrison. *Computer Ethics: Cautionary Tales and Ethical Dilemmas in Computing.* Cambridge, MA: The MIT P, 1992.

Fraser, Bruce. "Graphic Images." *Publish* Mar. 1996: 50–55.

Freeh, Louis J. *Opening Statement Before the Subcommittee on Terrorism, Technology, and Government Information, Committee on the Judiciary,* 19 Oct. 1995. FBI Online. Available at URL: http://www.fbi.gov/rubystat.htm#top

Fumento, Michael. "Will the U.S. Repeat Peru's Deadly Chlorine Folly?" Online. Available at URL: http://www.townhall.com/townhall/columnists/fumento /fume021496.html

Galluzzi, Michael C. "Post-Production Support Through Real-Time Information Software and Design Process Reform." Redstone Arsenal, AL: U.S. Army Missile Command, n.d.

Gerson, Steven M. "Commentary: Teaching Technical Writing in a Collaborative Computer Classroom." *Journal of Technical Writing and Communication* 23.1 (1993): 23–31.

Glossbrenner, Alfred. *The Little Online Book: A Gentle Introduction to Modems, Online Services, Electronic Bulletin Boards, and the Internet.* Berkeley: Peachpit, 1994.

Godnig, Edward C., and John S. Hacunda. *Computers and Visual Stress: Staying Healthy.* Grand Rapids, MI: Abacus, 1991.

Gookin, Dan, and Andy Rathbone. *PCs for Dummies.* 2nd ed. San Mateo, CA: IDG, 1994.

Gould, Stephen Jay. *The Mismeasure of Man.* New York: Norton, 1981.

Green, Barry. "At Your Service: OPI servers streamline printing process." *Publish* Mar. 1996: 83t.

Greenbaum, Sidney, Denis Baron, and Tom McArthur. "Grammar." *The Oxford Companion to the English Language.* Ed. Tom McArthur. New York: Oxford UP, 1992.

Hahn, Harley, and Rick Stout. *The Internet Complete Reference.* New York: Osborne, 1994.

Halloran, Michael. "Technical Writing and the Rhetoric of Science." *Journal of Technical Writing and Communication* 8.2 (1978): 77–88.

Harmon, Joseph. "Perturbations in the Scientific Literature." *Journal of Technical Writing and Communication* 16.4 (1986): 311–317.

Harris, John. "Metaphor in Technical Writing." *Technical Writing Teacher* 2.2 (1975): 9–13.

———. "The Naming of Parts: An Examination of the Origins of Technical and Scientific Vocabulary." *Journal of Technical Writing and Communication* 14.3 (1984): 183–191.

———. "Shape Imagery in Technical Terminology." *Journal of Technical Writing and Communication* 16:1/2 (1986): 55–61.

Hayakawa, S. I. "How to Attend a Conference." *The Use and Misuse of Language.* Ed. S. I. Hayakawa. Greenwich, CT: Fawcett, 1962.

Hays, Robert. *Principles of Technical Writing.* Reading, MA: Addison-Wesley, 1965.

———. "What is Technical Writing?" *Word Study* Apr. 1961: 1–4.

Hedtke, John. *Using Computer Bulletin Boards.* 2nd ed. New York: MIS, 1992.

Hickey, Dona. *Developing a Written Voice.* Mountain View, CA: Mayfield, 1993.

Holleran, Andrew. "Bedside Manners." *A Writer's Reader.* 7th ed. New York: HarperCollins, 1994, 226–231.

Holman, C. Hugh, and William Harmon. *A Handbook to Literature.* 6th ed. New York: Macmillan, 1992.

Houp, Kenneth W., Thomas E. Pearsall, and Elizabeth Tebeaux. *Reporting Technical Information.* 8th ed. Boston: Allyn & Bacon, 1995.

Howe, Irving, ed. *Orwell's* Nineteen Eighty-Four*: Text, Sources, Criticism.* 2nd ed. New York: Harcourt Brace Jovanovich, 1982.

Introducing Microsoft Windows 95 ®. Redmond, WA: Microsoft, 1995.

Jack, Judith. "Teaching Analytical Editing." *Technical Communication* 31.1 (1984): 9–11.

Johnston, David. "Report Shows FBI Officials Blocked Access to Documents." *The New York Times News Service.* Online. Available at URL: http://www.tribnet.com/~stories/news/oo768.htm

Journet, Debra. "Parallels in Scientific and Literary Discourse: Stephen Jay Gould and the Science of Form." *Journal of Technical Writing and Communication* 16.4 (1986): 299–310.

Karow, Peter. *Font Technology: Methods and Tools.* Berlin: Springer-Verlag, 1994. Trans. of *Schrifttechnologie.*

Kelley, Patrick, and Roger Masse. "A Definition of Technical Writing." *Technical Writing Teacher* 4 (1977): 94–97.

Kies, Daniel. "Some Stylistic Features of Business and Technical Writing: The Functions of Passive Voice, Nominalization, and Agency." *Journal of Technical Writing and Communication* 15.4 (1985): 299–308.

Killingsworth, Jimmie. *Information in Action.* Boston: Allyn & Bacon, 1996.

Kilpatrick, James. "Mrs. Malaprop's Mangled Prose Set a President." *Smithsonian* Jan. 1995: 82–87.

———. *The Writer's Art.* New York: Andrews, 1984.

Kolln, Martha. *Rhetorical Grammar: Grammatical Choices, Rhetorical Effects.* New York: Macmillan, 1991.

Krol, Ed. *The Whole Internet User's Guide and Catalog.* 2nd ed. Sebastopol, CA: O'Reilly, 1994.

Laib, Nevin. *Rhetoric and Style: Strategies for Advanced Writers.* Englewood Cliffs, NJ: Prentice-Hall, 1993.

Lakoff, Robin. *Talking Power: The Politics of Language in Our Lives.* New York: Basic Books, 1990.

Lamb, Linda. *Learning the vi Editor.* Sebastopol, CA: O'Reilly, 1994.

Lane, Karen, David Foster, and Heather Stuart. *LView Pro 1.6 User's Guide.* Merritt Island, FL: DKH Productions, 1995.

Lanham, Richard A. *Analyzing Prose.* New York: Scribner's, 1983.

———. *The Electronic Word: Democracy, Technology, and the Arts.* Chicago: University of Chicago Press, 1993.

———. *Revising Business Prose.* New York: Macmillan, 1992.

———. *Revising Prose.* 3rd ed. New York: Macmillan, 1992.

Lardner, George Jr. "Freeh Says Actions at Ruby Ridge Were 'Flawed.'" *Washington Post* 20 Oct. 1995. Online. Available at URL: http://the-tech.mit.edu/V115/N50/freeh.50w.html

Lay, Mary, Billie Wahlstrom, Stephen Doheny-Farina, Ann Hill Duin, Sherry Burgus Little, Carolyn Rude, Cynthia Selfe, and Jack Selzer. *Technical Communication.* Chicago: Irvin, 1995.

Legg, Cyndy. "The Different Faces of Multimedia." Unpublished essay, 1994.

Leonhard, Woody. *The Underground Guide to Word for Windows:™ Slightly Askew Advice from a WinWord Wizard.* Reading, MA: Addison-Wesley, 1994.

Levine, John, and Margaret Levine Young. *UNIX for Dummies.* San Mateo, CA: IDG, 1994.

Limerick, Patricia. "Dancing With Professors: The Trouble With Academic Prose." *New York Times Book Review* 31 Oct. 1993: 3+.

Lipsyte, Robert. "Boxing, Baseball, and 'Rejuberation.'" *New York Times Online* 19 Jan. 1996.

Locke, David. *Science As Writing.* New Haven: Yale University Press, 1992.

Loh, Sandra Tsing. "In L. A. They're Called Professional Girlfriends." *Cosmopolitan* Feb. 1996: 182–187.

Lutz, William. *Doublespeak: From "Revenue Enhancement" to "Terminal Living"—How Government, Business, Advertisers, and Others Use Language to Deceive You.* New York: Harper and Row, 1989.

Mancuso, Joseph. *Technical Editing.* Englewood Cliffs, NJ: Prentice-Hall, 1992.

"Marburg (Green Monkey) Virus Disease (African Hemorrhagic Fever)." *Review of Medical Microbiology.* 13th ed. Los Altos, CA: Lange Medical, 1978.

Marder, Daniel. "Technical Reporting is Technical Rhetoric." *Technical Communication* 25.4 (1978): 11–13.

McArthur, Tom, ed. *The Oxford Companion to the English Language.* New York: Oxford UP, 1992.

McCarthy, Mary. *Memories of a Catholic Girlhood.* San Diego: Harcourt Brace Jovanovich, 1985.

McCormick, John. *Create Your Own Multimedia System.* New York: Windcrest, 1995.

McFedries, Paul. *The Complete Idiot's Guide to Windows® 95.* Indianapolis: Que, 1995.

Merriam-Webster's Tenth Collegiate Dictionary and Thesaurus. Electronic Edition. CD-ROM. 1995.

Microsoft PowerPoint for Windows 95: Step by Step. Redmond, WA: Microsoft Press, 1995.

Miller, Carolyn R. "A Humanistic Rationale for Technical Writing." *College English* 40.6 (1979): 610–617.

Miller, Casey, and Kate Swift. *The Handbook of Nonsexist Writing.* 2nd ed. New York: Harper and Row Publishers, 1988.

———. "One Small Step for Genkind." *The Gender Reader.* Ed. Evelyn Ashton-Jones and Gary A. Olson. Boston: Allyn & Bacon, 1991. 247–259.

———. "One Small Step for Genkind." *Language Awareness.* Ed. Paul Eschholz, Alfred Rosa, and Virginia Clark. 5th ed. New York: St. Martin's Press, 1990. 307–319.

Mirshafiei, Mohsen. "Culture as an Element in Teaching Technical Writing." *Technical Communication* 41.2 (1994): 276–282.

Montaigne, Michel de. *The Complete Essays of Montaigne.* Trans. Donald M. Frame. Stanford: Stanford UP, 1989.

Montgomery, John. *The Underground Guide to UNIX.* Reading, MA: Addison-Wesley, 1994.

Morrison, Toni. *Beloved.* New York: Knopf, 1987.

Murphy, Kathleen. "The Bard: Hi-Tech Storytelling." *Film Comment* 31.4 Jul./Aug. 1995: 37–42.

Nash, Walter. *Jargon: Its Uses and Abuses.* Cambridge, MA: Blackwell, 1993.

Nelson, Stephen L. *Field Guide to Microsoft Windows® 95.* Redmond, WA: Microsoft, 1995.

Newman, Edwin. *Strictly Speaking.* New York: Warner, 1974.

Nilsen, Alleen Pace. "Sexism in English: A 1990s Update." *The Gender Reader.* Ed. Evelyn Ashton-Jones and Gary A. Olson. (Boston: Allyn & Bacon, 1991). 260–267.

Nunberg, Geoffrey. "The Decline of Grammar." *Atlantic Monthly* Dec. 1983: 31–46.

Oliver, Jim. "The Randy Weaver Case." *American Rifleman.* Nov. 1993. Online. Available at URL: http://eagle.tamn.edu/~carlp/Liberty/Weaver.Case.AR.html

Oriel, John. *NAWCTSD Technical Report 93–022. Engineering Specification Editing Tools. Final Report.* Orlando, FL: Naval Air Warfare Center Training Systems Div., Dec. 1993.

Partridge, Eric. *Usage and Abusage: A Guide to Good English.* New York: Norton, 1995.

Peck, Robert Newton. *Fiction is Folks.* Cincinnati: Writer's Digest, 1983.

Personal Computer Operation and Installation Guide. Rev. D. Text Block, 1992.

Pinker, Steven. *The Language Instinct.* New York: Morrow, 1994.

Pirsig, Robert. *Zen and the Art of Motorcycle Maintenance.* New York: Morrow, 1974.

Popper, Karl. *Conjectures and Refutations: The Growth of Scientific Knowledge.* New York: Harper and Row, 1963.

Pringle, Ian. Rev. of *Style: Ten Lessons in Clarity and Grace,* by Joseph Williams. *College Composition and Communication* 34.1 (1983): 91–98.

Prosise, Jeff. "Help: Working Smarter—Care and Feeding of Your Hard Disk." *PC Computing,* June 1995: 194–198.

Prudden, Bonnie. *Myotherapy: Bonnie Prudden's Complete Guide to Pain-Free Living.* New York: Ballantine, 1984.

Quantum ProDrive®, IDE-AT Interface, Hardware Installation Manual.

Raspberry, William. "What It Means to Be Black." *Language Awareness.* 5th ed. Ed. Paul Eschholz, Alfred Rosa, and Virginia Clark. New York: St. Martin's, 1990. 269–271.

Rawson, Hugh. *A Dictionary of Euphemism and Other Doubletalk: Being a Compilation of Linguistic Fig Leaves and Verbal Flourishes for Artful Users of the English Language.* New York: Crown, 1981.

Ray, Robert J. *The Weekend Novelist.* New York: Dell, 1994.

Redish, Janice. "Writing in Organizations." *Writing in the Business Professions.* Ed. Myra Kogen. Urbana, IL: NCTE/ABC, 1989. 97–124.

"Relational Databases." *Guide To Cube.* Orlando, FL: Newtrend, 1994.

Reynolds, Steve. "Can Consumer Online Be Master of Its Own Evolution?" *The Red Herring* Mar. 1995.

Rheingold, Howard. *The Virtual Community: Homesteading on the Electronic Frontier.* Reading, MA: HarperCollins, 1994.

Robin, Laura, "Hypertext Authoring Environment: A Critical Review." *Ejournal* 3.3 (Nov. 1993). Online. Available at http://carbon.cudenver.edu/~mryder/itcdatajournals.html

RoboHELP User's Guide. La Jolla, CA: Electron Image, 1994.

Rodman, Lilita. "Anticipatory It in Scientific Discourse." *Journal of Technical Writing and Communication* 21.1 (1991): 17–27.

Room, Adrian. *Room's Dictionary of Confusibles.* Boston: Routledge, 1979.

Rosner, Mary. "Style and Audience in Technical Writing: Advice from the Early Texts." *Technical Writing Teacher* 11.1 (1983): 38–45.

Rowell, Andrew. *The Exxon Valdez: A Case of Virtual Reality.* Ed. Andrea Goodall. Amsterdam: Greenpeace International, 1994.

Rubens, Philip, ed. *Science and Technical Writing: A Manual of Style.* New York: Holt Rinehart and Winston, 1992.

Rude, Carolyn D. *Technical Editing.* Belmont, CA: Wadsworth, 1991.

Rugg, Tom. *LANtastic Made Easy.* Berkeley: Osborne McGraw-Hill, 1992.

Ruggiero, Vincent Ryan. *Thinking Critically About Ethical Issues.* 3rd ed. Mountain View, CA: Mayfield, 1992.

Rutter, Russell. "History, Rhetoric, and Humanism: Toward a More Comprehensive Definition of Technical Communication." *Journal of Technical Writing and Communication* 21.2 (1991): 133–153.

Sachs, Harley. "Rhetoric, Persuasion, and the Technical Communicator." *Technical Communication* 25.4 (1978): 14–15.

Sacks, Oliver W. *A Leg to Stand On.* New York: Summit, 1984.

———. *The Man Who Mistook His Wife for a Hat and Other Clinical Tales.* New York: Summit, 1985.

Safire, William. *Coming to Terms.* New York: Holt Rinehart and Winston, 1991.

Sanders, Scott. "How Can Technical Writing Be Persuasive?" *Solving Problems in Technical Writing.* Ed. Lynne Beene and Peter White. New York: Oxford UP, 1988. 55–78.

Schindler, George E., Jr. "Why Engineers and Scientists Write As They Do—Twelve Characteristics of Their Prose." *IEEE Transactions on Professional Communication* 18.1 (1975): 5–10.

Shakespeare, William. *Hamlet.* Ed. Harold Jenkins. New York: Methuen, 1982.

Shirk, Henrietta Nickels. "Prologue to Teaching Software Documentation." *Perspectives on Software Documentation: Inquiries and Innovations.* Ed. Thomas T. Barker. Amityville, NY: Baywood, 1991.

Simon, John. *Paradigms Lost.* New York: Clarkson Potter, 1980.

Smith, Sally. "Getting There from Here: Total Immersion Editing." *Editorial Eye: Focusing on Publications Standards and Practices.* Online. Available http://www.eei-alex.com/eye/tarutz.html

Specter, Arlen, and Herb Kohl. *Executive Summary of Ruby Ridge Report of the Subcommittee on Terrorism, Technology, and Government Information of the Senate Judiciary Committee.* Online. Available at URL: http://www.nra.org/pub/general/96-02-02_senate_report_on_ruby_ridge

State of Alaska. Alaska Oil Commission. *Spill: The Wreck of the Exxon Valdez.* Final Rept., Feb. 1990.

STC Membership Directory 1995–1996 42.3A (Sept. 1995).

Stephens, Mitchell, and Nadyne Edison. "Coverage of Events at Three Mile Island." *Mass Comm Review* 7.3 (Fall 1980): 3–9.

Stent, Gunther S. Preface. *The Double Helix.* By James Watson. New York: Norton, 1980. ix-x.

Stern, Jerry. *Making Shapely Fiction.* New York: Norton, 1991.

Stott, Bill. *Write to the Point and Feel Better About Your Writing.* Garden City, NY: Anchor, 1984.

Strunk, William, and E. B. White. *The Elements of Style.* 3rd ed. New York: Macmillan, 1979.

STS and Associated Payloads: Glossary, Acronyms, and Abbreviations. NASA Reference Publication, 1981.

Suchan, James, and Robert Colucci. "An Analysis of Communication Efficiency Between High-Impact and Bureaucratic Written Communication." *Management Communication Quarterly* 2 (1989): 454–484.

Susskind, Charles. *Understanding Technology.* Baltimore: Johns Hopkins UP, 1973. Online. Available at URL: http://ei.cs.vt.edu:80/~cs3604/spring.95/Hippocr.Oath.html

Tarutz, Judith. *Technical Editing.* Boston: Addison-Wesley, 1992.

Tenner, Edward. *Tech Speak or How to Talk High Tech.* New York: Crown, 1986.

"The New Hacker's Dictionary—A Portrait of J. Random Hacker." Online. Available at URL: http://www.tpconsultants.com/tnhd/portrait.htm

Thomas, Lewis. *The Fragile Species.* New York: Collier, 1995.

———. *Late Night Thoughts on Listening to Mahler's Ninth Symphony.* New York: Bantam, 1984.

———. *Lives of a Cell: Notes of a Biology Watcher.* New York: Bantam, 1983.

———. *The Medusa and the Snail: More Notes of a Biology Watcher.* New York: Bantam, 1980.

———. "A Meliorist View of Disease and Dying." *Journal of Medicine and Philosophy* 1.3 (1976): 212–221.

Thoreau, Henry David. *Walden and Civil Disobedience.* Ed. Owen Thomas. New York: Norton, 1966.

Toulmin, Stephen, Richard Rieke, and Allan Janik. *An Introduction to Reasoning.* 2nd ed. New York: Macmillan, 1984.

Turk, Christopher, and John Kirkman. *Effective Writing: Improving Scientific, Technical, and Business Communication.* 2nd ed. New York: Spon, 1989.

T/V Exxon Valdez Oil Spill: Federal On Scene Coordinator's Report. Vol. 1. U.S. Coast Guard, Dept. of Transportation, 1993.

Twain, Mark [Samuel Langhorne Clemens]. *Adventures of Huckleberry Finn.* 2nd ed. Ed. Sculley Bradley, Richmond Croom Beatty, E. Hudson Long, and Thomas Colley. New York: Norton, 1977.

Tyman, Stephen T. "Ricoeur and the Problem of Evil." *The Philosophy of Paul Ricoeur.* Ed. Lewis Edwin Hahn. Open Court, 1995.

United States. Presidential Commission on the Space Shuttle *Challenger* Accident. *Report to the President.* William Rogers, chm. 5 vols. Washington: GPO, 1986.

———. President's Commission on the Accident at Three Mile Island. *Report of the President's Commission on the Accident at Three Mile Island: The Need for Change.* By John G. Kemeny. Washington: GPO, 1979.

User's Guide: Microsoft Word. Redmond, WA: Microsoft, 1994.

Van Buren, Robert, and Mary Fran Buehler. *The Levels of Edit.* 2nd ed. Pasadena, CA: Jet Propulsion Lab., 1980.

VanDeWeghe, Richard. "What is Technical Communication? A Rhetorical Analysis." *Technical Communication* 38.3 (1991): 295–299.

Vaughan, Tay. *Multimedia: Making It Work.* 2nd ed. New York: Osborne McGraw-Hill, 1994.

Wales, Katie. *A Dictionary of Stylistics.* New York: Longman, 1989.

Walter, John. "Usage and Style in Technical Writing: A Realistic Position." *The Practical Craft: Readings for Business and Technical Writers.* Eds. W. Keats Sparrow and Donald Cunningham. Boston: Houghton Mifflin, 1978.

Watson, James D. *The Double Helix.* Ed. Gunther S. Stent. New York: Norton, 1980.

Watson, J. D., and F. H. C. Crick. "A Structure for Deoxyribose Nucleic Acid." *Nature* 25 Apr. 1953: 737–38. Rpt. in *The Double Helix: A Personal Account of the Discovery of the Structure of DNA.* By James D. Watson. Ed. Gunther S. Stent. New York: Norton, 1980.

Webster's New World Dictionary. 2nd ed. Ed. David B. Gurlanik. New York: Simon and Schuster, 1982.

Webster's Tenth Collegiate Dictionary and Thesaurus. CD-ROM. 1995.

Weiss, Timothy. "Translation in a Borderless World." *Technical Communication Quarterly* 4.4 (1995): 407–425.

Whissen, Thomas. *A Way With Words: A Guide for Writers.* New York: Oxford UP, 1982.

Whitburn, Merrill. "Personality in Scientific and Technical Writing." *Journal of Technical Writing and Communication* 6.4 (1976): 299–306.

———. "The Plain Style in Scientific and Technical Writing." *Journal of Technical Writing and Communication* 8.4 (1978): 349–57.

White, Alex. *How to Spec Type.* New York: Watson-Guptill, 1987.

Whittington, Michael. *The Canadian Political System: Environment Structure, and Process.* 4th ed. Toronto: McGraw-Hill, 1987.

William, B. *United States Petroleum Strategies in the Decade of the Environment.* Tulsa, OK: Penn Well, 1991.

Williams, Joseph. *Style: Ten Lessons in Clarity and Grace.* 4th ed. New York: Harper and Row, 1994.

Winsor, Dorothy A. "The Construction of Knowledge in Organizations: Asking the Right Questions About the *Challenger.*" *Journal of Business and Technical Communication* 4.2 (1990): 7–20.

Yeoman, R. S. *A Guide Book to United States Coins.* 48th ed. Racine, WI: Whitman, 1995.

Young, Richard E., Alton L. Becker, and Kenneth L. Pike. *Rhetoric: Discovery and Change.* New York: Harcourt, 1970.

Younger, Irvin. "The English Language is Sex-Neutral." *Persuasive Writing.* Minnetonka, MN: Professional Education Group, 1990. 85–88.

Zinsser, William. *On Writing Well: An Informal Guide to Writing Nonfiction.* 5th ed. New York: Harper and Row, 1994.

Zuchero, John. "Computer-Based Training: Reinvent Yourself Through Multimedia." *Intercom* Mar. 1996: 18–20.

Index

abbreviations, 132, 133–134
   Latin, 101
absolutes
   for combining sentence elements, 152
   nominative, 146
abstracts and tone, 192–194
abstract words, 96
academic writing, 24, 46
   academese, 120–122
   advantages of, 24
   peer standards in, 25
accuracy, 52–53
   diction strategy, 91–92
   editing for, 257
accusation, tone of, 197–198
acronyms, 132, 133–134
action verbs, diction strategy, 92–93
active voice
   diction strategy, 92–93
   in Rogerian argument, 72
   in technical writing, 192
addition and deletion for combining sentence elements, 152–153
affected style, 51–52, 54
affective meaning. *See* connotation
ageism, 228
agreement, violations of, 106
Allan, Keith, 110, 225–226
Allen, Jo, 46
alliteration, 107
ambiguous language, 93–95, 243–245
analogy paragraphs, 172

analysis (classification) paragraph, 172
anecdotes
   in narrative paragraphs, 171
   in technical writing, 192
anger and tone, 195, 197–198
anthropomorphism, 107
   compared to personification, 108–109
anti-Semitism, 224
antithesis in sentence construction, 148–149
Apollo 13 (case study), 135–138
appearance of text and tone, 203
appositives for combining sentence elements, 151
appropriateness, 42–57. *See also* inappropriate language
   diction strategy, 93–94
   editing for, 260
   and inappropriateness, 46
   of technical terminology, 120
archaisms (barbarisms), 97–99
argot and cant, 130–131
argument
   Aristotelian, 73–74
   Rogerian, 71–74
   Toulmin, 74
Aristotle, 8
   persuasion, modes of, 73–74
arrangement (rhetorical canon), 68
artful prose, 156
artistry, editing for, 257
assonance, 107
attitude
   and bias, 219

and slang, 89–90
and tone, 189, 194–198
audience, 10, 37
    acronyms and initialisms, familiarity with, 133–134
    analysis, 37
    and discourse community, 25–26
    and humor, 206
    political considerations, 23
    relationship with, 198
    and technical terminology, 120
    and tone, 198, 202–203
August, Eugene, 219
authority, establishing, 199–200
awkward sentences, 158

Baker, Russell, 51
balanced sentences, 148–149
barbarisms, 98–99
Barry, John A., 51–52, 107, 118, 121–122, 133
Barth, John, 168
Barzun, Jacques, 1, 163, 164, 187
Becker, Alton, 71–72
Bernstein, Theodore, 95
bias, 215–231
    ageist, 228
    corporate, 222–224
    cultural, 226
    defining, 217–218
    editing for, 264
    ethnic, 225
    gender, 218–222
    levels of, 218
    national, 225–226
    philosophical, 224–225
    political, 225
    racial, 225–226
    religious, 224–225
    sexist, 218–222
    in technical writing, 216–218
blanket statements. *See* totality words
bloopers, 98–99
Boiarsky, Carolyn, 227
Bolinger, Dwight, 44
bottom-up editing, 256
brevity, 8–9, 95
Buehler, Mary Fran, 254, 257
bureaucratese, 23, 120–122, 127, 243–244
    and doublespeak, 131
Burgess, Anthony, 130

Burridge, Kate, 110, 225–226
buzzwords, 131

canons of rhetoric, 68–69
    and editing, 257
cant
    and argot, 130–131
    barbarism, 98–99
Carroll, Lewis, 105
case studies
    Apollo 13 and jargon, 135–138
    computer viruses and tone, 208–210
    the *Exxon Valdez*, deception and, 247–249
    the *Exxon Valdez* and persuasion, 75–79
    gender-neutral language, 230–231
    the Ruby Ridge incident and diction, 110–112
    the Space Shuttle *Challenger* and different discourse communities, 26–33
    Three Mile Island and ineffective communication, 12–14
catchwords and catchphrases, 99
cause-and-effect paragraphs, 174
C-B-S Theory of Style, 8–9
ceremonial style, 46
Challenger, Space Shuttle (case study), 26–33
Chapman, Robert, 89
character and tone, 199
Charrow, Veda, 52
choices
    stylistic, 71–74
    word, and appropriateness, 94
Ciardi, John, 126
clarity, 8–9, 46, 52–53
    diction strategy, 94–95
    editing for, 257
classification paragraphs, 172–173
clauses
    defining, 146
    forming dependent, 151
    independent, 145–146
cliches, 99
clipped words, 99
codes of conduct, 240–241
cognitive meaning. *See* denotation
coherence, 175–177
cohesion, 174–175
Coletta, W. John, 216
colloquial diction, 88–90
    jargon and slang, 130

colloquialisms, 97–98
colorism, 225
combining sentences, 150–153
comma splices, 156–157
communication, ineffective, Three Mile Island and (case study), 12–14
comparison paragraphs, 171
complement, object, 146
complex-compound sentences, 146–147
complexity, 96–97
complex sentences, 146–147
complex style, 45, 53–54
compound-complex sentences, 146–147
compound sentences, 146–147
computer
   ethics, 238–241
   jargon, 133
   viruses (*See* viruses, computer)
computerese, 120–122, 133
conciseness, 95
concrete words, 95–96
confidentiality and ethics, 240
confusibles, 100
confusing language (tech speak), 132
conjunctions, 150, 151
connotation, 90. *See also* denotation
   and political bias, 225
   and sexism, 220–221
consequences and ethics, 241–242
consistency, 96
consonance, 107–108
contractions, 99
contracts and tone, 192–194
contrast paragraphs, 171–172
control, establishing, 201
conventions, cultural and rhetorical, 226–228
conversational
   prose, 88–89
   writing, 47
conveyor-belt metaphor, 6, 63
Cook, Reginald, 187, 199
coordinating conjunctions, 150
coordination edit, 257
coordination of sentence elements, 150
copy-clarification edit, 258
copyediting, 255, 264–265
corporate bias, 222–224
corporate writing, 10
correspondence, legal, 201

credibility, 73, 199–200
Crick, Francis, 56, 67, 71, 122–123
criminal language, 130. *See also* cant
Cross, Donna, 225
Crystal, David, 93–94, 101, 122, 218–219, 222, 224, 229
cultural bias and conventions, 226–228
cumulative sentences, 149–150
cuteness and tone, 198

dangling modifiers, 106, 157–158
Day, Robert, 4, 64
dead metaphors, 108
deception
   the *Exxon Valdez* and (case study), 247–249
   and political bias, 225
deceptive language, 243–245
   and computer jargon, 133
declarative sentences, 147
deduction in science, 66
deference and tone, 201–202
definition paragraphs, 170
delicacy and tone, 202
delivery (rhetorical canon), 68
denotation, 90. *See also* connotation
   and discourse community, 25–26
dependent clauses
   defining, 146
   forming, 151
description paragraphs, 170
descriptive approaches, 43–44
design
   consistency in, 96
   and tone, 203
detail
   in paragraphs, 177–179
   and tone, 203
devices, literary, in technical writing, 106–109
dialect, 97–98
diction, 85–110
   colloquial, 98
   consistency, 96
   defining, 86
   editing for, 261
   faults, 97–106
   informal, 88
   levels of, 87–90
   Ruby Ridge (case study), 110–112
   slang, 89–90

specialized, 90–91
strategies in, 91–97
and tone, 203
dilemmas, moral, and ethics, 242
direct address, nouns of, 146
directness, 53
disabilities, discrimination, 228
disclaimers
and ethics, 245
tone in, 192–194
discourse community, 19–38
Apollo 13 (case study), 135–138
and audience analysis, 37
Challenger, Space Shuttle (case study), 26–33
complex audience, writing for, 25
and complex style, 54
corporate, 222–223
defining, 20
editing for, 259–260
examples of, 20
fields of study and, 20
influence on style, 22
on the Internet, 39
and jargon, 22
management versus engineering, 32–33
and technical terminology, 119
writing for, 25
discrimination
ageist, 228
in language, 106
domain of technical writing style, 7–9
Dombrowski, Paul, 32–33
doublespeak, 120–122, 131–132, 236, 243–245
and political bias, 225
Dragga, Sam, 257
dysphemisms, 109–110
and racial bias, 225–226

Eastman, Richard, 199
edit, levels of, 257–258
editing, 253–265
for accuracy, 257
approaches to, 256–258
guides, 258
misconceptions about, 255–256
and proofreading, 255
strategies, 259–264
technical, 254

editors
as advocate, 254
and writers, 255–256
effectiveness of prose, 10
Eiseley, Loren, 81
elegance in technical writing, 148, 156
elegancies, 108
e-mail and tone, 198
emotion
in Aristotelian argument, 74
and tone, 194–198
empathy and tone, 196
emphasis, 153–154
engineerese, 120–122
engineering writing, 23
enumeration paragraphs, 173
ethics, 236–249
computer, 238–241
defining, 237
editing for, 264
guidelines, 238–241
and jargon, 240
and language, 241–245
personal, 246
and the professions, 238–241
of style, 245–246
and technical writing, 238
ethnic bias, 225
ethos (Aristotelian mode of persuasion), 2, 73–74, 246
corporate, 223
and ethics, 237
and persona, 199
and tone, 199–202
euphemisms
and doublespeak, 131, 243
and dysphemisms, 109
as reassurance, 13
Ewing, David W., 163, 187, 196, 203
exaggeration, hyperbole, 109
example paragraphs, 167–168
exclamations, 106, 146
exclamatory sentences, 147
ex-jargon, 132
expletives, 105–106
expressions, parenthetical, 146
expressiveness in technical writing, 69, 70
Exxon Valdez
deception and (case study), 247–249
persuasion and (case study), 75–79

Fahey, Tom, 118, 128, 131
familiarity and tone, 198
familiar words, 95
faults in paragraphs, 179–180
Fielden, John S., 1, 47, 187
figures of speech, 108–109
flow in paragraphs, 174–177
forceful style, 47
foreign expressions in polite discussion, 98–99
Forester, Tom, 239
formal diction, 87
formality, levels of
    and discourse community, 25–26
    and tone, 191
formal tone in technical writing, 191–192
format consistency, 96
format edit, 258
fragments, sentence, 157
fraud in science, 218
friendliness and tone, 196
functions of sentences, 147
fused sentences, 157

gender
    bias, 218–222
    and stereotypes, 220–221
gender-neutral language, 221–222
    case study, 230–231
gender-related suffixes, 222
generalities and propaganda, 225
general-to-specific paragraphs, 168
generic "he," 106, 215, 219, 221–222, 230–231
genres and humor, 206
glossary, acronyms and initialisms in, 134
gobbledygook, 51–52, 126–129, 243–244
    disadvantages of, 129
    and doublespeak, 131
Gong, Gwendolyn, 257
Gould, Stephen Jay, 65, 66, 67, 215, 216
governmentese, 120–122
graciousness and tone, 196
grammar, 37
    defining, 43
    descriptive and prescriptive approaches to, 43–46
grammarians, pop, 44
grand style, 46, 47
group labels, 228
guidelines, ethical, 238–241
guides, editing, 258

hackers, 21–22
    and ethics, 239
Hahn, Harley, 52
Haig, Alexander, 244
Halloran, Michael, 65
Harmon, Joseph, 71
Harris, John, 56, 124
Hays, Robert, 85
headings in technical writing, 165–167
Hickey, Dona, 2, 190
high style, 46, 47
hobbyists and discourse communities, 20–22
homonyms, homographs, homophones, 100
honesty and ethics, 240
humor in technical writing, 204–207
hyperbole, 109

ideals and ethics, 241–242
idioms, 101
    barbarisms, 98–99
impartiality, 64
imperative sentences, 147
impersonal style
    in letter writing, 47
    in technical writing, 192–194
improprieties, 106
inappropriate language, 106. *See also* appropriateness
    jargon, 120–122
independent clauses and elements, 145–146
induction in science, 66
industrial writing, 46
ineffective communication, Three Mile Island and (case study), 12–14
inflated language, 244
    and doublespeak, 131
informal diction, 88
    slang, 89–90
informal tone in technical writing, 191–192
initialisms, 133–134
instructions, 5, 48
    action verbs in, 93
    negative, 104
    paragraph length for, 165
integrity edit, 258
intensifiers, 104
interjections, 146
international communication, rhetorical conventions in, 227

interpreter, technical writer as, 6
interrogative sentences, 147
invention (rhetorical canon), 68
"izing," 101–102

"Jabberwocky" (Lewis Carroll), 105
Jack, Judith, 57–58
jargon, 46, 120–122, 125–133
   Apollo 13 case study, 135–138
   barbarisms, 98–99
   computerese, 133
   disadvantages of, 129
   as doublespeak, 243
   editing for, 261–262
   ex-, 132
   and slang, 89–90, 130
   in technical writing, 35
   tech speak, 132
Jet Propulsion Lab levels of edit, 257–258
Johnson, Samuel, 130–131

Kilpatrick, James, 2, 109
Kirkman, John, 155
knowledge and credibility, 199–200
Kolln, Martha, 175
Kreicker, Kimberly, 220–221

labels, group, 228
Laib, Nevin, 2, 8
language
   criminal, 130
   deceptive, and computer jargon, 133
   and ethics, 242–245
   gender neutral, 221–222
   inappropriate, 106
   inflated, and doublespeak, 131, 244
   non-sexist, 221–222
   pretentious, and computer jargon, 133
   sexist, 106, 219
language edit, 258
Lanham, Richard, 8, 22, 47, 49, 118, 127, 148, 149, 192, 236, 245–246
   C-B-S Theory of Style, 8–9
Latin abbreviations, 101
layout and tone, 203
learned style, 55
legal
   correspondence, 201
   terms, 103–104

length
   of paragraphs, 164–165, 177–179
   of sentences
      in technical writing, 34
      variety in, 155
letter writing, 47
   and angry tone, 197–198
   and empathetic tone, 196–197
   employment, action verbs in, 92–93
   legal, 201
levels of diction, 87–90
levels of edit, 257–258
levels of formality
   and discourse community, 25–26
   and tone, 191
levels of headings in technical writing, 166–167
license agreements, tone in, 192–194
Limerick, Patricia, 24–25
lists, parallelism in, 149
literarisms, 108
literary devices, 54–56
   in scientific writing, 56, 71
   in technical writing, 106–109
litotes in scientific writing, 71
Locke, David, 5, 64, 65, 67, 70
logic trap words, 102–103
logos (Aristotelian mode of persuasion), 73–74
loose sentences, 149–150
low style, 46, 47
Lutz, William, 45, 131–132, 236
lying with language, 243–245

main clauses. *See* independent clauses
Mancuso, Joseph, 257
manuals (military), primer style in, 48
Marder, Daniel, 7
mavens, language, 44
Maverick, Maury, 126
McArthur, Tom, 100, 130, 163, 165, 218
meaningless words, 102–103, 105
meanings, double, 94
mechanical style edit, 258
mechanics, 37
mechanism, 107
memory (rhetorical canon), 68
Mencken, H. L., 2
mentally challenged, discrimination against, 228
metaphor, conveyor-belt, 6

metaphors
   literary devices, 108
   technical, 131
   in technical writing, 56
middle style, 46, 47
military
   training manuals, 48
   writing style, 23
Miller, Carolyn, 62, 65–66
Miller, Casey, 215, 219, 221–222, 224
Mirshafiei, Mohsen, 226–227
mixaphors. *See* mixed metaphors
mixed metaphors, 108
models in science, 67
modifiers, 104, 146
   dangling, 106, 157–158
   meaningless, 104
   strings of, 104
   in technical terminology, 122–123
Moely, Barbara, 220–221
Montaigne, Michel de, 201
mood and paragraph consistency, 180
moral dilemmas, 242
morphology, 43
Morrison, Perry, 239
Morrison, Toni, 16

narration paragraphs, 170–171
narrative voice, 190
Nash, Walter, 125
national bias, 225–226
negative instructions, 104
negatives, double, 100
negative words and phrases, 104
neologisms, 105
newspeak, 120–122, 125–126
Nilsen, Alleen, 220
nominalizations, 49–51, 101–102
nominal style, 49–51
nominative absolutes, 146
nonce words, 105
nonsense words, 105
non-sexist language, 221–222
noun
   adjectives, 49–51
   phrases, 104
   strings, 49–51, 104
   style, 49–51
   substitutes, for combining sentence elements, 152

nouns of direct address, 146
number and paragraph consistency, 180
Nunberg, Geoffrey, 44, 98–99

object complement, 146
objective style, 53
objectivity, 6, 63, 64, 68
   as rhetorical stance, 73
   in science, 74, 216–217
   and technical terminology, 122–123
obligations and ethics, 241–242
obscurity
   advantages of, 24
   in technical writing, 94–95
occupational slang, 131
office politics, 225
officialese, 120–122
Official Style, The. *See* The Official Style
organizing
   according to purpose, 11
   faults of, 103
ornate style, 53, 55
Orwell, George, 125–126

page layout and paragraph length, 165–167
paradigm
   new rhetoric of science, 65–66
   old and new, 6, 63–64, 123
   shift, 64, 216–217
paragraphs, 163–181
   details in, 177–179
   development of, 167–174
   editing, 263
   faults in, 179–180
   flow in, 174–177
   length of, 164–165, 177–179
paragraph types
   analogy, 172
   analysis (classification), 172
   cause-and-effect, 174
   classification, 172
   comparison, 171
   contrast, 171–172
   definition, 170
   description, 170
   enumeration, 173
   example, 167–168
   general-to-specific, 168
   narration, 170–171

question-to-answer, 169–170
specific-to-general, 168–169
parallelism
　faulty, 158
　in headings, 166–167
　in sentence construction, 148–149
parenthetical expressions, 146
participial phrases for combining sentence elements, 151, 152
Partridge, Eric, 98, 103, 105
passive style, 47, 48–49
　with nominalization, 102
passive verbs in Rogerian argument, 72
passive voice, 35, 93
pathos (Aristotelian mode of persuasion), 73–74
PC. *See* political correctness
pentagonese, 120–122
periodic sentences, 148
person
　and paragraph consistency, 180
　and tone, 188–190
persona, 2–3, 246
　and ethos, 199
　and tone, 188–190
personal
　style in letter writing, 47
　tone in technical writing, 192–194
personification, 108–109
persuasion, 62–79
　Aristotelian modes of, 199
　Exxon Valdez and (case study), 75–79
persuasiveness
　editing for, 260–261
　strategies for achieving, 71–74
　in technical writing, 6, 63, 70, 71–74
philosophical bias, 224–225
phrases, 146
　foreign, 101
　negative, 104
　participial, for combining sentence elements, 151
　prepositional, 35
physically challenged, discrimination against, 228
Pike, Kenneth, 71–72
Pinker, Steven, 42, 44
Pirsig, Robert, 68
plain style, 45, 46, 47, 52–53, 96–97
point of view, 6, 64
　and paragraph consistency, 180
　rhetorical, in science, 217

Rogerian argument, 71–74
　and tone, 188–190
policy edit, 257
political bias, 225
political correctness (PC), 228–229
politics
　influence on prose, 22
　office, 53, 225
pop grammarians, 44
Popper, Karl, 66
positivist view of science, 65
precision, diction strategy, 91–92
preconceptions and racial bias, 226
predicate, sentence, 145–146
prejudice. *See* bias
prepositional phrases, 35
　overuse of, 103–104
prepositions, functioning as connectors, 146
prescriptive approaches, 43–44, 45–46
pretentious language, 125, 130–131, 243–245
　and computer jargon, 133
pretentious style, 51–52
primer style, 47–48
principal clauses. *See* independent clauses
Pringle, Ian, 45
privacy and ethics, 239
professionalism and ethics, 240
professions and ethics, 238–241
pronouns
　relative, 151
　in technical writing, 192
proofreading and editing, 255
propaganda, 225
propriety, editing for, 257
prose
　science and engineering, 23 (*See also* science writing)
　technical (*See* technical writing)
punctuation, 37, 96
purpose, 36–37
　and humor, 206

quality and ethics, 240
question-to-answer paragraphs, 169–170

racial bias, 225–226
Radford levels of edit, 257
Raspberry, William, 226
Rawson, Hugh, 109

reader-friendly style, 52–53
rearrangement for combining sentence elements, 153
Redish, Janice, 10
redundancy, 105–106
reference books for technical terminology, 123
regionalisms. *See* dialect
relative clauses for combining sentence elements, 151
relative pronouns in dependent clauses, 151
relative words, 94
religion and ethics, 241–242
religious bias, 224–225
repetition, 105–106
   of key words and ideas, 176–177
reports, tone in, 192–194
respect and ethics, 241–242
responsives, 146
Rheingold, Howard, 200
rhetoric
   as basis of technical writing, 7
   canons of, 68–69
      and editing, 257
   choosing a strategy, 69
   conventions of, 227
   decisions in, 70
   defining, 8, 64
   devices of, 69, 71
   of science, 64–71
   theories of argument, 71–74
   VanDeWeghe's four kinds, 8
rhythm in sentences, 155
Rogerian argument, 71–74
Room, Adrian, 100
Rosner, Mary, 19
Rubens, Philip, 91
Ruby Ridge incident and diction (case study), 110–112
Rude, Carolyn, 2, 254, 256
Ruggiero, Vincent, 237, 241–242
run-on sentences, 157
Rutter, Russell, 62

Sachs, Harley, 62
Sacks, Oliver, 40, 55–56
Safire, William, 44
sales jargon, 128–129
Sanders, Scott, 73–74
Schindler, George, 23–24
science
   as point of view, 65
   rhetoric of, 64–71

   as social phenomenon, 67
science writing, 23
   audience for, 4
   claim to neutrality of, 64
   renaissance of, 6, 192
scientific method, 64
scientific objectivity, 216–217
   myth of, 74
scientific writing
   audience for, 4
   expressiveness of, 5
   literary devices in, 56, 71
scientist as interpreter, 68
scope in technical documentation, 37
screening edit, 258
segments in technical writing, 179
self-editing, 259
sentences, 144–159
   arrangement, 148–150
   awkward, 158
   balanced, 148–149
   combining techniques for, 150–153
   defining, 145
   editing, 262–263
   emphasis, 153–154
   emphatic endings, 153–154
   endings, 153–154
   faults, 156–158
   forms, 146–147
   fragments, 157
   functions of, 147
   fused, 157
   length, 34, 36
      variety in, 155
   parts of, 145–146
   periodic, 148
   rhythm, 155
   run-on, 157
   variety in, 155
seriousness and tone, 195
sexism, 106, 218–222
sexist language, 106
   case study, 230–231
sexual stereotypes, 220–221
Shakespeare, William, 85
shamans, 44
shape imagery in technical writing, 56
shop talk, 133
   Apollo 13 case study, 135–138

similes, 108
simple sentences, 146–147
simplicity, 8–9, 52–53
    as diction strategy, 96–97
sincerity, 52–53
    as diction strategy, 97
    and establishing trust, 200–201
    and ethos, 74
    and jargon, 129
slang, 89–90
    and jargon, 130
    occupational, 131
slanting, 218
slogans, corporate, and bias, 224
Society for Technical Communication (STC) ethical
        guidelines, 239–240
solecisms, 106
specialized languages, 119–133
specifications, meaningless words in,
        102–103
specific-to-general paragraphs, 168–169
speech, figures of, 108–109
spelling
    consistency, 96
    as stylistic choice, 37
Sprat, Thomas, 53
standards and ethics, 241–242
Stent, Gunther S., 65, 66
stereotypes and gender, 220–221
Stott, Bill, 2, 164–165
Stout, Rick, 52
strategies
    for editing, 259–264
    persuasive, 71–74
strings, modifier, 104
structure of sentences, 145–146
Strunk, William, 2, 221
study, fields of, 20
style
    choices, 71–74
    as communication, 9
    consistency in, 96
    controversial elements of, 46
    defining, 2, 3
    elements of, 34
    good and bad, 45–46
    guides, 3
    and mechanics
        choices in, 37

    making unconventional choices, 45
    misconceptions about, 3
    play and competition in, 9
    and sentence arrangement, 148–150
    in technical writing, 4
style (rhetorical canon), 68
stylistics, 5
subheadings in technical writing, 165–166
subject
    complexity of, 34
    of sentence, 145
subordinate clauses. *See* dependent clauses
subordinating conjunctions, 151
subordination for combining sentence
        elements, 151
substantive editing, 255, 258
suffixes, gender related, 222
suggestibles, 100
Swift, Kate, 215, 219, 221–222, 224
symmetry in sentence construction, 148–149
synonyms, 100
    and consistency, 96
syntax, 43

taboo words, 98–99
tact and tone, 202
Tarutz, Judith, 254
tautology, 105–106
technical editing, 254
    guides for, 258
technical language
    level of, 34
    strategies for using, 134
technical metaphors, 131
technical prose. *See* technical writing
technical support and jargon, 129
technical terminology, 118–138
    Apollo 13 case study, 135–138
    and audience, 25–26
    challenges of, 91, 122–124
    editing for, 261–262
    learning, 91
    placement of, 154
    reference books for, 123
    using, 119–124
technical writing
    active voice in, 192
    anecdotes in, 192
    audience for, 4–5

technical writing *(continued)*
  bias in, 216–218
  characteristics of, 7
  contractions in, 99
  as craft, 11
  creativity in, 6
  and culture, 227
  defining, 69
  domain of style in, 7–9
  and ethics, 238
  expressiveness in, 69, 70
  failure of, 10
  humor in, 204–207
  jargon in, 35
  levels of formality in, 191–192
  literary style in, 55–56, 106–109
  misunderstandings about, 7
  old and new paradigms, 6
  paradigm shift in, 216–217
  paragraph patterns in, 7
  persuasion in, 6, 63–64, 71–74
  purpose of, 7
  and sentence length, 34
  sentence types in, 7
  as shorthand, 23
  as social activity, 22
  specialized vocabulary in, 7
  style, domain of, 7–9
  style in, 1–14
  tone in, 191–194
  and voice, 190
technobabble, 51–52, 133
tech speak and technospeak, 120–122, 132
telegraphic style, 48, 57–58
Tenner, Edward, 132
terminology
  defining, 120
  professional or technical, 131
terms, defining, 120
The Official Style, 47, 127–129, 245–246
  tone in, 192
Thomas, Lewis, 54, 68, 93, 171, 181–182, 184, 191–192
Thoreau, Henry David, 55, 144, 156, 200
Three Mile Island (case study), 12–14
tone, 187–210
  and attitude, 194–198
  balance of, and information, 192
  computer viruses (case study), 208–210
  defining, 188
  editing for, 263–264
  and emotion, 194–198
  and ethos, 199–202
  and paragraph consistency, 180
  in technical writing, 69
top-down editing, 256
topic
  development, 167–174
  limiting for flow, 175–177
  sentences, in technical writing, 164
totality words, 102–103
Toulmin, Stephen, 74
Toulmin logic, 74
trades as discourse community, 20
transitional words and coherence, 175–176
transitions and repetition, 105–106
translation and cultural conventions, 227
transparent style, 4, 5
trust and ethos, 200–201
Turk, Christopher, 155
Twain, Mark, 85, 113

understatements, 56, 109
  as reassurance, 13
  in scientific writing, 71
unmarked usage, 230–231
unusual words, 95

vagueness, 102–103
Van Buren, Robert, 254, 257
VanDeWeghe, Richard, 8
variety in sentences, 155
verbizing, 99, 101–102
verbs
  active, diction strategy, 92–93
  in technical terminology, 122–123
verb style, 49
verb tense and paragraph consistency, 180
viruses, computer
  and ethics, 239
  and tone (case study), 208–210
vocabulary, specialized, 120–121
vogue words, 99
voice
  active, in technical writing, 192
  as diction strategy, 92–93
  and paragraph consistency, 180
  passive, 35, 93

and style, 190
　　and tone, 188–190
vulgarisms, 98–99

Wales, Katie, 5, 89, 189
Walter, John, 1
warmth and tone, 196
warranties, 245
　　tone in, 192–194
Watson, James D., 56, 67, 71, 122–123
Weiss, Timothy, 215, 227
Whissen, Thomas, 90
Whitburn, Merrill, 53, 68, 69
White, E.B., 2, 221
Williams, Joseph, 2, 45, 144, 156, 174–175, 236
window pane theory of language, 65
Winsor, Dorothy, 32
-wise words, 101–102
wooliness, 102–103
word choice and appropriateness, 94
wordiness, 95

words
　　archaic, 97–98
　　defining, 120
　　foreign, 101
　　inoffensive, 109
　　meaningless, 105
　　negative, 104
　　nonce, 105
　　nonsense, 105
　　pretentious, 125
　　transitional, 175–176
workplace politics and style, 53
writers
　　and editors, 255–256
　　as interpreters, 6
writing, good and bad, 44–45

Young, Richard, 71–72
Younger, Irving, 230–231

Zinsser, William, 2, 190, 192, 199–200, 204, 255–256

This page consitutes a continuation of the copyright page.

John Barth. From "Teacher," copyright © John Barth, originally appeared in *Harper's* and reprinted by permission of The Wylie Agency, Inc.

David Crystal. From *The Cambridge Encyclopedia of the English Language,* Cambridge University Press © 1995. Reprinted with the permission of Cambridge University Press.

Stephen Jay Gould. From *The Mismeasure of Man* by Stephen Jay Gould. Copyright © 1981 by Stephen Jay Gould. Reprinted by permission of W. W. Norton & Company, Inc.

Harley Hahn and Rick Stout. From *The Internet Complete Reference,* copyright © 1994 by Osborne McGraw-Hill. Reproduced with permission of The McGraw-Hill Companies.

Ed Krol. Reprinted with permission from *The Whole Internet User's Guide and Catalog.* Copyright © 1994 O'Reilly and Associates, Inc. For orders and information call 800-998-9938.

Richard Lanham. From *Revising Prose* © 1992. All rights reserved. Reprinted by permission of Allyn and Bacon.

David Locke. From *Science as Writing.* Copyright © 1992. Reprinted with permission of Yale University Press.

William Lutz. From *Doublespeak* by William Lutz. Copyright © 1989 by Blonde Bear, Inc. Reprinted by permission of HarperCollins Publishers, Inc.

Stephen L. Nelson. From *Field Guide to Windows 95.* Copyright © 1995 by Stephen L. Nelson, Inc. Reprinted with permission of Microsoft Press.

Stephen L. Nelson. From *Field Guide to Windows 3.1.* Copyright © 1994 by Stephen L. Nelson, Inc. Reprinted with permission of Microsoft Press.

Vincent Ryan Ruggiero. From *Thinking Critically About Ethical Issues,* Third Edition by Vincent Ryan Ruggiero. Copyright © 1992 by Mayfield Publishing Company. Reprinted by permission of the publisher.

Oliver Sacks. Reprinted with the permission of Simon & Schuster from *The Man Who Mistook His Wife for a Hat* by Oliver Sacks. Copyright © 1970, 1981, 1983, 1985 by Oliver Sacks.

Lewis Thomas. "Things Unflattened by Science," Copyright © 1983 by Lewis Thomas, "Humanities and Science," Copyright © 1983 by Lewis Thomas, "On Smell," Copyright © 1983 by Lewis Thomas, from *Late Night Thoughts on Listening to Mahler's Ninth* by Lewis Thomas. Used by permission of Viking Penguin, a division of Penguin Books USA, Inc.

Lewis Thomas. "On Warts," copyright © 1979 by the *New England Journal of Medicine,* "The Hazards of Science," copyright © 1977 by Lewis Thomas, "On Natural Death," copyright © 1979 by Lewis Thomas, From *The Medusa and the Snail* by Lewis Thomas. Used by permission of Viking Penguin, a division of Penguin Books USA Inc.

Lewis Thomas. Reprinted with the permission of Scribner, a division of Simon & Schuster from *The Fragile Species* by Lewis Thomas. Copyright © 1992 by Lewis Thomas.

Tay Vaughan. From *Multimedia: Making It Work,* Second Edition. Copyright © 1994. Reproduced with permission of The McGraw-Hill Companies.

Irving Younger. Reprinted with permission from *Persuasive Writing,* a collection of 16 columns originally penned for the *ABA Journal* by Irving Younger and republished in book form (102 pgs.) by The Professional Education Group, 12401 Minnetonka Boulevard, Minnetonka, MN 55305/800-229-2531 [ISBN 094-338-0022].

William K. Zinsser. From *On Writing Well,* Fifth Edition. Copyright © 1976, 1980, 1985, 1988, 1990, 1994 by HarperCollins. Reprinted by permission of the author.